*The Most Successful Way of JEE Preparation*

# DPP

## Daily Practice Problems

# JEE Main & Advanced

# Atomic Structure & Chemical Bonding

## CHEMISTRY Vol-I

**Dr. RK GUPTA**
*Retd. Associate Professor*
*KGPG College, Simbhawali, GZB*



arihant
ARIHANT PRAKASHAN (SERIES), MEERUT

DPP
Daily Practice Problems

**arihant**

# ARIHANT PRAKASHAN (SERIES), MEERUT



**Administrative & Production Offices**

**Corporate Office** 'Ramchhaya' 4577/15, Agarwal Road, Darya Ganj, New Delhi -110002
Tele: 011- 47630600, 43518550; Fax: 011- 23280316

**Head Office** Kalindi, TP Nagar, Meerut (UP) - 250002
Tel: 0121-2401479, 2512970, 4004199; Fax: 0121-2401648

**Sales & Support Offices**

Agra, Ahmedabad, Bengaluru, Bareilly, Chennai, Delhi, Guwahati, Hyderabad, Jaipur, Jhansi, Kolkata, Lucknow, Meerut, Nagpur & Pune

**ISBN :** 978-93-13193-40-1

**Price :** ₹210

Published by Arihant Publications (I) Ltd.

**Production Team**

| | | | |
|---|---|---|---|
| *Publishing Managers* | Mahendra Singh Rawat & Keshav Mohan | *Inner Designer* | Mazher Chaudhary |
| *Project Head* | Sorabh Chaudhary | *Page Layouting* | Sunil Sharma, Hemant |
| *Cover Designer* | Shanu Mahad | *Proof Reading* | Salman |

For further information about the products from Arihant, log on to www.arihantbooks.com or email to info@arihantbooks.com

/arihantpub /@arihantpub Arihant Publications /arihantpub

# ABOUT DPP

## *Daily Practice Problems*

Indian Institutes of Technology better known as IITs have always been the dream destination of each science group students who want to pursue their career as Engineers. These are the institutes which have got very high value not even in India but in the whole world. To get the admission in such institutes one need to face entrance Examinations; JEE Main & JEE Advanced, out of which JEE Main is considered as the preliminary exam & JEE Advanced is the final exam to get into the IITs. Approximately per year 12 lakh students enroll themselves for JEE Main and only the top 2.25 lakh in JEE Main merit can attempt for JEE Advanced, the fact clearly express the degree of toughness of JEE Main & JEE Advanced.

To have success in such an exam one needs to be very planned & systematic in its preparation. A very popular wise saying 'Rome was not built in a day' suggests that to achieve a big target like IITs we need to work continuously. For continuous work, one need to have proper planning that how much work he/she has to do on regular basis, i.e., on daily basis. Each student preparing for JEE has many books but unfortunately none of them gives the students a plan for daily studies, which is the most effective success key for JEE.

Keeping the daily practice needs of the students in mind we have come up with DPP Series (Daily Practice Problems) having approximately 26 books. The primary focus of this series is on achieving success through practice & proper planning. For that we have divided the whole syllabus of JEE Main & JEE Advanced; Physics, Chemistry & Mathematics, into different volumes (around 8 volumes of Mathematics & 9 volumes each of Physics & Chemistry), each volume has Daily Practice Problems Sheets having specific questions on various Topics of the individual chapter of the volume ensuring the complete Practice & Assessment of the topic.

**DPP**

*Daily Practice Problems*

In this way we have tried to give you an effective daily planning for JEE keeping in mind the daily hours that a student can devote to his practice, you are just required to practice one sheet of each subject on daily basis. This habit of daily practice will keep you away from exam stress and thoughts like; how much I have prepared/am I preparing properly or not/ do I need to study more on daily basis or what I am doing is ok, etc., which regularly tease your mind and can give you panic at any stage of your preparation.

Some special features that we have tried to incorporate in our Daily Practice Problems (DPP) series are discussed below

1. Micro Level coverage of all the topics of a Chapter/ Unit given in JEE Main & Advanced Syllabus.
2. Each DPP has various types of questions; Subjective Questions, Single/ Multiple Correct Options, Statements type, Comprehension based, Integer Value Answer type, Matching, etc., making you practice & ready for all the question formats of JEE.
3. New & innovative questions in each DPP challenging & improving your problem solving skills.
4. Along with Topical Coverage, Revisal DPPs for JEE Main & JEE Advanced with each Chapter, covering multidisciplinary questions proving comprehensive practice.
5. JEE Main & JEE Advanced Archive (collection of last 13 years' exams questions) with each chapter.
6. Complete Solutions for each DPP.

We hope that DPP series will prove to be very beneficial for the students preparing for JEE, by giving them daily practice in a systematic & planned manner. The suggestions/criticism from the students, teachers & parents about DPP series will be welcomed whole heartedly and we promise to update this series accordingly from time to time.

DPP
Daily Practice Problems

# DAILY TASKS...

## 1 STRUCTURE OF ATOM



## 2 REDOX REACTIONS



## 3 CHEMICAL BONDING

DPP
Daily Practice Problems



## 4 STOICHIOMETRY

# 1. Structure of Atom



DPP
DAILY PRACTICE PROBLEMS

# DPP-1 Electromagnetic Radiations

## Subjective Questions

Working Space

**Direction** *(Q. Nos. 1-4) These questions are subjective in nature, need to be solved completely on notebook.*

**1** Calculate the wavelength, frequency and wave number of a light whose period is $2.0\times 10^{-10}$ s.

**2** What is the number of photons of light with a wavelength of 4000 pm that provides 1 J of energy?

**3** Electromagnetic radiation of wavelength 242 nm is just sufficient to ionise sodium atom. Calculate the ionisation energy of sodium in kJ $mol^{-1}$.

**4** Calculate energy of radiation whose wave number is 1.0 $cm^{-1}$ in terms of

i. J per photon

ii. kJ $mol^{-1}$ of photons

iii. kcal $mol^{-1}$ of photons

## Only One Option Correct Type

**Direction** *(Q. Nos. 5-12) This section contains 8 multiple choice questions. Each question has four choices (a), (b), (c) and (d), out of which ONLY ONE option is correct.*

**5** FM radio broadcasts at 900 kHz. What wavelength does this corresponds to?

a. 333 m  b. $3.03\times 10^{-3}$ m

c. 330 cm  d. 3300 m

**6** A wavelength of 400 nm corresponds to

a. frequency $(\nu)= 7.5\times 10^{14}$ Hz

b. wave number $(\bar{\nu}) = 2.5\times 10^{-6}\ m^{-1}$

c. momentum $(mv) = 1.66\times 10^{-27}\ kg\ ms^{-1}$

d. All of the above

**7** Moseley's equation is represented as $\sqrt{\nu} = a(Z - b)$ where, $a$ and $b$ are constants.

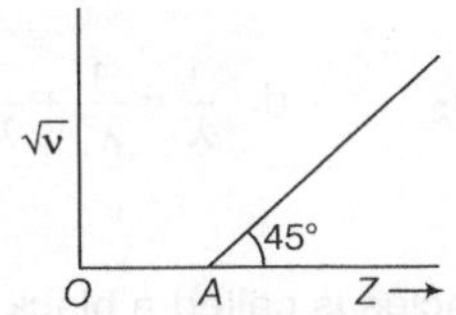


If $OA = 1$, then atomic number of the element showing frequency of 400 Hz is

a. 21  b. 11

c. 401  d. 19

*Working Space*

**8** A gas absorbs photon of wavelength 355 nm and emits at two wavelengths. If one of the emissions is at 680 nm, the other is at **[AIEEE 2011]**

a. 1035 nm
b. 325 nm
c. 743 nm
d. 518 nm

**9** The longest wavelength of light capable of breaking a single (Cl—Cl) bond in $Cl_2$ is if **[AIEEE 2010]**

$Cl—Cl\,(g) \longrightarrow 2Cl\,(g)\,;\ \Delta H = 242\ \text{kJ mol}^{-1}$

a. 494 pm
b. 494 nm
c. 494 Å
d. 247 nm

**10** Which of the following conclusions could not be derived from Rutherford's α-particle scattering experiment?

a. Electrons and nucleus are held together by electrostatic forces of attraction
b. The radius of the atom is larger than that of the nucleus
c. Most of the space in the atom is empty
d. Electrons move in a circular path of fixed energy. Circular paths are called orbits

**11** Select the incorrect statement about cathode rays.

a. They travel from cathode to anode
b. They travel in a straight line in the absence of an external electrical or magnetic field
c. Their nature is independent of the materials of cathode
d. Their nature is dependent on the nature of the gas present

**12** Energy of a mole of radio wave photons with a frequency of 909 kHz is

a. $6.02 \times 10^{-28}$ J
b. $3.62 \times 10^{-4}$ J
c. $1.00 \times 10^{-4}$ J
d. $6.02 \times 10^{-31}$ J

## One or More than One Options Correct Type

***Direction** (Q. Nos. 13-15) This section contains 3 multiple choice questions. Each question has four choices (a), (b), (c) and (d), out of which ONE or MORE THAN ONE are correct.*

**13** Select the correct statement(s).

a. Electromagnetic radiation is a form of energy consisting of oscillating electric and magnetic fields
b. Visible light is a form of electromagnetic radiation
c. The electromagnetic spectrum of sunlight received at the Earth's surface differs from that emitted by the Sun
d. Ultraviolet radiation is absorbed in the stratosphere by ozone and oxygen

**14** Consider the following transitions $n = 1$ to $n = 2$ and then, $n = 2$ to $n = 3$.

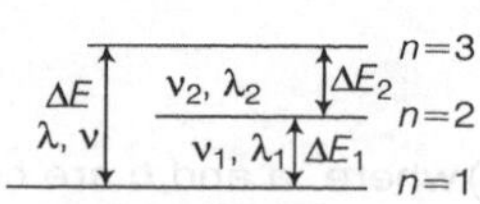


Then

a. $\Delta E = \Delta E_1 + \Delta E_2$
b. $\lambda = \lambda_1 + \lambda_2$
c. $\nu = \nu_1 + \nu_2$
d. $\dfrac{1}{\lambda} = \dfrac{1}{\lambda_1} + \dfrac{1}{\lambda_2}$

**15** Select the correct statement(s).

a. The ideal body, which emits and absorbs all frequencies is called a black body
b. The exact frequency distribution of the emitted radiation from a black body depends upon its temperature
c. The radiation emitted goes from a lower frequency to higher frequency as the temperature increases
d. In vacuum, all types of electromagnetic radiation travel at the same speed

## Comprehension Type

*Direction* (Q. Nos. 16 and 17) *This section contains a paragraph, describing theory, experiments, data, etc. Two questions related to the paragraph have been given. Each question has only one correct answer among the four given options (a), (b), (c) and (d).*

A hypothetical electromagnetic wave is pictured here.

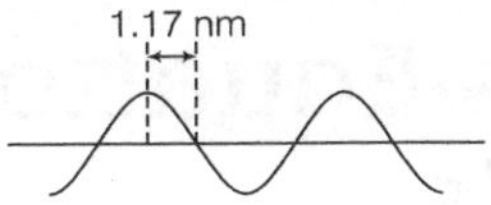


**16** Energy associated with this wave is

a. $4.24 \times 10^{-19}$ J  b. $2.12 \times 10^{-19}$ J
c. $1.06 \times 10^{-19}$ J  d. $8.49 \times 10^{-19}$ J

**17** A standing wave in a string 42 cm long has a total of six nodes (including those at the ends). Wavelength of the standing wave is

a. 7.0 cm  b. 16.8 cm
c. 10.5 cm  d. 5.25 cm

## Matching List Type

*Direction* (Q. No. 18) *Choices for the correct combination of elements from Column I and Column II are given as options (a), (b), (c) and (d) out of which one is correct.*

**18** The colours of visible light are at different wavelength. Match the wavelengths in Column I with their colours in Column II.

| | Column I | | Column II |
|---|---|---|---|
| i. | 700 nm | p. | Violet |
| ii. | 600 nm | q. | Blue |
| iii. | 580 nm | r. | Green |
| iv. | 500 nm | s. | Yellow |
| v. | 460 nm | t. | Orange |
| vi. | 400 nm | u. | Red |

**Codes**

| | i | ii | iii | iv | v | vi |
|---|---|---|---|---|---|---|
| a. | p | q | r | s | t | u |
| b. | u | t | s | r | q | p |
| c. | u | t | r | q | p | s |
| d. | t | u | r | p | s | q |

## One Integer Value Correct Type

*Direction* (Q. Nos. 19 and 20) *This section contains 2 questions. When worked out will result in an integer from 0 to 9 (both inclusive).*

**19** In Millikan's experiment, static electric charge on the oil drops has been obtained by shining X-rays. If the static electric charge on the oil drop is $-1.282 \times 10^{-18}$ C, calculate the number of electrons present on it.

**20** Energy change associated per mole of atoms with an atomic transition giving rise to radiations of 1 Hz is $x \times 10^{-10}$ J mol$^{-1}$. What is the value of $x$?

Working Space

# DPP-2 Dual Nature of Electron (de-Broglie's Equation), Heisenberg's Uncertainty Principle and Photoelectric Effect

## Subjective Questions

*Direction* (Q. Nos. 1-4) *These questions are subjective in nature, need to be solved completely on notebook.*

Working Space

**1** A photon is moving with a velocity of light and emit radiations of wavelength 3.6 Å. What is mass of the photon?

**2** i. Kinetic energy associated with the electron is $3.0 \times 10^{-25}$ J. Calculate its wavelength.

ii. Nitrogen laser produces a radiation at a wavelength of 337.1 nm. If the number of photons emitted is $5.6 \times 10^{24}$ $s^{-1}$, calculate the power of this laser.

**3** Electrons are emitted with zero velocity from a metal surface when it is exposed to radiation of wavelength 6800 Å. Calculate

i. threshold frequency ($\nu_0$) and

ii. work function ($\phi$) of the metal.

**4** The ejection of the photoelectron from the silver metal in the photoelectric effect experiment can be stopped by applying the voltage of 0.35 V when the radiation of 256 nm is used. Calculate the work function for silver metal.

## Only One Option Correct Type

*Direction* (Q. Nos. 5-14) *This section contains* 10 *multiple choice questions. Each question has four choices (a), (b), (c) and (d), out of which ONLY ONE option is correct.*

**5** Given,

I. Radiations for microwave oven

II. Amber light from traffic signals

III. Radiations from FM radio

IV. Cosmic rays from outerspace

V. X-rays

Increasing order of their energies is

a. I < II < III < IV < V

b. V < IV < III < II < I

c. I < III < V < IV < II

d. III < I < II < V < IV

Working Space

**6** The threshold frequency $\nu_0$ for a metal is $7.0 \times 10^{14}\ s^{-1}$. Radiation of frequency $\nu = 1.0 \times 10^{15}\ s^{-1}$ hits the metal. Kinetic energy of the emitted electron is

a. $2.0 \times 10^{-18}$ J
b. $1.60 \times 10^{-17}$ J
c. $1.60 \times 10^{-19}$ J
d. $2.0 \times 10^{-19}$ J

**7** Which of the following rules/principles is responsible to rule out the existence of definite paths or trajectories of electrons?

a. Aufbau (ascending energy) rule
b. Hund's rule of maximum multiples
c. Pauli's exclusion principle
d. Heisenberg's uncertainty principle

**8** In an atom, an electron is moving with a speed of 600 $ms^{-1}$ with an accuracy of 0.005%. Certainty with which the position of the electron can be located is

$[h = 6.6 \times 10^{-34}$ Js, $m_e = 9.1 \times 10^{-31}$ kg] **[AIEEE 2009]**

a. $1.52 \times 10^{-4}$ m
b. $5.10 \times 10^{-3}$ m
c. $1.92 \times 10^{-3}$ m
d. $3.84 \times 10^{-3}$ m

**9** The wavelength (in nanometer) associated with a proton (mass = $1.67 \times 10^{-27}$ kg $atom^{-1}$) at the velocity of $1.0 \times 10^{-3}\ ms^{-1}$ is

a. 0.032 nm
b. 0.400 nm
c. 2.500 nm
d. 14.000 nm

**10** **Statement I** It is impossible to determine position and momentum of the moving electron simultaneously with same accuracy.

**Statement II** The path of an electron in an atom is clearly defined.

a. Both Statement I and Statement II are correct and Statement II is the correct explanation of Statement I
b. Both Statement I and Statement II are correct and Statement II is not the correct explanation of Statement I
c. Statement I is correct but Statement II is incorrect
d. Statement II is correct but Statement I is incorrect

**11** If the photon of the wavelength 150 pm strikes an atom and one of its inner bound electrons is ejected out with a velocity of $1.5 \times 10^{7}\ ms^{-1}$, then binding energy by which electron is bound to nucleus is

a. $1.223 \times 10^{-15}$ J
b. $-1.223 \times 10^{-15}$ J
c. $2.345 \times 10^{-15}$ J
d. $1.428 \times 10^{-15}$ J

**12** The position of both, an electron and a helium atom is known within 1.0 mm. Further more the momentum of the electron is known within $5.0 \times 10^{-26}$ kg $ms^{-1}$. The minimum uncertainty in the measurement of the momentum of the helium atom is

a. 50.0 kg $ms^{-1}$
b. 80.0 kg $ms^{-1}$
c. $80.0 \times 10^{-26}$ kg $ms^{-1}$
d. $5.0 \times 10^{-26}$ kg $ms^{-1}$

Working Space

**13** The wavelength of a neutron with a translatory kinetic energy equal to $kT$ at 300 K is

a. 178 pm b. 200 pm c. 17.8 pm d. 20.0 pm

**14** Frequency of a matter wave is equal to

a. $\frac{(KE)}{2h}$ b. $\frac{2(KE)}{h}$ c. $\frac{(KE)}{h}$ d. $\lambda$

## Comprehension Type

***Direction*** *(Q. Nos. 15 and 16) This section contains a paragraph, describing theory, experiments, data, etc. Two questions related to the paragraph have been given. Each question has only one correct answer among the four given options (a), (b), (c) and (d).*

Photoelectric effect can be expressed in terms of the following graph

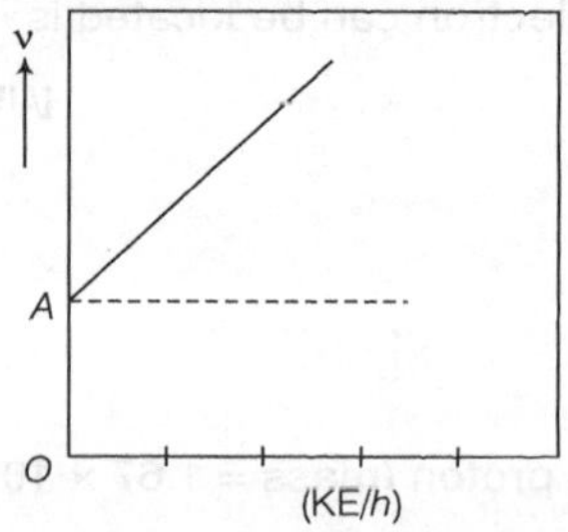


$OA = 5 \times 10^{14}\ s^{-14}$

**15** What is work function for this photoelectric emission of electrons?

a. 199.262 kJ $mol^{-1}$ b. 199.262 J $mol^{-1}$

c. 3.3 kJ $mol^{-1}$ d. $3.3 \times 10^{-19}$ kJ $mol^{-1}$

**16** Kinetic energy imparted to the moving electron at a wavelength of $2.5 \times 10^{-7}$ m is

a. −278.97 kJ $mol^{-1}$ b. 278.97 kJ $mol^{-1}$

c. 168.23 kJ $mol^{-1}$ d. −168.23 kJ $mol^{-1}$

## Matching List Type

***Direction*** *(Q. No. 17) Choices for the correct combination of elements from Column I and Column II are given as options (a), (b), (c) and (d), out of which one is correct.*

**17** Neon gas is generally used in sign boards. It emits radiations at wavelength 616 nm.

Match the values derived given in Column II with the corresponding parameter given in Column I.

| | Column I | | Column II |
|---|---|---|---|
| i. | Frequency (in $s^{-1}$) of emission | p. | $6.2 \times 10^{18}$ |
| ii. | Distance travelled (in km) by it in 30 s | q. | $3.22 \times 10^{-22}$ |
| iii. | Energy of quantum (in kJ) | r. | $9.0 \times 10^{6}$ |
| iv. | Number of quanta present if it produces 2 J of energy | s. | $4.87 \times 10^{14}$ |

**Codes**

| | i | ii | iii | iv | | i | ii | iii | iv |
|---|---|---|---|---|---|---|---|---|---|
| a. | p | r | s | q | b. | s | r | q | p |
| c. | s | q | r | p | d. | p | s | r | q |

Working Space

## One Integer Value Correct Type

*Direction* (*Q. Nos.* 18 -20) *This section contains* 3 *questions. When worked out will result in an integer from* 0 *to* 9 *(both inclusive).*

**18** The work function ($\phi$) of some metals is listed below. The number of metals which will show photoelectric effect when light of 300 nm wavelength falls on the metal is ..........

| Metal | $\phi$ (eV) |
|---|---|
| Li | 2.4 |
| Na | 2.3 |
| K | 2.2 |
| Mg | 3.7 |
| Cu | 4.8 |
| Ag | 4.3 |
| Fe | 4.7 |
| Pt | 6.3 |
| W | 4.75 |

**[IIT JEE 2011]**

**19** If the uncertainties in the measurements of position and momentum are equal, calculate the uncertainty in the measurement of velocity (in $ms^{-1}$) of particle of mass $1.21 \times 10^{-18}$ kg.

**20** Wavelength associated with an electron having KE $3.0 \times 10^{-25}$ J is $x \times 10^{-7}$ m. What is the value of $x$?

# DPP-3 Bohr's Theory of H-Atom

## Subjective Questions

Working Space

**Direction** (Q. Nos. 1-5) *These questions are subjective in nature, need to be solved completely on notebook.*

**1** The energy of the electron in the second and third Bohr's orbit of the hydrogen atom is $-5.42 \times 10^{-12}$ erg and $-2.42 \times 10^{-12}$ erg respectively. Calculate the wavelength of the emitted light when the electron drops from the third to the second orbit.

**2** Calculate the wavelength of the photon that is emitted when an electron in the Bohr's orbit, $n = 2$, returns to the orbit, $n = 1$ in the hydrogen atom. The ionisation potential of the ground state hydrogen atom is $2.17 \times 10^{-11}$ erg per atom.

**3** i. Calculate the energy required for the process

$$He^{+}(g) \longrightarrow He^{2+}(g) + e^{-}$$

The ionisation energy for the H-atom in the ground state is $2.18 \times 10^{-18}$ J atom$^{-1}$.

ii. Calculate the wave number for the longest wavelength transition in the Balmer series of atomic hydrogen.

**4** i. What transition in the hydrogen spectrum would have the same wavelength as the Balmer transition $n = 4$ to $n = 2$ of $He^{+}$ spectrum?

ii. The Lyman series of the hydrogen spectrum can be represented by the equation

$$v = 3.2881 \times 10^{15}\ s^{-1}\left[\frac{1}{1^2} - \frac{1}{n^2}\right]$$

Calculate the maximum and minimum frequency of lines in this series.

**5** Energy of the electron in H-atom in ground state is 13.6 eV. What is its energy in first excited state?

## Only One Option Correct Type

**Direction** (Q. Nos. 6-16) *This section contains 11 multiple choice questions. Each question has four choices (a), (b), (c) and (d) out of which ONLY ONE option is correct.*

**6** Given, $\Delta E = \overline{R}_H Z^2 \left[\frac{1}{n_1^2} - \frac{1}{n_2^2}\right]$

$\overline{R}_H$ (Rydberg constant) for H-atom is $x$ m$^{-1}$. Corresponding value for this constant for $He^{+}$ ion is

a. $100x$ cm$^{-1}$ b. $400x$ cm$^{-1}$
c. $4x$ cm$^{-1}$ d. $0.04x$ cm$^{-1}$

**7** 'Hartree' is the atomic unit of energy and is equal to

a. $-13.6$ eV atom$^{-1}$ b. $-27.2$ eV atom$^{-1}$
c. $+13.6$ eV atom$^{-1}$ d. $+27.2$ eV atom$^{-1}$

Working Space

**8** Wave number of a spectral line for a given transition is $x$ cm$^{-1}$ for $He^+$, then its value for $Be^{3+}$ (isoelectronic of $He^+$) for the same transition is

a. $x$ cm$^{-1}$ b. $4x$ cm$^{-1}$ c. $\frac{x}{4}$ cm$^{-1}$ d. $2x$ cm$^{-1}$

**9** Which of the following electronic transitions requires that the greatest quantity of energy be absorbed by a hydrogen atom?

a. $n = 1$ to $n = 2$ b. $n = 2$ to $n = 4$ c. $n = 3$ to $n = 6$ d. $n = 3$ to $n = \infty$

**10** An electron in H-atom in its ground state absorbs 1.5 times as much as energy as the minimum required for its escape from the atom

$$H(g) \longrightarrow H^+(g) + e^-; \qquad \Delta H = 13.6 \text{ eV atom}^{-1}$$

Thus, kinetic energy given to the emitted electron is

a. 20.4 eV b. 34.0 eV c. –20.4 eV d. –34.0 eV

**11** Ionisation energy of $He^+$ is $19.6 \times 10^{-18}$ J atom$^{-1}$. The energy of the first stationary state ($n = 1$) of $Li^{2+}$ is **[AIEEE 2010]**

a. $4.41 \times 10^{-16}$ J atom$^{-1}$ b. $-4.41 \times 10^{-17}$ J atom$^{-1}$

c. $-2.20 \times 10^{-15}$ J atom$^{-1}$ d. $8.82 \times 10^{-17}$ J atom$^{-1}$

**12** The kinetic energy of an electron in the second Bohr orbit of a hydrogen atom is ($a_0$ is Bohr radius) **[AIEEE 2012]**

a. $\frac{h^2}{4\pi m a_0^2}$ b. $\frac{h^2}{16\pi^2 m a_0^2}$ c. $\frac{h^2}{32\pi^2 m a_0^2}$ d. $\frac{h^2}{64\pi^2 m a_0^2}$

**13** Energy of the electron in $n$th orbit is given by $E = -2.178 \times 10^{-18} \text{ J}\left(\frac{Z^2}{n^2}\right)$

Wavelength of light required to excite an electron in an H-atom from level $n = 1$ to $n = 2$ will be ($h = 6.62 \times 10^{-34}$ J s; $c = 3.0 \times 10^8$ ms$^{-1}$) **[AIEEE 2012]**

a. $1.216 \times 10^{-7}$ m b. $2.816 \times 10^{-7}$ m c. $6.500 \times 10^{-7}$ m d. $8.500 \times 10^{-7}$ m

**14** The potential energy of an electron in the second Bohr's orbit of the $He^+$ ion is

a. –54.4 eV b. –27.2 eV c. –108.8 eV d. –13.6 eV

**15** In Lyman series, shortest wavelength of H-atom appears at $x$ m, then longest wavelength in Balmer series of $He^+$ appear at

a. $\frac{9x}{5}$ m b. $\frac{36x}{5}$ m c. $\frac{x}{4}$ m d. $\frac{5x}{9}$

**16** If the radius of the first Bohr orbit is $x$, then de-Broglie wavelength of the electron in the third orbit is nearly

a. $2\pi x$ b. $6\pi x$ c. $9x$ d. $\frac{x}{3}$

## Comprehension Type

**Direction** *(Q. Nos. 17 and 18) This section contains a paragraph, describing theory, experiments, data, etc. Two questions related to the paragraph have been given. Each question has only one correct answer among the four given options (a), (b), (c) and (d).*

Radius of Bohr's orbit of H-atom is 52.9 pm. An emission in H-atom starts from the orbit having radius 1.3225 nm and ends at 211.6 pm.

**17** Wavelength (in nm) associated with this emission is

a. 434.17 nm b. 230.20 nm c. 144.70 nm d. 289.40 nm

**18** Spectral line appears in .......... region.

a. UV b. visible c. near IR d. far IR

## Matching List Type

Working Space

***Direction*** *(Q. Nos. 19 and 20) Choices for the correct combination of elements from Column I and Column II are given as options (a), (b), (c) and (d) out of which one is correct.*

**19** Match the equation in Column I with the name type in Column II.

| | Column I | | Column II |
|---|---|---|---|
| i. | $\Delta x \Delta v \geq \frac{h}{4\pi m}$ | p. | Photoelectric effect |
| ii. | $mvr = \frac{nh}{2\pi}$ | q. | de-Broglie equation |
| iii. | $r_1 = \frac{h^2}{4\pi^2 Me^2}$ | r. | Ritz combination principle |
| iv. | $\bar{\nu} = R_H \left[\frac{1}{n_1^2} - \frac{1}{n_2^2}\right]$ | s. | Heisenberg's uncertainty principle |
| v. | $h\Delta\nu = \frac{1}{2}mv^2$ | t. | Bohr's radius of H–atom |
| vi. | $\lambda = \frac{h}{\sqrt{2mKE}}$ | u. | Angular momentum is quantised |

**Codes**

| | i | ii | iii | iv | v | vi |
|---|---|---|---|---|---|---|
| a. | s | u | t | r | p | q |
| b. | s | u | r | t | q | p |
| c. | u | s | t | r | q | p |
| d. | p | q | t | u | r | s |

**20** If $E_n$ = total energy, $K_n$ = kinetic energy, $V_n$ = potential energy and $r_n$ = radius of the $n$th orbit, then based on Bohr's theory, match the parameter in Column I with the values in Column II.

| | Column I | | Column II |
|---|---|---|---|
| i. | $V_n / K_n$ | p. | 0 |
| ii. | $r_n \propto E_n^x, x = ?$ | q. | –1 |
| iii. | Angular momentum in the lowest orbital | r. | –2 |
| iv. | $r_n \propto \frac{1}{Z^n}, n = ?$ | s. | +1 |

**Codes** **[IITJEE 2006]**

| | i | ii | iii | iv | | i | ii | iii | iv |
|---|---|---|---|---|---|---|---|---|---|
| a. | p | q | r | s | b. | r | q | p | s |
| c. | r | p | q | s | d. | s | r | q | p |

## One Integer Value Correct Type

***Direction*** *(Q. Nos. 21-24) This section contains 4 questions. When worked out will result in an integer from 0 to 9 (both inclusive).*

**21** The energy of an electron in the first Bohr orbit of H-atom is –13.6 eV. What is the possible value of quantum number for the excited state to have energy –3.4 eV?

**22** An emission is $Be^{3+}$ in observed at 2.116 A°. In which orbit is it placed?

**23** At what minimum atomic number, a transition from $n = 2$ to $n = 1$ energy level results in the emission of X-rays with wavelength $3.0 \times 10^{-8}$ m?

**24** Find the number of waves made by a Bohr's electron in one complete revolution in its 3rd orbit.

# DPP-4 Quantum Numbers and Electronic Configuration

## Subjective Questions

*Working Space*

**Direction** (Q. Nos. 1-4) *These questions are subjective in nature, need to be solved completely on notebook.*

**1** A compound of vanadium has magnetic moment of 1.73 BM for vanadium.

i. Determine unpaired electrons in vanadium ion.

ii. Write its electronic configuration.

**2** What is the lowest value of $n$ that allows $g$ orbitals to exist? Derive values of suborbits and orbitals in it.

**3** Given below are several electronic configuration that may be correct for oxygen atom. Indicate whether each of these represents the ground state or an excited state or is a forbidden configuration for oxygen atom ($Z = 8$).

| | 1s | 2s | 2p | | | 3s |
|---|---|---|---|---|---|---|
| i. | ↑↓ | ↑↓ | ↑↓ | ↑ | – | ↓ |
| ii. | ↑↓ | ↑↓ | ↓ | ↓ | ↑↓ | – |
| iii. | ↑↓ | ↑↓ | ↓ | ↓ | ↓ | ↓ |
| iv. | ↑↓ | ↑↓ | ↑↓ | ↑↓ | – | – |
| v. | ↑↓ | ↑↓ | ↓ | ↓ | ↓↓ | – |
| vi. | ↑↓ | ↑↓ | ↑ | ↑↓ | ↑ | – |

**4** How many elements would be in the second period of the periodic table if the spin quantum number $m_s$ could have the value $-\frac{1}{2}, 0, +\frac{1}{2}$?

## Only One Option Correct Type

**Direction** (Q. Nos. 5-14) *This section contains 10 multiple choice questions. Each question has four choices (a), (b), (c) and (d), out of which ONLY ONE option is correct.*

**5** Which is a possible set of quantum numbers for a valence unpaired electrons in ground state atom of phosphorus ($Z = 15$)?

| | $n$ | $l$ | $m_l$ | $m_s$ |
|---|---|---|---|---|
| a. | 2 | 1 | 0 | $+\frac{1}{2}$ |
| b. | 3 | 2 | 0 | $+\frac{1}{2}$ |
| c. | 3 | 0 | 0 | $+\frac{1}{2}$ |
| d. | 3 | 1 | 0 | $-\frac{1}{2}$ |

Working Space

6 For a multi-electron atom, set of quantum numbers is given as

$$2, \ 0, \ 0, \ \frac{1}{2} \qquad 2, \ 0, \ 0, \ -\frac{1}{2}$$

Thus, the next higher allowed set of $n$ and $l$ quantum numbers for this atom in its ground state is

a. $\frac{n}{3}, \frac{l}{0}$ b $\frac{n}{3}, \frac{l}{1}$ c. $\frac{n}{2}, \frac{l}{1}$ d. $\frac{n}{2}, \frac{l}{0}$

7 Magnetic moments of the following isoelectronic species (24 electrons) are in order

$$Mn^{+}, Cr, Fe^{2+}, Co^{3+}$$

a. $Fe^{2+} < Co^{3+} < Mn^{+} < Cr$
b. $Cr < Mn^{+} < Fe^{2+} < Co^{3+}$
c. $Co^{3+} = Fe^{2+} < Mn^{+} = Cr$
d. $Co^{3+} < Fe^{2+} < Mn^{+} < Cr$

8 The orbital angular momentum for a $d$-orbital electron is given by

a. $\sqrt{6}\left(\frac{h}{2\pi}\right)$ b. $\sqrt{2}\left(\frac{h}{2\pi}\right)$ c. $\left(\frac{h}{2\pi}\right)$ d. $2\left(\frac{h}{2\pi}\right)$

9 The number of radial nodes in $3s$ and $2p$ respectively are

a. 2 and 0
b. 0 and 2
c. 1 and 2
d. 2 and 1

10 A hydrogen like specics in fourth orbit has radius 1.5 times that of Bohr's orbit. In neutral state, its valence electron is in

a. $1s$
b. $2p$
c. $3s$
d. $3d$

11 Following suborbits with values of $n$ and $l$ are given

| | $n$ | $l$ |
|---|---|---|
| I. | 4 | 1 |
| II. | 4 | 0 |
| III. | 3 | 2 |
| IV. | 3 | 1 |

Increasing order of energy of these suborbits is

a. IV < III < II < I
b. IV< III < I < II
c. IV < II < III < I
d. I < III < II < IV

12 The correct set of four quantum numbers for the valence electron of rubidium atom ($Z = 37$) is **[JEE Main 2013]**

a. $5, 0, 0, +\frac{1}{2}$
b. $5, 1, 0, +\frac{1}{2}$
c. $5, 1, 1, +\frac{1}{2}$
d. $5, 0, 1, +\frac{1}{2}$

13 For Cr (24), (EC) : [Ar] $4s^1 3d^5$ The number of electrons with $l = 1$ to $l = 2$ are respectively

a. 12 and 4
b. 12 and 5
c. 16 and 4
d. 16 and 5

14 Last filling electron in lanthanides find place in $4f$-orbital. Which of the following sets of quantum number is correct for an electron in $4f$ orbital?

| | $n$ | $l$ | $m_l$ | $m_s$ |
|---|---|---|---|---|
| a. | 4 | 3 | +4 | $+\frac{1}{2}$ |
| b. | 4 | 4 | −4 | $-\frac{1}{2}$ |
| c. | 4 | 3 | +1 | $+\frac{1}{2}$ |
| d. | 4 | 2 | −2 | $+\frac{1}{2}$ |

## One or More than One Options Correct Type

**Direction** *(Q. Nos. 15-19) This section contains 5 multiple choice questions. Each question has four choices (a), (b), (c) and (d), out of which ONE or MORE THAN ONE are correct.*

**15** Quantum numbers have been matched with their characteristics. Select correct matching(s).

| | Quantum numbers | Characteristics |
|---|---|---|
| a. | $l$ | Determines three-dimensional shape of the orbital |
| b. | $n$ | Orientation and energy of the orbital |
| c. | $m_l$ | Size of the orbital |
| d. | $m_s$ | Orientation of the spin of the electron relative to chosen axis. |

**16** In the following pair, each has two orbitals I and II. Select the ones in which II experiences larger effective nuclear charge than I

| | I | II | | I | II |
|---|---|---|---|---|---|
| a. | $2s$ | $3s$ | b. | $4d$ | $4f$ |
| c. | $3d$ | $3p$ | d. | $3d$ | $4s$ |

**17** In a multi-electron atom, which of the following orbitals described by the three quantum numbers will have the same energy in the absence of magnetic and electric fields?

| | $n$ | $l$ | $m_l$ | | $n$ | $l$ | $m_l$ |
|---|---|---|---|---|---|---|---|
| a. | 2 | 0 | 0 | b. | 2 | 1 | 1 |
| c. | 3 | 2 | 1 | d. | 3 | 2 | 0 |

**18** Which of the following statements is/are correct?

a. The electronic configuration Cr is [Ar] $3d^5, 4s^1$

b. Spin quantum number of K (19) can have value $-\frac{1}{2}$

c. A $p$-orbital can have maximum of two electrons

d. A $s$-orbital can have maximum of two electron

**19** Select the correct configuration(s) of the given species.
If Aufbau rule is not followed

a. $M^{2+}$ : (electrons = 24) : [Ar] $4s^1 3d^5$

b. $M$ (electrons = 24) : [Ar] $4s^1 3d^5$

c. $M^+$ (electrons = 24) : [Ar] $4s^1 3d^5$

d. $M$ (electrons = 19) : [Ar] $3d^1$

## Comprehension Type

**Direction** *(Q. Nos. 20 and 21) This section contains a paragraph, describing theory, experiments, data, etc. Two questions related to the paragraph have been given. Each question has only one correct answer among the four given options (a), (b), (c) and (d).*

Assume that azimuthal quantum number could have value of $n$ also [in addition to normal value 0, 1, 2, ..., $(n-1)$].

**20** Electronic configuration of vanadium ($Z = 23$) under the above condition would have been

a. $1s^2, 2s^2, 2p^6, 3s^2, 3p^6, 4s^2, 3d^3$

b. $1s^2, 1p^6, 2s^2, 2p^6, 2d^7$

c. $1s^2, 1p^6, 2s^2, 2p^6, 3s^2, 2d^5$

d. $1s^2, 1p^6, 2s^2, 2p^6, 3s^1, 2d^6$

**21** Under the condition given above, magnetic moment of Ti ($Z = 22$) would have been

a. $\sqrt{24}$ BM  b. $\sqrt{48}$ BM  c. $\sqrt{35}$ BM  d. 0

## Matching List Type

*Working Space*

***Direction** (Q. No. 22) Choices for the correct combination of elements from Column I and Column II are given as options (a), (b), (c) and (d), out of which one is correct.*

**22** Match the entries in Column I with correctly related quantum number(s) in Column II **[IITJEE 2008]**

| | Column I | | Column II |
|---|---|---|---|
| i. | Orbital angular momentum of the electron in a hydrogen-like atomic orbital | p. | Principal quantum number ($n$) |
| ii. | A hydrogen-like one electron wave function obeying Pauli's exclusion principle. | q. | Azimuthal quantum number ($l$) |
| iii. | Shape, size and orient tation of hydrogen-like atomic orbitals. | r. | Magnetic quantum number ($m$) |
| iv. | Probability density of electron at the nucleus in hydrogen like atom | s. | Electron spin quantum number ($s$) |

**Codes**

| | i | ii | iii | iv |
|---|---|---|---|---|
| a. | p,r | s | p,q | p |
| b. | q | p,q,r,s | p,q,r | p,q,r |
| c. | p,r,s | p,s | q,r,s | r,s |
| d. | p,q | p,r | r,s | p,s |

## One Integer Value Correct Type

***Direction** (Q. Nos. 23 and 24) This section contains 2 questions. When worked out will result in an integer from 0 to 9 (both inclusive).*

**23** In an atom, the total number of electrons having quantum number $n = 4, |m_l| = 1$ and $m_s = -\frac{1}{2}$, is .......... **[JEE Advanced 2014]**

**24** The maximum number of electrons that can have principal quantum number, $n = 3$ and spin quantum number, $m_s = -\frac{1}{2}$, is ........ . **[IIT JEE 2011]**

# DPP-5 Quantum Mechanical Model of an Atom

## Only One Option Correct Type

***Direction** (Q. Nos. 1-12) This section contains 12 multiple choice questions. Each question has four choices (a), (b), (c) and (d), out of which ONLY ONE option is correct.*

*Working Space*

**1** Radial wave functions ($R$) of different orbitals are plotted. Which is/are correct graphs?

a. 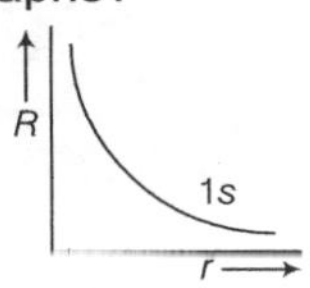


b. 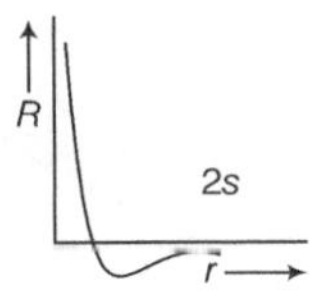


c. 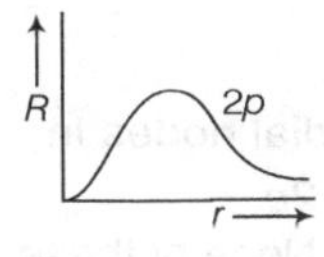


d. All of these

**2** For 2$s$-orbital electron, radial probability density $R^2$ as function of $r$ (distance) is given by

a. 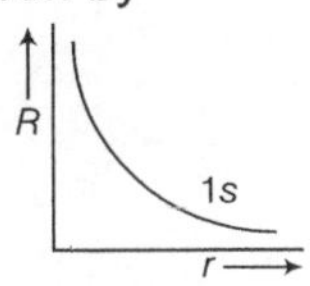


b. 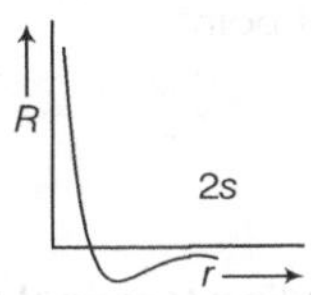


c. 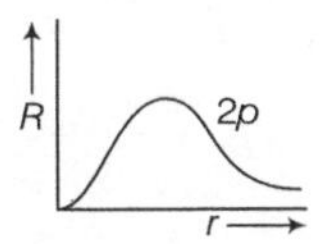


d. All of these

**3** For an electron in 2$p$-orbital, radial probability function $4\pi r^2 R^2$ as a function of $r$ is given by

a. 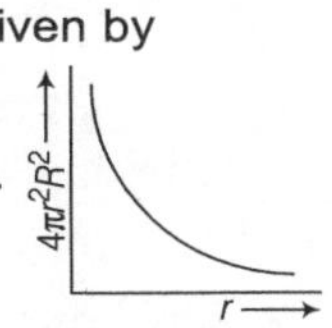


b. 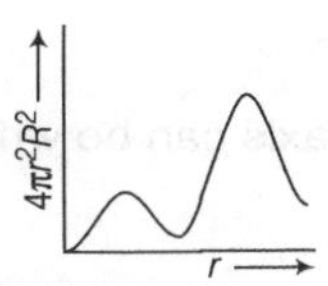


c. 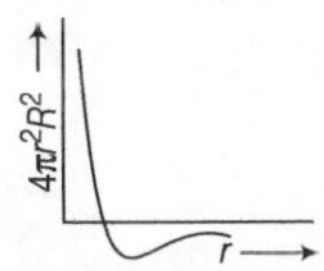


d. 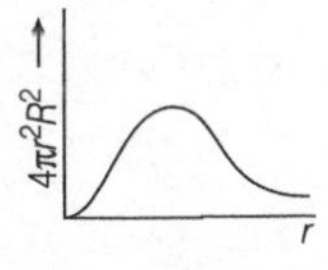


**4** Consider the following figures $A$ and $B$ indicating distribution of charge density (electron probability $\psi^2$) with distance $r$

*A.* 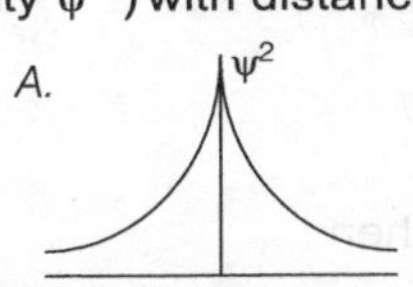


*B.* 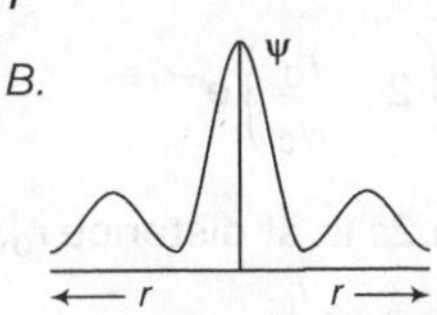


Select the correct statement(s).

a. Both $A$ and $B$ are for 1$s$

b. Both $A$ and $B$ are for 2$s$

c. $A$ is for 2$s$ and $B$ is for 1$s$

d. $A$ is for 1$s$ and $B$ is for 2$s$

Working Space

5 Select the correct statement(s).

a. $|\psi^2|$ measures the electron probability density at point in an atom

b. $\psi$ and $|\psi^2|$ vary as a function of the three coordinates $r$ (radial), $\theta$ and $\phi$ (angular part)

c. Angular wave function '$\theta\phi$' depends an only $l$ and $m_l$ and is independent of $n$

d. All of the above are correct statements

6 Which orbital gives an electron, a greater probability being found close to the nucleus?

a. 3*p* b. 3*d*

c. 3*s* d. Equal

7 Hamiltonian operator $\hat{H}$ is the sum of two energy operators. These are

a. mechanical and potential

b. kinetic and mechanical

c. kinetic and potential

d. thermal and potential

8 Angular nodes in 4*s*- suborbit is equal to radial nodes in

a. 3*d* b. 2*p*

c. Both of (a) and (b) d. None of these

9 $\psi^2$ at any point is proportional to

a. the probability of finding the particle at that point

b. energy of the electron

c. radius of the orbital

d. $(\text{radius})^{-1}$ of the orbital

10 There is formation of in all five nodes (including terminal nodes) in a string of 12 cm. Thus, wavelength of the waves formed is

a. 6.0 cm b. 4.8 cm

c. 2.4 cm d. 1.2 cm

11 For *x*-axis, wave function ($\psi$) can be written as

$$\psi = A\sin\left(\frac{2\pi x}{\lambda}\right)$$

Thus, Schrodinger wave equation along *x*-axis can be written as :

a. $\frac{d^2\psi}{dx^2} - \frac{4\pi^2\psi}{\lambda^2} = 0$

b. $\frac{d^2\psi}{dx^2} + \frac{2\pi\psi}{\lambda} = 0$

c. $\frac{d^2\psi}{dx^2} + \frac{4\pi^2\psi}{\lambda^2} = 0$

d. None of the above

12 The Schrodinger wave equation for H-atom is

$$\psi_{2s} = \frac{1}{4(2\pi)^{1/2}}\left(\frac{1}{a_0}\right)^{3/2}\left(2 - \frac{r_0}{a_0}\right)e^{-r/a_0}$$

where, $a_0$ is Bohr's radius. If radial node in 2*s* is at distance $r_0$, then

a. $a_0 = r_0$ b. $a_0 = \frac{r_0}{2}$

c. $r_0 = \frac{a_0}{2}$ d. $r_0^2 = a_0$

Working Space

## One or More than One Options Correct Type

*Direction (Q. Nos. 13-15) This section contains 3 multiple choice questions. Each question has four choices (a), (b), (c) and (d), out of which ONE or MORE THAN ONE are correct.*

**13** Which of the following properties can be described by wave function $\psi$?

a. Type of orbital occupied by electron
b. Energy of the electron in the orbital
c. Shape of the orbital
d. Probability of finding the electron in any particular region within the orbital

**14** As compared to 1*s* electron of H-atom in ground state, which of the following properties appear(s) in the radial probability density of 2*s* electron of H-atom in first excited state?

a. Spherical node appears
b. Electron charge density is highest in the vicinity of the nucleus
c. Electron density drops to zero after maximum probability is reached
d. Electron density rises to second highest value

**15** Radial probability density in the occupied orbital of a hydrogen atom in the ground state (1*s*) is given below

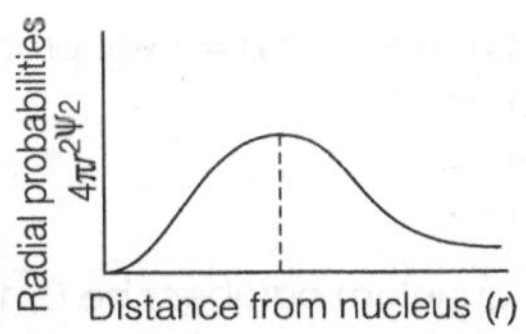


Select the correct alternate(s).

a. Radial probability is maximum when $(r = a_0)$
b. Radial probability is maximum when $\left(r = \frac{1}{a_0}\right)$
c. Radial probability almost falls to zero when, $r > 1$Å
d. Radial probability is independent of $r$

SET - 1

# Revisal Problems for JEE Main

## Only One Option Correct Type

Working Space

***Direction*** *(Q. Nos. 1-24) This section contains 24 multiple choice questions. Each question has four choices (a), (b), (c) and (d), out of which ONLY ONE option is correct.*

**1** Wavelength associated with water molecule moving with a velocity of $500 \text{ ms}^{-1}$ is

a. 44.3 pm
b. 443 pm
c. 0.0443 pm
d. 0.443 pm

**2** Assume H-atom ($r_H = 0.0529$ nm) and proton ($r_p = 1.5 \times 10^{-15}$ m) to be spherical. Fraction of the space in an atom of hydrogen that is occupied by the nucleus is

a. $2.3 \times 10^{-14}$
b. $1.6 \times 10^{-14}$
c. $2.3 \times 10^{-12}$
d. $1.6 \times 10^{-12}$

**3** What electronic transition in a hydrogen atom, starting from the orbit $n = 7$ will produce infrared light of wavelength 2170 nm? ($\bar{R}_H = 1.09677 \times 10^7 \text{ m}^{-1}$)

a. $n = 7$ to $n = 5$
b. $n = 7$ to $n = 4$
c. $n = 7$ to $n = 3$
d. $n = 7$ to $n = 2$

**4** If for a given value of $n$, perimissible values of azimuthal quantum number $l$ be 0, 1, 2, ..., $n$, then magnetic moment of chromium (24) would be

a. $\sqrt{15}$ BM
b. 0
c. $\sqrt{48}$ BM
d. $\sqrt{24}$ BM

**5** Velocity of the electron in H-atom in second orbit is

a. $\frac{4\pi m a_0}{h}$
b. $\frac{h}{\pi m a_0}$
c. $\frac{h}{4\pi m a_0}$
d. $\frac{4h}{\pi m a_0}$

**6** If uncertainties in the measurements of position and momentum are equal, then uncertainty in the measurement of velocity is

a. $\frac{1}{2m}\sqrt{\frac{h}{\pi}}$
b. $\frac{1}{4m}\sqrt{\frac{h}{\pi}}$
c. $\frac{1}{2m}\sqrt{\frac{\pi}{h}}$
d. $\frac{1}{4m}\sqrt{\frac{\pi}{h}}$

**7** For the two moving particles $A$ and $B$, following values are given

| Particle | Velocity | Mass |
|---|---|---|
| $A$ | $0.1 \text{ ms}^{-1}$ | $m$ kg |
| $B$ | $0.05 \text{ ms}^{-1}$ | $5m$ kg |

Ratio of de-Broglie wavelength associated with the particles $A$ and $B$ is

a. 2 : 5
b. 3 : 4
c. 6 : 4
d. 5 : 2

**8** The wave number of the spectral line in the emission spectrum of H-atom will be equal to $\frac{8}{9}$ times the Rydberg's constant if the electron jumps from

a. $n = 3$ to $n = 1$
b. $n = 10$ to $n = 1$
c. $n = 9$ to $n = 1$
d. $n = 2$ to $n = 1$

**9** Which one of the following constitutes a group of the isoelectronic species?

a. $C_2^{2-}$, $O_2^-$, CO, NO
b. $NO^+$, $C_2^{2-}$, $CN^-$, $N_2$
c. $CN^-$, $N_2$, $O_2^{2-}$, $CO_3^{2-}$
d. $N_2$, $O_2^-$, $NO^+$, CO

*Working Space*

**10** In an experiment reproducing the measurements of Rutherford and his co-workers, $22 \times 10^{-3}$ mg of He gas was collected in one year from a sample of radium. This sample was observed to emit $1.06 \times 10^{11}$ $\alpha$-particles per second. Thus, Avogadro's number is

a. $5.92 \times 10^{23}$ mol$^{-1}$  b. $6.92 \times 10^{23}$ mol$^{-1}$
c. $6.08 \times 10^{23}$ mol$^{-1}$  d. $6.58 \times 10^{-24}$ mol$^{-1}$

**11** Medical experts generally consider a lead level of 30 $\mu$g Pb per (dL) of blood to pose a significant health risk (Pb = 208). Number of lead atoms per cm$^3$ of blood is

a. $8.68 \times 10^{16}$  b. $8.68 \times 10^{14}$
c. $8.68 \times 10^{12}$  d. $8.68 \times 10^{10}$

**12** In Schrodinger wave equation given below

$$\nabla^2\psi + \frac{8\pi^2 m}{h^2}(E - U)\psi = 0$$

$\nabla^2$ and $U$ are

| | $\nabla^2$ | $U$ |
|---|---|---|
| a. | $\frac{d^2}{dx^2} + \frac{d^2}{dy^2} + \frac{d^2}{dz^2}$ | $\frac{Ze^2}{r}$ |
| b. | $\frac{d^2}{dx^2} + \frac{d^2}{dy^2} + \frac{d^2}{dz^2}$ | $-\frac{Ze^2}{r}$ |
| c. | $\frac{d}{dx} + \frac{d}{dy} + \frac{d}{dz}$ | $-\frac{Ze^2}{r}$ |

d. None of the above is correct

**13** Which hydrogen like species will have same radius as that of Bohr's orbit of hydrogen atom?

a. $n = 2$, $Li^{2+}$  b. $n = 2$, $Be^{3+}$
c. $n = 2$, $He^{+}$  d. $n = 3$, $Li^{2+}$

**14** An atom for which the electron probability distribution is spherically symmetric about the nucleus is

a. $2p$  b. $3s$
c. $3p$  d. $3d$

**15** Which of the following statements in relation to the hydrogen atom is correct?

a. $3s$, $3p$ and $3d$ orbitals all have the same energy
b. $3s$ and $3p$ orbitals are of lower energy than $3d$ orbital
c. $3p$ orbital is lower in energy than $3d$ orbital
d. $3s$ orbital is lower in energy than $3s$ orbital

**16** Which of the following options does not represent ground state EC of an atom?

a. [Ar] $4s^2 3d^8$  b. [Ar] $4s^2 3d^9$
c. [Ar] $4s^1 3d^{10}$  d. [Ar] $4s^1 3d^5$

**17** **Statement I** The electronic configuration of N-atom is represented as :

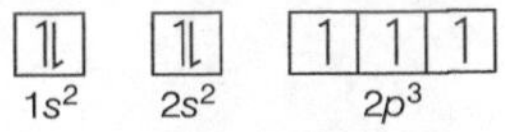

and not as 

**Statement II** The electronic configuration of the ground state of an atom is the one which has the greatest multiplicity.

a. Both Statement I and Statement II are correct and Statement II is the correct explanation of Statement I
b. Both Statement I and Statement II are correct and Statement II is not the correct explanation of Statement I
c. Statement I is correct but Statement II is incorrect
d. Statement II is correct but Statement I is incorrect

**Working Space**

**18** If $n$ and $l$ are respectively the principal and azimuthal quantum numbers, then the expression for calculating the total number of electrons in any energy level is

a. $\sum_{l=1}^{l=n} 2(2l+1)$
b. $\sum_{l=1}^{l=n-1} 2(2l+1)$
c. $\sum_{l=0}^{l=n+1} 2(2l+1)$
d. $\sum_{l=0}^{l=(n-1)} 2(2l+1)$

**19** 3$d$-suborbit first occurs in fourth orbit. It has

a. two angular nodes
b. zero radial nodes
c. two total nodes
d. All of the above are correct

**20** How fast is an electron moving if it has a wavelength equal to the distance it travels in one second?

a. $\sqrt{\frac{h}{m}}$
b. $\sqrt{\frac{m}{h}}$
c. $\sqrt{\frac{h}{p}}$
d. $\sqrt{\frac{h}{2\,KE}}$

**21** In the presence of magnetic field, $d$-suborbit is

a. 5-fold degenerate
b. 3-fold degenerate
c. 7-fold degenerate
d. non-degenerate

**22** Electrons are used to determine the structure of crystal surfaces. To have diffraction, the wavelength $\lambda$, of the electrons should be of the order of the lattice constant, which is typically 0.30 nm. What energy do such electrons have?

a. $2.67 \times 10^{-18}$ J
b. 16.72 eV
c. $2.67 \times 10^{-21}$ kJ
d. All of the above are correct

**23** The number of $d$-electrons in $Fe^{2+}$ (24 electrons) is not equal to

a. $p$-electrons in Ne (10)
b. $s$-electrons in Mg (12)
c. $d$-electrons in Fe (26)
d. $p$-electrons in Cl (17)

**24** Bohr's theory is applicable to the following set of species

a. $H^+$, $Li^{2+}$, $He^+$, $Be^{3+}$
b. $H^+$, $Li^+$, $He^+$, $Be^+$
c. H, $He^+$, $Li^{2+}$, $Be^{3+}$
d. For all these sets

SET - 2

# Revisal Problems
## for JEE Advanced

### Only One Option Correct Type

Working Space

**Direction** *(Q. Nos. 1-10) This section contains 10 multiple choice questions. Each question has four choices (a), (b), (c) and (d), out of which ONLY ONE option is correct.*

**1** Energy for $7.25 \times 10^{15}$ photons of $5.37 \times 10^{14}\ s^{-1}$ frequency in Einstein unit is

a. $1.20 \times 10^{-8}$
b. $2.58 \times 10^{-3}$
c. $3.56 \times 10^{-19}$
d. $8.33 \times 10^{7}$

**2** What is the minimum amount of time required to measure the kinetic energy of an electron with a speed of $(100 \pm 1)\ ms^{-1}$

a. 3.0 ms
b. 3.0 ns
c. 3.0 μs
d. 0.3 μs

**3** When a certain metal was irradiated with light of frequency $1.6 \times 10^{16}$ Hz, the photoelectrons emitted had twice the kinetic energy as did photoelectrons emitted when the same metal was irradiated with light of frequency $1.0 \times 10^{16}$ Hz. Thus, threshold frequency for the metal is

a. $1.0 \times 10^{15}$ Hz
b. $2.0 \times 10^{5}$ Hz
c. $4.0 \times 10^{15}$ Hz
d. $6.0 \times 10^{15}$ Hz

**4** Consider the following dissociation of $O_2$ (dissociation energy 498 kJ $mol^{-1}$)

$$O_2 \xrightarrow{h\nu} O + O^*$$

$O^*$ is more energetic than normal oxygen atom (O) by 1.967 eV. Thus, maximum wavelength for photochemical dissociation is

a. 174 nm
b. 398 nm
c. 174 pm
d. 398 pm

**5** Which of the following pairs of orbitals have electron density along the axis?

a. $d_{yz}, d_{xz}$
b. $d_{x^2-y^2}, d_{z^2}$
c. $d_{xy}, d_{yz}$
d. $d_{xy}, d_{z^2}$

**6** The normalised wave function of 1s orbital is $\psi = \sqrt{N}\, e^{-zr/a_0}$ and the radial distribution function $= 4\pi r^2 \psi^2$ where, $N = \frac{1}{\pi}\left(\frac{Z}{a_0}\right)^3$

Most probable distance at which the 1s electron of H-like atom with atomic number $Z$ is to be found, is

a. $a_0$
b. $\frac{a_0}{Z}$
c. $\frac{Z}{a_0}$
d. $\frac{a_0^2}{Z}$

**7** For any axis (say x-axis) Schrodinger wave equation is represented as $\frac{d^2\psi}{dx^2} + \frac{4\pi^2}{\lambda^2}\psi = 0$.

Its solution can be

a. $\psi = A \sin\left(\frac{2\pi x}{\lambda}\right)$
b. $\psi = Ae^{-2\pi x/\lambda}$
c. Both a and b
d. None of these

Working Space

**8** If a 1.00 g body is travelling along the X-axis at $100 \text{ cm s}^{-1}$ within $1 \text{ cm s}^{-1}$, then uncertainty in its position is

a. $2.636 \times 10^{-30}$ m  b. $5.272 \times 10^{-30}$ m
c. $1.318 \times 10^{-30}$ m  d. $6.590 \times 10^{-31}$ m

**9** Iodine molecule dissociates into atoms after absorbing light of 4500 Å. If one quantum of radiation is absorbed by each molecule, kinetic energy of each iodine atom is (Bond energy of $I_2 = 240 \text{ kJ mol}^{-1}$)

a. $4.26 \times 10^{-20}$ J  b. $2.130 \times 10^{-20}$ J
c. $3.984 \times 10^{-19}$ J  d. $1.992 \times 10^{-9}$ J

**10** Quantum numbers of the electrons are given

I. $n = 4, m_s = -\frac{1}{2}$  II. $n = 3, l = 0$
III. $n = 2, l = 1$  IV. $n = 2, m_l = 0$

How many electrons in an atom may have these quantum numbers?

| | I | II | III | IV |
|---|---|---|---|---|
| a. | 16 | 2 | 6 | 8 |
| b. | 32 | 18 | 8 | 8 |
| c. | 16 | 6 | 8 | 6 |
| d. | 16 | 2 | 6 | 4 |

## Statement Type

***Direction*** *(Q. Nos. 11-13) This section is based on Statement I and Statement II. Select the correct answer from the codes given below.*

a. Both Statement I and Statement II are correct and Statement II is the correct explanation of Statement I
b. Both Statement I and Statement II are correct and Statement II is not the correct explanation of Statement I
c. Statement I is correct but Statement II is incorrect
d. Statement II is correct but Statement I is incorrect

**11** **Statement I** On heating a metal for a longer time, radiations become white and then blue as the temperature increases.
**Statement II** Radiations emitted go from a lower frequency to higher frequency as the temperature increases.

**12** **Statement I** Black body is an ideal body that emits and absorbs radiations of all frequencies.
**Statement II** The frequency of radiations emitted by a body goes from a lower frequency to higher frequency with an increase in temperature.

**13** **Statement I** Electrons are negatively charged.
**Statement II** The application of electric and magnetic field deflected the rays in the discharge tube towards the positive plate.

## One or More than One Options Correct Type

***Direction*** *(Q. Nos. 14-16) This section contains 3 multiple choice questions. Each question has four choices (a), (b), (c) and (d), out of which ONE or MORE THAN ONE are correct.*

**14** Spectral series of H-atom given $\overline{\nu}_{min}$ and $\overline{\nu}_{max}$ depending on the types of transition. Which is/are correct matchings?

| | Series | $\overline{\nu}_{max}$ | $\overline{\nu}_{min}$ |
|---|---|---|---|
| a. | Lyman | $\overline{R}_H$ | $\frac{3}{4}\overline{R}_H$ |
| b. | Balmer | $\overline{R}_H / 4$ | $\frac{5}{36}\overline{R}_H$ |
| c. | Paschen | $\overline{R}_H / 9$ | $\frac{7}{144}\overline{R}_H$ |
| d. | Brackett | $\overline{R}_H / 16$ | $\frac{9}{400}\overline{R}_H$ |

Working Space

**15** Velocity of the electron in third orbit of $Li^{2+}$ is equal to velocity of the electron in

a. second orbit of $He^+$
b. first orbit of H
c. third orbit of $Be^{3+}$
d. first orbit of $He^+$

**16** If the quantum number $l$ could have the value of $n$ also then

a. Sc ($Z = 21$) would have configuration as $1s^2 1p^6 2s^2 2p^6 2d^3 3s^2$
b. Fe ($Z = 26$) would have two unpaired electrons
c. $Fe^{2+}$ and $Fe^{3+}$ would be coloured ion
d. $Cu^+$ would be coloured and paramagnetic

## Comprehension Type

**Direction** *(Q. Nos. 17-22) This section contains 3 paragraphs, each describing theory, experiments, data etc. Six questions related to the paragraphs have been given. Each question has only one correct answer among the four given options (a), (b), (c) and (d).*

### Passage I

The ionisation energy of hydrogen atom is 13.6 eV and the first ionisation energy of helium is 24.6 eV.

**17** What is the first ionisation energy of $He^+$ (isoelectronic of H-atom)?

a. 13.6 eV
b. 24.6 eV
c. 38.2 eV
d. 54.4 eV

**18** What is the enthalpy change of the following process

$$He^{2+}(g) + 2e^- \longrightarrow He(g)$$

a. −13.6 eV
b. −24.6 eV
c. −79.0 eV
d. −38.2 eV

### Passage II

The position and momentum of electron of energy 0.5 keV are simultaneously determined. Its position is located within 0.2 nm.

**19** Uncertainty in momentum is

a. $1.3 \times 10^{-28}$ kg ms$^{-1}$
b. $1.3 \times 10^{-26}$ kg ms$^{-1}$
c. $2.6 \times 10^{-26}$ kg ms$^{-1}$
d. $2.6 \times 10^{-28}$ kg ms$^{-1}$

**20** Percentage uncertainty in momentum is

a. 2.2%
b. 0.022%
c. 1.1%
d. 0.011%

### Passage III

Yellow lamp of 100 W emits radiations of wavelength 560 nm.

**21** Number of photons emitted by this lamp in 10 s is

a. $2.82 \times 10^{21}$
b. $2.82 \times 10^{19}$
c. $1.12 \times 10^{21}$
d. $2.22 \times 10^{21}$

**22** Time taken to produce 1 mole of the photons by this lamp is

a. 40 min
b. 36 min
c. 12 min
d. 340 min

## One Integer Value Correct Type

**Direction** (Q. No. 23-26) *This section contains 4 question. When worked out will result in an integer from 0 to 9 (both inclusive).*

**23** What is the maximum number of emission lines when the excited electron of a H-atom in $n = 4$ drops to the ground state?

**24** Ionisation energy of H-atom in the ground state is 25 times the ionisation energy when it is in $n$th state. What is value of $n$?

**Working Space**

**25** Emission transitions in the Paschen series end at orbit $n = 3$ and start from orbit $n$ and can be represented as
$\nu$ (frequency) $= 3.29 \times 10^{15}\ s^{-1}\left[\frac{1}{3^2} - \frac{1}{n^2}\right]$.
Calculate value of $n$ if, the transition is observed at 1285 nm.

**26** i. The atomic mass of He and Ne are 4 and 20 amu, respectively. The value of the de-Broglie wavelength of He gas at −73°C is '$m$' times that of the de-Broglie wavelength of Ne at 727°C.$m$ is ...... **[JEE Advanced 2013]**

ii. The value of the de-Broglie wavelength of He gas at 100 K is 5 times that of the de-Broglie wavelength gas $X$, at 1250 K. What is atomic mass of $X$?

## Matching List Type

***Direction** (Q. Nos. 27 and 28) Choices for the correct combination of elements from Column I and Column II are given as options (a), (b), (c) and (d), out of which one is correct.*

**27** Match the name of units given in Column I with their relation (in terms of physical constants) given in Column II.

| | Column I | | Column II |
|---|---|---|---|
| i. | Hartree (atomic unit) | p. | $e^2 / 8\pi ch\varepsilon_0 a_0$ |
| ii. | Molar atomic unit | q. | $e / 4\pi\varepsilon_0 a_0$ |
| iii. | Atomic unit (Hertz) | r. | $e^2 / 4\pi h\varepsilon_0 a_0$ |
| iv. | Atomic unit (eV) | s. | $N_0 e^2 / 4\pi h\varepsilon_0 a_0$ |
| v. | Rydberg constant | t. | $e^2 / 4\pi\varepsilon_0 a_0$ |

**Codes**

| | i | ii | iii | iv | v |
|---|---|---|---|---|---|
| a. | t | s | r | q | p |
| b. | t | s | q | r | p |
| c. | s | t | r | q | p |
| d. | p | q | r | t | s |

**28** Match the species in Column I with their related properties in Column II.

| | Column I | | Column II |
|---|---|---|---|
| i. | N(7) | p. | Paramagnetic |
| ii. | $Co^{2+}$ (27) | q. | Coloured ion |
| III. | $Mn^{2+}$ (25) | r. | Sum of spin 1.5 |
| iv. | $Fe^{2+}$ (26) | s. | Magnetic moment $\sqrt{24}$ BM |

**Codes**

| | i | ii | iii | iv |
|---|---|---|---|---|
| a. | p,r | q,s | r,s | p,r,s |
| b. | p | q,r | p,s | s |
| c. | p,r | p,q,r | p,q | p,q,s |
| d. | r | s | r,s | p,s |

# JEE Main & AIEEE Archive

## (Compilation of Last 13 Years Questions)

**1** The correct set of four quantum numbers for the valence electrons of rubidium atom $[Z = 37]$ is **(JEE Main 2014)**

a. $5, 0, 0, +\frac{1}{2}$ b. $5, 1, 0, +\frac{1}{2}$

c. $5, 1, 1, +\frac{1}{2}$ d. $5, 0, 1, +\frac{1}{2}$

**2** Energy of an electron is given by $E = -2.178 \times 10^{-18}\ \text{J}\left(\frac{Z^2}{n^2}\right)$. Wavelength of light required to excite an electron in an hydrogen atom from level $n = 1$ to $n = 2$ will be ($h = 6.62 \times 10^{-34}$ Js and $c = 3.0 \times 10^8\ \text{ms}^{-1}$) **(JEE Main 2013)**

a. $1.214 \times 10^{-7}$ m b. $2.816 \times 10^{-7}$ m

c. $6.500 \times 10^{-7}$ m d. $8.500 \times 10^{-7}$ m

**3** The electrons identified by quantum numbers $n$ and $l$

(1) $n = 4, l = 1$ (2) $n = 4, l = 0$

(3) $n = 3, l = 2$ (4) $n = 3, l = 1$

can be placed in the order of increasing energy as **(AIEEE 2012)**

a. (3) < (4) < (2) < (1)

b. (4) < (2) < (3) < (1)

c. (2) < (4) < (1) < (3)

d. (1) < (3) < (2) < (4)

**4** A gas absorbs photon of 355 nm and emits at two wavelengths. If one of the emission is at 680 nm, the other is at **(AIEEE 2011)**

a. 1035 nm b. 325 nm

c. 743 nm d. 518 nm

**5** The frequency of light emitted for the transition $n = 4$ to $n = 2$ of $He^+$ is equal to the transition in H atom corresponding to which of the following? **(AIEEE 2011)**

a. $n = 3$ to $n = 1$ b. $n = 2$ to $n = 1$

c. $n = 3$ to $n = 2$ d. $n = 4$ to $n = 3$

**6** The energy required to break one mole of Cl—Cl bonds in $Cl_2$ is 242 kJ mol$^{-1}$. The longest wavelength of light capable of breaking a single Cl—Cl bond is **(AIEEE 2010)**

a. 594 nm b. 640 nm

c. 700 nm d. 494 nm

**7** Ionisation energy of $He^+$ is $19.6 \times 10^{-18}$ J atom$^{-1}$. The energy of the first stationary state ($n = 1$) of $Li^{2+}$ is **(AIEEE 2010)**

a. $4.41 \times 10^{-16}$ J atom$^{-1}$

b. $-4.41 \times 10^{-17}$ J atom$^{-1}$

c. $-2.2 \times 10^{-15}$ J atom$^{-1}$

d. $8.82 \times 10^{-17}$ J atom$^{-1}$

**8** In an atom, an electron is moving with a speed of 600 m/s with an accuracy of 0.005%. Certainity with which the position of the electron can be located is ($h = 6.6 \times 10^{-34}$ kg m$^2$ s$^{-1}$, mass of electron, $e_m = 9.1 \times 10^{-31}$ kg) **(AIEEE 2009)**

a. $1.52 \times 10^{-4}$ m b. $5.10 \times 10^{-3}$ m

c. $1.92 \times 10^{-3}$ m d. $3.84 \times 10^{-3}$ m

**9** Calculate the wavelength (in nanometer) associated with a proton moving at $1.0 \times 10^3$ ms$^{-1}$ (Mass of proton = $1.67 \times 10^{-27}$ kg and $h = 6.63 \times 10^{-34}$ Js) **(AIEEE 2009)**

a. 0.032 nm b. 0.40 nm

c. 2.5 nm d. 14.0 nm

**10** The ionisation enthalpy of hydrogen atom is $1.312 \times 10^6$ J mol$^{-1}$. The energy required to excite the electron in the atom from $n_1 = 1$ to $n = 2$ is **(AIEEE 2008)**

a. $8.51 \times 10^5$ J mol$^{-1}$ b. $6.56 \times 10^5$ J mol$^{-1}$

c. $7.56 \times 10^5$ J mol$^{-1}$ d. $9.84 \times 10^5$ J mol$^{-1}$

**11** Which of the following sets of quantum numbers represents the highest energy of an atom? **(AIEEE 2007)**

a. $n = 3, l = 1, m = 1, s = +1/2$

b. $n = 3, l = 2, m = 1, s = +1/2$

c. $n = 4, l = 0, m = 0, s = +1/2$

d. $n = 3, l = 0, m = 0, s = +1/2$

**12** Which of the following nuclear reactions will generate an isotope? **(AIEEE 2007)**

a. Neutron particle emission

b. Positron emission

c. $\alpha$-particle emission

d. $\beta$-particle emission

**13** Uncertainty in the position of an electron (mass = $9.1 \times 10^{-31}$ kg) moving with a velocity 300 m s$^{-1}$, accurate upto 0.001% will be ($h = 6.63 \times 10^{-34}$ Js) **(AIEEE 2006)**

a. $19.2 \times 10^{-2}$ m b. $5.76 \times 10^{-2}$ m

c. $1.92 \times 10^{-2}$ m d. $3.84 \times 10^{-2}$ m

**14** According to Bohr's theory, the angular momentum of an electron in 5th orbit is **(AIEEE 2006)**

a. $25\frac{h}{\pi}$ b. $1.0\frac{h}{\pi}$

c. $10\frac{h}{\pi}$ d. $2.5\frac{h}{\pi}$

**15** Which of the following statements in relation to the hydrogen atom is correct? **(AIEEE 2005)**

a. 3s, 3p and 3d orbitals all have the same energy

b. 3s and 3p orbitals are of lower energy than 3d orbital

c. 3p orbital is lower in energy than 3d orbital

d. 3s orbital is lower in energy than 3p orbital

**16** In a multielectron atom, which of the following orbitals described by the three quantum numbers will have the same energy in the absence of magnetic and electric fields? **(AIEEE 2005)**

(A) $n = 1, l = 0, m = 0$
(B) $n = 2, l = 0, m = 0$
(C) $n = 2, l = 1, m = 1$
(D) $n = 3, l = 2, m = 1$
(E) $n = 3, l = 2, m = 0$

a. (D) and (E) b. (C) and (D)
c. (B) and (C) d. (A) and (B)

**17** Of the following sets, which one does not contain isoelectronic species? **(AIEEE 2005)**

a. $BO_3^{3-}, CO_3^{2-}, NO_3^-$ b. $SO_3^{2-}, CO_3^{2-}, NO_3^-$
c. $CN^-, N_2, C_2^{2-}$ d. $PO_4^{3-}, SO_4^{2-}, ClO_4^-$

**18** The wavelength of the radiation emitted, when in a hydrogen atom electron falls from infinity to stationary state 1, would be (Rydberg constant $= 1.097 \times 10^7 m^{-1}$) **(AIEEE 2004)**

a. 91 nm b. 192 nm
c. 406 nm d. $9.1 \times 10^{-8}$ nm

**19** Which of the following sets of quantum numbers is correct for an electron in 4*f* orbital? **(AIEEE 2004)**

a. $n = 4, l = 3, m = +4, s = +1/2$
b. $n = 4, l = 4, m = -4, s = -1/2$
c. $n = 4, l = 3, m = +1, s = +1/2$
d. $n = 3, l = 2, m = -2, s = +1/2$

**20** Consider the ground state of Cr atom ($Z = 24$). The numbers of electrons with the azimuthal quantum numbers, $l = 1$ and 2 are, respectively **(AIEEE 2004)**

a. 12 and 4 b. 12 and 5
c. 16 and 4 d. 16 and 5

**21** Which one of the following sets of ions represents the collection of isoelectronic species? **(AIEEE 2004)**

(Atomic numbers of F = 9, Cl = 17, Na = 11, Mg = 12, Al = 13, K = 19, Ca = 20, Sc = 21)

a. $K^+, Ca^{2+}, Sc^{3+}, Cl^-$
b. $Na^+, Ca^{2+}, Sc^{3+}, F^-$
c. $K^+, Cl^-, Mg^{2+}, Sc^{3+}$
d. $Na^+, Mg^{2+}, Al^{3+}, Cl^-$

**22** The de-Broglie wavelength of a tennis ball of mass 60 g moving with a velocity of 10 m/s is approximately **(AIEEE 2003)**

(Planck's constant, $h = 6.63 \times 10^{-34}$ Js)

a. $10^{-33}$ m b. $10^{-31}$ m
c. $10^{-16}$ m d. $10^{-25}$ m

**23** In Bohr series of lines of hydrogen spectrum, the third line from the red end corresponds to which one of the following inner-orbit jumps of the electron for Bohr orbits in an atom of hydrogen? **(AIEEE 2003)**

a. $3 \rightarrow 2$ b. $5 \rightarrow 2$
c. $4 \rightarrow 1$ d. $2 \rightarrow 5$

**24** The number of *d*-electrons retained in $Fe^{2+}$ (atomic number Fe = 26) ion is **(AIEEE 2003)**

a. 3 b. 4
c. 5 d. 6

**25** Which one of the following groupings represents a collection of isoelectronic species? **(AIEEE 2003)**

(Atomic numbers of Cs = 55, Br = 35)

a. $Na^+, Ca^{2+}, Mg^{2+}$ b. $N^{3-}, F^-, Na^+$
c. $Be, Al^{3+}, Cl^-$ d. $Ca^{2+}, Cs^+, Br^-$

**26** The orbital angular momentum for an electron revolving in an orbit is given by $\sqrt{l(l+1)}\,\frac{h}{2\pi}$. This momentum for *s*-electron will be given by **(AIEEE 2003)**

a. $+\frac{1}{2}\cdot\frac{h}{2\pi}$ b. zero
c. $\frac{h}{2\pi}$ d. $\sqrt{2}\cdot\frac{h}{2\pi}$

**27** Uncertainty in position of a particle of 25 g in space is $10^{-5}$ m. Hence, uncertainty in velocity ($ms^{-1}$) is
(Planck's constant, $h = 6.6 \times 10^{-34}$ Js) **(AIEEE 2002)**

a. $2.1 \times 10^{-28}$
b. $2.1 \times 10^{-34}$
c. $0.5 \times 10^{-34}$
d. $5.0 \times 10^{-24}$

**28** Energy of H-atom in the ground state is –13.6 eV hence energy in the second excited state is **(AIEEE 2002)**

a. – 6.8 eV b. – 3.4 eV
c. – 1.51 eV d. – 4.53 eV

# JEE Advanced & IIT JEE Archive

(Compilation of Last 10 Years Questions)

**1** In an atom, the total number of electrons having quantum numbers $n = 4, |m_l| = 1$ and $m_s = -\frac{1}{2}$ is **(2014 Adv., Integer Type)**

**2** The atomic masses of He and Ne are 4 and 20 amu, respectively. The value of the de-Broglie wavelength of He gas at –73°C is "$M$" times that of the de-Broglie wavelength of Ne at 727°C. $M$ is **(2013 Adv., Integer Type)**

**3** The kinetic energy of an electron in the second Bohr orbit of a hydrogen atom is [$a_0$ is Bohr radius] **(2012, Single Option Correct Type)**

a. $\frac{h^2}{4\pi^2 m a_0^2}$ b. $\frac{h^2}{16\pi^2 m a_0^2}$

c. $\frac{h^2}{32\pi^2 m a_0^2}$ d. $\frac{h^2}{64\pi^2 m a_0^2}$

**4.** The maximum number of electrons that can have principal quantum number, $n = 3$ and spin quantum number, $m_s = -1/2$, is **(2011, Integer Type)**

**5.** The work function ($\Phi$) of some metals is listed below. The number of metals which will show photoelectric effect when light of 300 nm wavelength falls on the metal is **(2011, Integer Type)**

| Metal | $\phi$ (eV) |
|---|---|
| Li | 2.4 |
| Na | 2.3 |
| K | 2.2 |
| Mg | 3.7 |
| Cu | 4.8 |
| Ag | 4.3 |
| Fe | 4.7 |
| Pt | 6.3 |
| W | 4.75 |

**Passage for Q. Nos. (6-8)**

The hydrogen-like species $Li^{2+}$ is in a spherically symmetric state $S_1$ with one radial node. Upon absorbing light the ion undergoes transition to a state $S_2$. The state $S_2$ has one radial node and its energy is equal to the ground state energy of the hydrogen atom. **(2010, Comprehension Type)**

**6** The state $S_1$ is

a. 1s b. 2s c. 2p d. 3s

**7.** Energy of the state $S_1$ in units of the hydrogen atom ground state energy is

a. 0.75 b. 1.50 c. 2.25 d. 4.50

**8** The orbtial angular momentum quantum number of the state $S_2$ is

a. 0 b. 1 c. 2 d. 3

**9** Match the entries in Column I with the correctly related quantum number(s) in Column II. **(2008, Matching Type)**

| | Column I | | Column II |
|---|---|---|---|
| i. | Orbital angular momentum of the electron in a hydrogen-like atomic orbital | p. | Principal quantum number |
| ii. | A hydrogen-like one-electron wave function obeying Pauli's principle | q. | Azimuthal quantum number |
| iii. | Shape, size and orientation of hydrogen-like atomic orbitals | r. | Magnetic quantum number |
| iv. | Probability density of electron at the nucleus in hydrogen-like atom | s. | Electron spin quantum number |

**Codes**

| | i | ii | iii | iv |
|---|---|---|---|---|
| a. | q,p | r | s | p |
| b. | q | s | p,q,r | p,q,r |
| c. | p | q | r,s | s |
| d. | q | p | p,s | r |

**10** According to Bohr's theory

$E_n$ = Total energy, $K_n$ = Kinetic energy

$V_n$ = Potential energy, $r^n$ = Radius of $n$th orbit

Match the following : **(2006, Matching Type)**

| | Column I | | Column II |
|---|---|---|---|
| i. | $V_n/K_n = ?$ | p. | 0 |
| ii. | If radius of $n$th orbit $\propto E_n^x, x = ?$ | q. | – 1 |
| iii. | Angular momentum in lowest orbital | r. | – 2 |
| iv. | $\frac{1}{r^n} \propto Z^y, y = ?$ | s. | 1 |

**Codes**

| | I | ii | iii | iv |
|---|---|---|---|---|
| a. | r | q | p | s |
| b. | p | q | s | r |
| c. | r | s | p | p |
| d. | q | s | r | p |

**11** The number of radial nodes in 3s and 2p respectively are **(2005, Single Option Correct Type)**

a. 2 and 0 b. 0 and 2

c. 1 and 2 d. 2 and 1

**12** i. Calculate velocity of electron in first Bohr orbit of hydrogen atom (Given, $r = a_0$).

ii. Find de-Broglie wavelength of the electron in first Bohr orbit

iii. Find the orbital angular momentum of 2p-orbital in terms of $h/2\pi$ units. **(2005, Subjective Type)**

# 2. Redox Reactions



DPP
DAILY PRACTICE PROBLEMS

# DPP-1 Oxidation Number

## Subjective Questions

Working Space

*Direction* (Q. Nos. 1-5) *These questions are subjective in nature, need to be solved completely on notebook.*

**1** Assign oxidation number to the underlined elements in each of the following species

i. $H_4\underline{P}_2O_7$ ii. $Ca\underline{O}_2$
iii. $H_2\underline{S}_2O_7$ iv. $KAl(\underline{S}O_4)_2$
v. $\underline{Fe}_3O_4$ vi. $\underline{C}H_3\underline{C}H_2OH$
vii. $\underline{C}H_3\underline{C}OOH$

**2** Thiosulphuric acid (a dibasic acid, $H_2S_2O_3$) contains one sulphur and three oxygen atoms directly joined to central sulphur atom. What is the oxidation number of sulphur atom?

**3** What is the oxidation number of chlorine atom in

i. $Ca(ClO_2)_2$ and ii. $CaOCl_2$?

**4** The tetrathionate anion $S_4O_6^{2-}$, is a sulphur oxoanion derived from the compound tetrathionic acid ($H_2S_4O_6$). It can be viewed as the adduct formed from the reaction between $S_2^{2-}$ and $SO_3$.

$$S_2^{2-} + 2SO_3 \longrightarrow S_4O_6^{2-}$$

Derive the value of oxidation number of sulphur atom.

**5** i. Write the various compounds of nitrogen of different oxidation number varying between −3 to + 5.

ii. Determine the change in oxidation number of the underlined atom.

a. $\underline{Cl_2} \rightarrow Cl^-$ b. $\underline{Cl_2} \rightarrow ClO_3^-$
c. $P\overline{Cl}_5 \rightarrow PCl_3 + Cl_2$ d. $\underline{N}H_4\underline{N}O_2 \rightarrow N_2 + 2H_2O$
e. $\underline{C}O_2 + 2H_2O \rightarrow HCO_3^- + H_3O^+$

## Only One Option Correct Type

*Direction* (Q. Nos. 6-17) *This section contains 12 multiple choice questions. Each question has four choices (a), (b), (c) and (d), out of which ONLY ONE option is correct.*

**6** Consider the following reactions.

I. $Zn + \text{dil.}\, H_2SO_4 \longrightarrow ZnSO_4 + H_2$
II. $Zn + \text{conc.}\, H_2SO_4 \longrightarrow ZnSO_4 + SO_2 + H_2O$
III. $Zn + HNO_3 \longrightarrow Zn(NO_3)_2 + H_2O + NH_4NO_3$

Oxidation number of H changes in

a. I, II and III b. I and II
c. II and III d. Only I

Working Space

7 In which case change of oxidation number of V is maximum?

a. $VO^{2+} \rightarrow VO_2^+$ b. $VO_2^+ \rightarrow VO^+$

c. $VO^{2+} \rightarrow V^{3+}$ d. Equal in each case

8 In which of the following set of compounds oxidation number of oxygen is not (−2)?

a. $KO_2$, $OF_2$, $H_2O_2$ b. $S_2O_8^{2-}$, $SO_4^{2-}$, $S_2O_3^{2-}$

c. $CO_2$, $CrO_3$, $P_2O_3$ d. $HNO_2$, $HNO_3$, $N_2O$

9 Which is chlorate (I) ion?

a. $ClO_2^-$ b. $ClO_3^-$ c. $ClO_4^-$ d. $ClO^-$

10 Consider the following conjugate bases of oxyacids of chlorine.

I. $ClO^-$ II. $ClO_2^-$

III. $ClO_3^-$ IV. $ClO_4^-$

Arrange these in the increasing strength of conjugate bases

a. I < II < III < IV b. IV < III < II < I

c. I = II = III = IV d. These are not conjugate bases

11 Which of the following atoms has been assigned only single oxidation number?

a. H b. O c. N d. F

12 The oxidation number of P in $Ba(H_2PO_2)_2$, $Ba(H_2PO_3)_2$ and $Ba(H_2PO_4)_2$ are respectively

a. + 1, + 2, + 3 b. + 3, + 2, + 1

c. + 1, + 3, + 5 d. + 2 in all

13 Oxidation number of sulphur in $S_8$, $S_2F_2$, $H_2S$ and $H_2SO_5$ are respectively

a. 0, + 1, − 2, + 6 b. 0, + 1, − 2, + 8

c. 0, − 1, − 2, + 8 d. 0, + 1, + 2, + 6

14 In which of the following reactions oxidation number of chromium has been affected?

a. $2CrO_4^{2-} + 2H^+ \longrightarrow Cr_2O_7^{2-} + H_2O$

b. $Cr_2O_7^{2-} + 2OH^- \longrightarrow 2CrO_4^{2-} + H_2O$

c. $(NH_4)_2Cr_2O_7 \xrightarrow{\Delta} N_2 + Cr_2O_3 + 3H_2O$

d. $CrO_2Cl_2 + 2OH^- \longrightarrow CrO_4^{2-} + 2HCl$

15 Prussian blue is represented by $KFe^x[Fe(CN)_6]$. Value of $x$ is

a. zero b. + 2 c. + 3 d. − 1

16 In the following reaction, $NH_4NO_2 \xrightarrow{\Delta} N_2 + 2H_2O$, oxidation number of

a. N changes from −2 to +2

b. N changes from −2 to 0

c. N in $NH_4^+$ changes from −3 to 0 and that in $NO_2^-$ changes from +3 to 0

d. N does not change

17 Select the set of compounds having fractional oxidation number in one or more atoms.

a. FeO, $Fe_2O_3$, $Fe_3O_4$, $C_3O_2$

b. $S_4O_6^{2-}$, $Pb_3O_4$, $C_3O_2$, $Br_3O_8$

c. $NO_2$, $N_2O_4$, $N_2O_5$, $CO_3^{2-}$, $S_2O_3^{2-}$

d. $CO_3^{2-}$, $PCl_5$, $P_2O_5$, $C_3O_2$, $Fe_3O_4$

Working Space

## One or More than One Options Correct Type

*Direction (Q. Nos. 18 and 19) This section contains 2 multiple choice questions. Each question has four choices (a), (b), (c) and (d), out of which ONE or MORE THAN ONE are correct.*

**18** Which of the following species has/have oxidation number of the metal as + 6?

a. $MnO_4^{2-}$ b. $CrO_2Cl_2$ c. $CrO_4^{2-}$ d. $MnO_4^-$

**19** Set of compounds with oxidation number of the underlined atoms + 6.

a. $\underline{S}O_3$, $\underline{Cr}O_3$, $\underline{Cr}O_5$
b. $\underline{S}_2O_8^{2-}$, $\underline{S}O_5^{2-}$, $\underline{S}O_4^{2-}$
c. $\underline{S}_2O_3^{2-}$, $\underline{S}_4O_6^{2-}$, $\underline{S}_2O_8^{2-}$
d. $\underline{Cr}O_4^{2-}$, $\underline{Mn}O_4^{2-}$, $\underline{Cr}O_5$

## Comprehension Type

*Direction (Q. Nos. 20 and 21) This section contains a paragraph, each describing theory, experiments, data, etc. Two questions related to the paragraph have been given. Each question has only one correct answer among the four given options (a), (b), (c) and (d).*

The complex $[Fe(H_2O)_5NO]^{2+}$ is formed in the **ring-test** for nitrate ion ($NO_3^-$) when freshly prepared $FeSO_4$ solution is added to aqueous solution of $NO_3^-$ followed by the addition of conc. $H_2SO_4$. NO exists as $NO^+$ (nitrosyl).

**20** Oxidation number of Fe in the ring is

a. + 1 b. + 2 c. + 3 d. + 4

**21** Magnetic moment ($= \sqrt{N(N+2)}$ for $N$ unpaired electrons) of Fe in the ring is

a. zero BM b. $\sqrt{8}$ BM c. $\sqrt{15}$ BM d. $\sqrt{35}$ BM

## Matching List Type

*Direction (Q. No. 22) Choices for the correct combination of elements from Column I and Column II are given*

**22** Match the compounds/ions having underlined atoms of different oxidation number (in Column I) with values (in Column II).

| | Column I | | Column II |
|---|---|---|---|
| i. | $CaO\underline{Cl}_2$ | p. | + 6 |
| ii. | $Na_2\underline{S}_2O_3$ | q. | + 3 |
| iii. | $\underline{N}_2H_4O_2$ | r. | + 2 |
| iv. | $[\underline{Fe}_2(CN)_6]^-$ | s. | + 1 |
| v. | $\underline{Cr}O_5$ | t. | − 1 |
| | | u. | − 2 |
| | | v. | − 3 |

Codes

| | i | ii | iii | iv | v |
|---|---|---|---|---|---|
| a. | p | u | s,r | q,s | p,u |
| b. | s,t | p,u | q,v | q,r | t,u |
| c. | r,u | p,q,s | r,s | q,t | q,u |
| d. | p,s,u | r,t | r,u | p,s | s,u |

## One Integer Value Correct Type

*Direction (Q. Nos. 23 and 24) This section contains 2 questions. When worked out will result in an integer from 0 to 9 (both inclusive).*

**23** $S_2O_3^{2-}$ has two types of sulphur atoms. What is the difference in the oxidation states of two types of sulphur atoms?

**24** The difference in the oxidation number of the two types of sulphur atoms in $Na_2S_4O_6$ is .......... **[IIT JEE 2011]**

# DPP-2 Oxidation-Reduction and Disproportionation Reactions

## Subjective Questions

Working Space

**Direction** *(Q. Nos. 1 and 2) These questions are subjective in nature, need to be solved completely on notebook.*

**1** i. Balance the following reaction in acidic medium

$$CrO_7^{2-} + SO_3^{2-} \longrightarrow SO_4^{2-} + Cr^{3+}$$

ii. Balance the following reaction in basic medium

$$Cl_2 \longrightarrow ClO_3^- + Cl^-$$

**2** Select the disproportionation reaction out of

i. $(CN)_2 + 2OH^- \longrightarrow CN^- + CNO^- + H_2O$

ii. $2Mn^{3+} + 2H_2O \longrightarrow Mn^{2+} + MnO_2 + 4H^+$

iii. $P_4 + 3OH^- + 3H_2O \longrightarrow PH_3 + 3H_2PO_2^-$

iv. $\begin{matrix} CHO \\ | \\ CH_2OH \end{matrix} + OH^- \longrightarrow \begin{matrix} CH_2OH \\ | \\ CH_2OH \end{matrix} + \begin{matrix} COO^- \\ | \\ CH_2OH \end{matrix}$

## Only One Option Correct Type

**Direction** *(Q. Nos. 3-20) This section contains 18 multiple choice questions. Each question has four choices (a), (b), (c) and (d), out of which ONLY ONE option is correct.*

**3** Oxidation can be defined as the terms

I. gain of electron and hydrogen
II. gain of oxygen and loss of electron
III. increase in oxidation number
IV. decrease in oxidation number

Select the correct terms

a. I and II  b. II and III  c. I and III  d. I and IV

**4** Reduction is defined in terms of

I. electronation and hydrogenation
II. deelectronation and gain of oxygen
III. increase in oxidation number
IV. decrease in oxidation number

Select the correct terms

a. I and IV  b. I and III  c. II and II  d. I and II

**5** In balancing the half-reaction, $CN^- \longrightarrow CNO^-$

The number of electrons that must be added is

a. 0  b. 4 on the right  c. 4 on the left  d. 2 on the right

Working Space

**6** Select the set of compounds with oxidation-reduction duality.

a. $Cl_2, H_3PO_4, HCHO, HNO_2$
b. $Cl_2, H_3PO_3, C_6H_5CHO, H_2O_2$
c. $Br_2, H_3PO_2, CH_3CHO, H_3PO_4$
d. $CrO_2Cl_2, KMnO_4, SO_3, CO_2$

**7** In the following balanced reaction,

$$4O_2^x + 2H_2O \longrightarrow 4OH^- + 3O_2$$

a. $x = -1$ and species is superoxide
b. $x = -2$ and species is peroxide
c. $x = 0$ and species is oxygen
d. $x = -4$ and species is oxide

**8** $I^-$ reduces $IO_3^-$ to $I_2$ and itself oxidised to $I_2$ in acidic medium. Thus, final reaction is

a. $I^- + IO_3^- + 6H^+ \longrightarrow I_2 + 3H_2O$
b. $I^- + IO_3 \longrightarrow I_2 + O_3$
c. $5I^- + IO_3^- + 6H^+ \longrightarrow 3I_2 + 3H_2O$
d. None of the above is correct

**9** $H_2O$ can be oxidised to

a. $H_2$ and $O_2$
b. $O^{2-}$
c. $OH^-$
d. $O_2$

**10** Consider the following reactions,

I. $Zn + \text{dil.}\, H_2SO_4 \longrightarrow ZnSO_4 + H_2$
II. $Zn + \text{conc.}\, H_2SO_4 \longrightarrow ZnSO_4 + SO_2 + H_2O$

Oxidising agents in I and II are

| | I | II |
|---|---|---|
| a. | $H^+$ | $SO_4^{2-}$ |
| b. | $SO_4^{2-}$ | $H^+$ |
| c. | $H^+$ | $H^+$ |
| d. | $SO_4^{2-}$ | $SO_4^{2-}$ |

**11** In the following reactions except in one, oxygen is the reducing agent. Exceptional reaction is

a. $2Ag_2O \longrightarrow 4Ag + O_2$
b. $4NH_3 + 3O_2 \longrightarrow 2N_2 + 6H_2O$
c. $2F_2 + 2H_2O \longrightarrow 4HF + O_2$
d. $2AgNO_3 + H_2O_2 \longrightarrow 2Ag + 2HNO_3 + O_2$

**12** In the following reaction,

$$2I^- + Cr_2O_7^{2-} + 14H^+ \longrightarrow I_2 + 2Cr^{3+} + 7H_2O$$

Unbalanced parts are

a. $I^-, I_2$
b. $Cr_2O_7^{2-}, Cr^{3+}$
c. $H^+, H_2O$
d. None of them are balanced

**13** $MnO_4^-$ oxidises $H_2O_2$ to $O_2$ in acidic medium

$$xMnO_4^- + yH_2O_2 + zH^+ \longrightarrow Mn^{2+} + O_2 + H_2O$$

Coefficient $x$, $y$ and $z$ are respectively

| | x | y | z |
|---|---|---|---|
| a. | 1 | 2 | 3 |
| b. | 3 | 2 | 5 |
| c. | 2 | 5 | 6 |
| d. | 2 | 6 | 5 |

Working Space

**14** Which of the following species does not show disproportionation reaction?

a. $ClO^-$  b. $ClO_2^-$

c. $ClO_3^-$  d. $ClO_4^-$

**15** Which is the intramolecular oxidation-reduction reaction?

a. $2KClO_3 \longrightarrow 2KCl + 3O_2$

b. $(NH_4)_2Cr_2O_7 \longrightarrow N_2 + Cr_2O_3 + 4H_2O$

c. $PCl_5 \longrightarrow PCl_3 + Cl_2$

d. All of the above

**16** Consider the following experimental facts,

I. When $Cl_2$ gas is passed into KI solution containing $CHCl_3$, violet colour appears in $CHCl_3$ layer.

II. When $Cl_2$ gas is passed into KBr solution containing $CHCl_3$, orange colour appears in $CHCl_3$ layer.

III. When $Cl_2$ gas is passed into a solution containing KBr, KI and KCl, containing $CHCl_3$, violet colour appears in $CHCl_3$ layer.

Select the correct experimental facts.

a. I and II  b. I and III

c. II and III  d. I, II and III

**17** Based on the following reaction,

$XeO_6^{4-}(aq) + 2F^-(aq) + 6H^+(aq) \longrightarrow XeO_3(g) + F_2(g) + 3H_2O(l) (\Delta G^\circ < 0)$

It can be concluded that

a. oxidising power of $F^-$ is greater than that of $XeO_6^{4-}$

b. oxidising power of $XeO_6^{4-}$ is greater than that of $F^-$

c. it is a disproportionation reaction

d. it is not a redox reaction

**18** For the redox reaction,

$$xFe^{2+} + yCr_2O_7^{2-} + zH^+ \longrightarrow Fe^{3+} + Cr^{3+} + H_2O$$

*x*, *y* and *z* are

| | *x* | *y* | *z* |
|---|---|---|---|
| a. | 3 | 1 | 14 |
| b. | 6 | 1 | 7 |
| c. | 6 | 1 | 14 |
| d. | 6 | 2 | 14 |

**19** In the following conversion of chromite into soluble chromium salt

$$4Fe(CrO_2)_2 + 8Na_2CO_3 + 7O_2 \longrightarrow 2Fe_2O_3 + 8Na_2CrO_4 + 8CO_2$$

There is

a. oxidation of Fe and reduction of Cr

b. oxidation of Fe and Cr both and reduction of O

c. oxidation of Fe and Cr both and reduction of carbon

d. No redox reaction

**20** In which of the following reactions $H_2O_2$ acts as a reducing agent? **[JEE Main 2014]**

I. $H_2O_2 + 2H^+ + 2e^- \longrightarrow 2H_2O$

II. $H_2O_2 - 2e^- \longrightarrow O_2 + 2H^+$

III. $H_2O_2 + 2e^- \longrightarrow 2OH^-$

IV. $H_2O_2 + 2OH^- - 2e^- \longrightarrow O_2 + 2H_2O$

a. I and II  b. III and IV

c. I and III  d. II and IV

Working Space

## One or More than One Options Correct Type

*Direction* *(Q. Nos. 21-25) This section contains 5 multiple choice questions. Each question has four choices (a), (b), (c) and (d) out of which ONE or MORE THAN ONE are correct.*

**21** Consider the following reaction,

$(NH_4)_2Cr_2O_7 \xrightarrow{\Delta} N_2 + Cr_2O_3 + 4H_2O$

a. it is an example of intramolecular redox reaction
b. it is an example of intermolecular redox reaction
c. $NH_4^+$ ion is a reducing agent and $Cr_2O_7^{2-}$ is an oxidising agent
d. it is an example of disproportionation reaction

**22** Following reaction, $2[Au(CN)_2]^- + Zn \longrightarrow 2Au + [Zn(CN)_4]^{2-}$ is an example of

a. oxidation of Zn
b. reduction of $Au^+$
c. displacement of Au(*s*) from solution by an active metal Zn
d. in which $\Delta G° < 0$

**23** Select the correct statement(s).

a. Compound of lower oxidation state is formed, if the reducing agent is in excess
b. Compound of higher oxidation state is formed, if the oxidising agent is in excess
c. Compound of higher oxidation state is formed, if the reducing agent is in excess
d. Compound of lower oxidation state is formed, if the oxidising agent is in excess

**24** For the reaction, $I^- + ClO_3^- + H_2SO_4 \rightarrow Cl^- + HSO_4^- + I_2$

The correct statement(s) in the balanced equation is/are **[JEE Advanced 2014]**

a. stoichiometric coefficient of $HSO_4^-$ is 6
b. iodide is oxidised
c. sulphuric acid is reduced
d. $H_2O$ is one of the products

**25** Identify intramolecular oxidation-reduction reaction.

a. $PCl_5 \longrightarrow PCl_3 + Cl_2$
b. $NH_4NO_3 \longrightarrow N_2O + 2H_2O$
c. $(NH_4)_2Cr_2O_7 \longrightarrow N_2 + Cr_2O_3 + 4H_2O$
d. 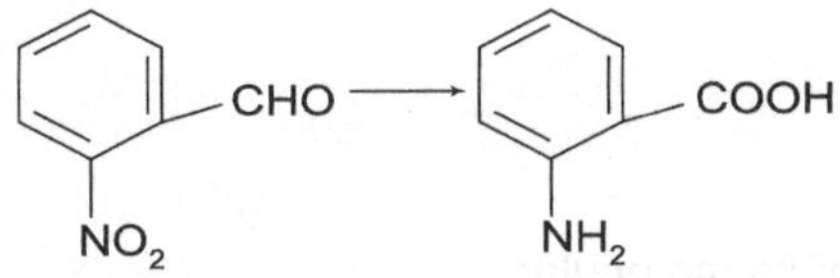

# DPP-3 Equivalent in Terms of Redox Reactions (Simple Calculations)

## Only One Option Correct Type

*Working Space*

**Direction** *(Q. Nos. 1-15) This section contains 15 multiple choice questions. Each question has four choices (a), (b), (c) and (d), out of which ONLY ONE option is correct.*

**1** $As_2S_3$ (yellow precipitate) is soluble in conc. $HNO_3$ as given in unbalanced equation

$$As_2S_3(s) + H^+ + NO_3^- \longrightarrow NO + H_2O + AsO_4^{3-} + SO_4^{2-}$$

Equivalent weight in terms of molar mass $M$ of $As_2S_3$ is

a. $\frac{M}{2}$ b. $\frac{M}{4}$

c. $\frac{M}{28}$ d. $\frac{M}{24}$

**2** In the following reaction,

$$2H_2S\,(g) + SO_2(g) \longrightarrow 3S\,(s) + 2H_2O\,(l)$$

One equivalent of $H_2S\,(g)$ will reduce

a. 1 mole $SO_2$ b. 0.5 mole $SO_2$

c. 0.25 mole $SO_2$ d. 2 moles $SO_2$

**3** Which has maximum number of equivalents per mole of the oxidant?

a. $Zn\,(s) + VO^{2+}(aq) \longrightarrow V^{3+}(aq) + Zn^{2+}(aq)$

b. $Ag\,(s) + NO_3^-(aq) \longrightarrow NO_2(g) + Ag^+(aq)$

c. $Mg\,(s) + VO_4^{3-}(aq) \longrightarrow Mg^{2+}(aq) + V^{2+}(aq)$

d. $I^-(aq) + IO_3^-(aq) \longrightarrow I_3^-(aq)$

**4** Equivalent weight of $H_3PO_2$ in a reaction is found to be half of its molecular weight. It can be due to its

a. reaction of two $H^+$ ions b. oxidation to $H_3PO_3$

c. oxidation to $H_3PO_4$ d. reduction to $PH_3$

**5** $KMnO_4$ (purple) is reduced to $K_2MnO_4$ (green) by $SO_3^{2-}$ in basic medium. 1 mole of $KMnO_4$ is reduced by

a. 1 mole of $SO_3^{2-}$ b. 2 moles of $SO_3^{2-}$

c. 0.5 mole of $SO_3^{2-}$ d. 1.5 mole of $SO_3^{2-}$

**6** Equivalent weight of $K_2Cr_2O_7$ in the following reaction is

$$Cr_2O_7^{2-} + Fe^{2+} \longrightarrow Fe^{3+} + Cr^{3+}$$

($M$ = molar mass of $K_2Cr_2O_7$)

a. $\frac{M}{3}$ b. $\frac{M}{4}$ c. $\frac{M}{5}$ d. $\frac{M}{6}$

Working Space

**7** In the following disproportionation of $Cl_2$ in basic medium

$$Cl_2 + 2KOH \longrightarrow KCl + KClO + H_2O$$

equivalent mass of $Cl_2$ is

a. 35.50 b. 11.83
c. 47.33 d. 71.00

**8** One mole of $CaOCl_2$ is dissolved in water and excess of KI added. Hypo ($Na_2S_2O_3$) required to react oxidised part completely is

a. 1 mole b. 1.5 mole
c. 2.0 moles d. 2.5 moles

**9** Consider conversion of $MnCl_2$ into

I. $MnCl_3$ II. $MnO_2$
III. $MnO_4^{2-}$ IV. $MnO_4^-$

Increasing order of equivalent weights of $MnCl_2$ in these conversions is

a. I < II < III < IV
b. IV < III < II < I
c. IV = III < II = I
d. equivalent weight is independent of oxidation number

**10** The equivalent weight of $MnSO_4$ is half of its molecular weight when it converts to

a. $Mn_2O_3$ b. $MnO_2$
c. $MnO_4^-$ d. $MnO_4^{2-}$

**11** The number of moles of $KMnO_4$ that will be needed to react with one mole of ferrous sulphite in acidic solution is

a. 0.4 b. 0.6
c. 0.8 d. 1.0

**12** The number of moles of $K_2Cr_2O_7$ that will be needed to react completely with one mole of ferric sulphite in acidic medium is

a. 0.4 b. 0.6
c. 0.8 d. 1.0

**13** Photosynthesis of carbohydrates in plants takes place as

$$6CO_2 + 12H_2O \xrightarrow{\text{Sunlight}} C_6H_{12}O_6 + 6O_2 + 6H_2O$$

Equivalent weight of $CO_2$ and $C_6H_{12}O_6$ respectively are

a. 11, 7.5 b. 44, 180
c. 22, 15 d. 44, 90

**14** Equivalent weight of $C_6H_5CHO$ is equal to molar mass in the following reaction.

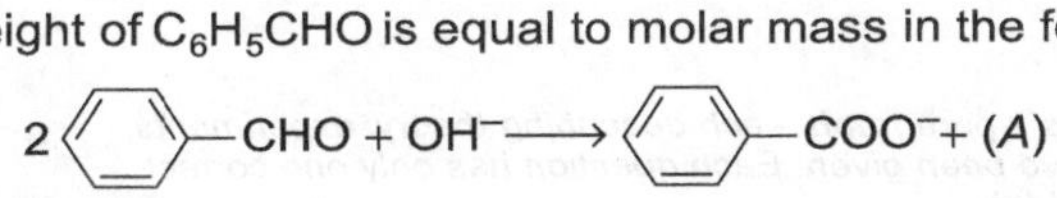


Thus, species (*A*) is

a. 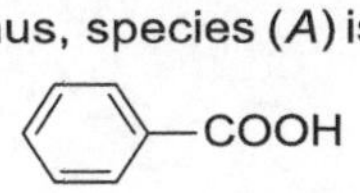


b. 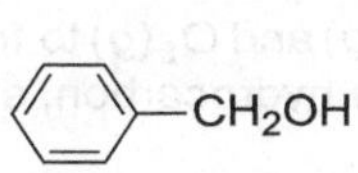


c. 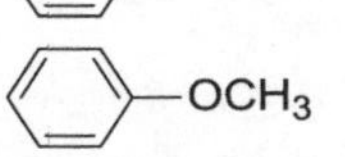


d. 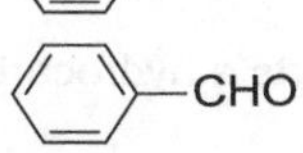

Working Space

**15** Equivalent weight of $KMnO_4$ and its reduced species in different mediums are given

| Medium | Equivalent weight | Reduced species |
|---|---|---|
| Acidic | $\frac{M}{5}$ | I |
| Neutral | $\frac{M}{1}$ | II |
| Concentrated Alkaline | $\frac{M}{3}$ | III |

I, II and III are

| | I | II | III |
|---|---|---|---|
| a. | $MnO_2$ | $MnO_4^{2-}$ | $Mn^{2+}$ |
| b. | $MnO_2$ | $Mn^{2+}$ | $MnO_4^{2-}$ |
| c. | $Mn^{2+}$ | $MnO_4^{2-}$ | $MnO_2$ |
| d. | $Mn^{2+}$ | $MnO_2$ | $MnO_4^{2-}$ |

## One or More than One Options Correct Type

*Direction (Q. Nos. 16-18) This section contains 3 multiple choice questions. Each question has four choices (a), (b), (c) and (d), out of which ONE or MORE THAN ONE are correct.*

**16** Sulphate (IV) ion reduces chromate (VI) to green coloured salt. Thus,

a. Cr (III) is formed

b. Equivalent mass of chromate (VI) $= \frac{M}{3}$

c. Equivalent mass of sulphate (IV) $= \frac{M}{2}$

d. Oxidation product is sulphate (VI)

**17** In which of the following reactions, equivalent mass of the underlined is equal to molar mass?

a. $\underline{Cl_2} + 2OH^- \longrightarrow Cl^- + ClO^- + H_2O$

b. $2\underline{HCHO} + OH^- \longrightarrow HCOO^- + CH_3OH$

c. $2\underline{MnO_4^{2-}} \longrightarrow MnO_2 + MnO_4^-$

d. $2\underline{KClO_3} \longrightarrow 2KCl + 3O_2$

**18** $NaHC_2O_4$ is neutralised by NaOH and can also be oxidised by $KMnO_4$ (in acidic medium). Equivalent weight is related to molecular weight ($M$) of $NaHC_2O_4$ in these two reactions as

a. $M, M$ b. $M, \frac{M}{2}$ c. $\frac{M}{2}, \frac{M}{5}$ d. $\frac{M}{5}, \frac{M}{2}$

## Comprehension Type

*Direction (Q. Nos. 19 and 20) This section contains a paragraph, each describing theory, experiments, data, etc. Two questions related to the paragraph have been given. Each question has only one correct answer among the four given options (a), (b), (c) and (d).*

Current fuel cells use the reaction of $H_2(g)$ and $O_2(g)$ to form $H_2O(l)$. Often the $H_2(g)$ is obtained by steam reforming of a hydrocarbon, such as

I. $C_3H_8(g) + 3H_2O \longrightarrow 3CO(g) + 7H_2(g)$

A future possibility is a fuel cell that converts a hydrocarbon such as propane directly to $CO_2(g)$ and $H_2O(l)$

II. $C_3H_8(g) + 5O_2(g) \longrightarrow 3CO_2(g) + 4H_2O(l)$

**19** Difference in equivalent weights of $C_3H_8$ based on I and II is

a. zero b. 0.94

c. −0.94 d. None is correct

Working Space

**20** Select the correct statements about reaction II taking place in acidic medium.

a. $C_3H_8$ has been oxidised and $O_2$ has been reduced
b. Twenty electrons are exchanged in this redox reaction
c. Equivalent mass of $C_3H_8$ is 2.2 and that of $O_2$ is 8
d. All of the above are correct statements

## Matching List Type

**Direction** *(Q. No. 21) Choices for the correct combination of elements from Column I and Column II are given as options (a), (b), (c) and (d), out of which one is correct.*

**21** Match the reactions of $H_3PO_3$ (molar mass = $M$) in Column I with the corresponding equivalent mass of $H_3PO_3$ in Column II.

| | Column I | | Column II |
|---|---|---|---|
| i. | $4\,H_3PO_3 \longrightarrow 3\,H_3PO_4 + PH_3$ | p. | $M$ |
| ii. | $H_3PO_3 + NaOH \longrightarrow NaH_2PO_3 + H_2O$ | q. | $M/2$ |
| iii. | $H_3PO_3 + Ca(OH)_2 \longrightarrow CaHPO_3 + 2H_2O$ | r. | $M/3$ |
| iv. | $H_3PO_3 + 2AgNO_3\ H_2O \longrightarrow H_3PO_4 + 2HNO_3 + 2Ag$ | s. | $2M/3$ |

**Codes**

| | i | ii | iii | iv |
|---|---|---|---|---|
| a. | p | r | s | q |
| b. | s | r | q | p |
| c. | s | p | q | q |
| d. | s | p | q | r |

## One Integer Value Correct Type

**Direction** *(Q. Nos. 22 and 23) This section contains 2 questions. When worked out will result in an integer from 0 to 9 (both inclusive).*

**22** How many moles of $SO_4^{2-}$ are formed per mole of $Mn^{2+}$ reacted in the following reaction?

$$Mn^{2+}(aq) + S_2O_8^{2-}(aq) \longrightarrow MnO_4^-(aq) + SO_4^{2-}(aq)$$

**23** In certain laboratory process, it is necessary to remove $NH_4^+$ ions from solution by heating with $HNO_3$. Calculate the moles of $HNO_3$ required by 1 mole of $(NH_4)_2SO_4$.

SET - 1

# Revisal Problems for JEE Main

## Only One Option Correct Type

*Working Space*

***Direction** (Q. Nos. 1-17) This section contains 17 multiple choice questions. Each question has four choices (a), (b), (c) and (d), out of which ONLY ONE option is correct.*

**1** $S_4O_6^{2-}$ is the oxidation product of $S_2O_3^{2-}$. Oxidation number of sulphur atoms in $S_4O_6^{2-}$ from left to right respectively are,

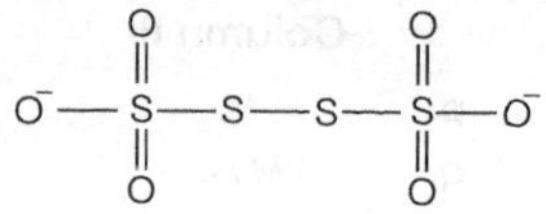

a. $-6, 0, 0, +6$
b. $+3, +1, +1, 3$
c. $+5, 0, 0, +5$
d. $+4, +1, +1, +4$

**2** $PbO_2$ oxidises MnO to purple coloured compound (*X*). In the balanced reaction $PbO_2$ and (*X*) appear in ratio as

a. 1 : 1
b. 2 : 5
c. 5 : 2
d. 1 : 3

**3** In the following redox reaction

$$xS_2O_4^{2-} + yCrO_4^{2-} + 4H_2O \longrightarrow 6SO_3^{2-} + zCr(OH)_3 + 2H^+$$

*x*, *y* and *z* are

| | *x* | *y* | *z* |
|---|---|---|---|
| a. | 3 | 2 | 2 |
| b. | 1 | 2 | 2 |
| c. | 2 | 1 | 1 |
| d. | 3 | 1 | 2 |

**4** One mole of $KMnO_4$ will oxidise *x* moles of Iron (II) oxalate in acidic medium. *x* is

a. 1/3
b. 2/3
c. 3/3
d. 5/3

**5** Which has the least number of equivalents per mole of the reactant?

a. $MnO_4^-$ changes to $MnO_2$
b. $MnO_4^-$ changes to $Mn^{2+}$
c. $MnO_4^-$ changes to $MnO_4^{2-}$
d. $MnO_4^-$ changes to $Mn_2O_3$

**6** Select the correct reaction.

a. $Pb_3O_4 + 8HCl \longrightarrow 3PbCl_2 + Cl_2 + 4H_2O$
b. $Pb_3O_4 + 4HNO_3 \longrightarrow 2Pb(NO_3)_2 + PbO_2 + 2H_2O$
c. Both (a) and (b) are correct reactions
d. None of the above is the correct reaction

**7** In a reduction of $HNO_3$, 4 moles of electrons are transferred per mole of $HNO_3$. The possible product is

a. 0.4 mol $N_2$
b. 0.5 mol $N_2O$
c. 1 mol $NO_2$
d. 1 mol $NH_3$

**8** Following reactions are feasible at 1400 K

I. $2C + O_2 \longrightarrow 2CO; \Delta G° = -460 \text{ kJ mol}^{-1}$

II. $2Zn + O_2 \longrightarrow 2ZnO; \Delta G° = -360 \text{ kJ mol}^{-1}$

Thus,

a. zinc can be oxidised by CO
b. zinc oxide can be reduced by carbon
c. Both (a) and (b) are correct
d. None of the above is correct

Working Space

9 A compound is made of Ba, Xe and O. Given, oxidation number of Ba = + 2, oxidation number of Xe = 8. Also Xe and O are present in atomic ratio of 1 : 6. Thus, compound is

a. $Ba_2XeO_3$
b. $Ba_2XeO_4$
c. $Ba_2XeO_6$
d. $BaXeO_6$

10 Following compounds of chromium (with oxidation number shown in roman digit) are known, all oxygen being in peroxy form

I. $(NH_3)_3\overset{IV}{Cr}O_x$

II. $K_3\overset{V}{Cr}O_y$

Thus, $x$ and $y$ are respectively

| | $x$ | $y$ | | $x$ | $y$ |
|---|---|---|---|---|---|
| a. | 4 | 4 | b. | 2 | 4 |
| c. | 4 | 2 | d. | 4 | 8 |

11 Select the correct statement (s).

a. When $NH_3$ reacts with excess of $Cl_2$ a product of nitrogen with oxidation number (+ 3) is formed by oxidation
b. When $Cl_2$ reacts with excess of $NH_3$ a product of nitrogen with oxidation number (zero) is formed by oxidation
c. Both (a) and (b) are correct
d. None of the abvoe is correct

12 A compound is formed with three elements *A*, *B* and *C*.

| Element | Oxidation number |
|---|---|
| *A* | +2 |
| *B* | +5 |
| *C* | −2 |

Thus, possible formula of the compound is

a. $A_2(BC_3)_2$
b. $A_3(BC_4)_2$
c. $A_3(B_4C)_2$
d. $ABC_2$

13 In the complex $[\overset{III}{Cr}(O_2)(NH_3)_4(H_2O)]Cl_2$, $O_2$ will be in the form of

a. dioxo
b. peroxo
c. superoxo
d. oxo

14 Chromite ($FeCr_2O_4$) is the source of chromium metal. Oxidation numbers of Fe and Cr in this mineral respectively are,

a. +2, +2
b. +3, +3
c. +2, +3
d. +3, +2

15 Oxidising behaviour of $H_2SO_4$ is depicted in the reaction

a. $2HI + H_2SO_4 \longrightarrow 2H_2O + I_2 + SO_2$
b. $Ca(OH)_2 + H_2SO_4 \longrightarrow CaSO_4 + 2H_2O$
c. $NaCl + H_2SO_4 \longrightarrow NaHSO_4 + HCl$
d. $PCl_5 + H_2SO_4 \longrightarrow 2POCl_3 + 2HCl + SO_2Cl_2$

16 Oxidation number and covalency of sulphur in $S_8$ are

a. 0 and 2
b. 0 and 8
c. +6 and 8
d. +6 and 2

17 In which of the following reactions $H_2O_2$ acts as a reducing agent? **[JEE Main 2014]**

I. $H_2O_2 + 2H^+ + 2e^- \longrightarrow 2H_2O$

II. $H_2O_2 - 2e^- \longrightarrow O_2 + 2H^+$

III. $H_2O_2 + 2e^- \longrightarrow 2OH^-$

IV. $H_2O_2 + 2OH^- - 2e^- \longrightarrow O_2 + 2H_2O$

a. I and II
b. III and IV
c. I and III
d. II and IV

## Statement Type

Working Space

***Direction*** *(Q. Nos. 18-20) This section is based on Statement I and Statement II. Select the correct answer from the codes given below.*

a. Both Statement I and Statement II are correct and Statement II is the correct explanation of Statement I
b. Both Statement I and Statement II are correct and Statement II is not the correct explanation of Statement I
c. Statement I is correct but Statement II is incorrect
d. Statement I is incorrect but Statement II is correct

**18** **Statement I** $SO_2$ and $H_2O_2$ can act as oxidising agents as well as reducing agents in their reactions.

**Statement II** Both are bleaching agents.

**19** **Statement I** Synthesis of carbohydrate is written as:

$$6CO_2(g) + 12H_2O\,(l) \longrightarrow C_6H_{12}O_6(aq) + 6H_2O\,(l) + 6O_2(g)$$

and not as $6CO_2(g) + 6H_2O\,(l) \longrightarrow C_6H_{12}O_6(aq) + 6O_2(g)$

**Statement II** By isotopic labelling of oxygen of $H_2O$ in reactant, there is no isotopic oxygen in other products. It is only in $O_2$.

**20** **Statement I** $Pb_3O_4$ is a double oxide of lead.

**Statement II** They appear on reaction with $HNO_3$.

# Revisal Problems for JEE Advanced

## Only One Option Correct Type

Working Space

**Direction** *(Q. Nos. 1-8) This section contains 8 multiple choice questions. Each question has four choices (a), (b), (c) and (d) out of which ONLY ONE option is correct.*

**1** Out of the following reactions, disproportionation is not observed in

a. $NH_4NO_3 \longrightarrow N_2O + 2H_2O$
b. $Cu_2O \longrightarrow CuO + Cu$
c. $2H_2O_2 \longrightarrow 2H_2O + O_2$
d. $6S + 12KOH \longrightarrow 2K_2SO_3 + 4K_2S + 6H_2O$

**2** In the following redox reaction.

$$Zn(s) + NO_3^-(aq) + H^+(aq) \longrightarrow Zn^{2+}(aq) + NH_4^+(aq)$$

$Zn(s)$ and $NO_3(aq)$ respectively are

a. oxidant, reductant
b. reductant, oxidant
c. Both oxidant
d. Both reductant

**3** Equivalent weight of $Br_2$ is 96 in the following disproportionation reaction

$$Br_2 + OH^- \longrightarrow Br^- + (X) + H_2O$$

Species $(X)$ is

a. $BrO^-$  b. $BrO_2^-$  c. $BrO_3^-$  d. $BrO_4^-$

**4** Insoluble $As_2S_3$ is dissolved in conc. $HNO_3$ when

a. $As^{3+}$ and $S^{2-}$ both are oxidised and $HNO_3$ is reduced
b. $As^{3+}$ is oxidised and $HNO_3$ is reduced
c. $S^{2-}$ is oxidised and $HNO_3$ is reduced
d. $HNO_3$ is oxidised and $S^{2-}$ is reduced

**5** 1 mole of $N_2H_4$ loses 10 moles of electrons to form a new compound Y, having same number of N-atoms as in $N_2H_4$. There is no change in the oxidation number of H. Oxidation state of nitrogen in the compound $Y$ is

a. −1  b. −3
c. +3  d. +5

**6** Blister copper is formed by the following reaction

$$Cu_2S + O_2(g) \longrightarrow Cu(l) + SO_2(g)$$

In the reaction

a. $O_2$ oxidises S of $Cu_2S$ to $SO_2$ and $Cu^+$ of $Cu_2S$ to Cu
b. $O_2$ oxidises S of $Cu_2S$ to $SO_2$ and reduces $Cu^+$ of $Cu_2S$ to Cu
c. $Cu_2S$ first changes to $Cu_2O$ by $O_2$ and then $Cu_2S$ changes to Cu and $SO_2$
d. All the above are correct statements

**7.** A sample of pure $KHC_2O_4 \cdot H_2C_2O_4 \cdot 2H_2O$ with three replaceable $H^+$ requires 4.62 moles of NaOH for the complete reaction. Thus, the given sample will require $KMnO_4$ in acidic medium

a. 4.62 mol  b. 2.31 mol
c. 0.924 mol  d. 1.232 mol

**8** 0.01 mole of $NH_2OH$ reduces 0.02 mole of $FeCl_3$ to $FeCl_2$ in acidic medium. Thus, nitrogen containing compound is

a. NO  b. $HNO_2$
c. $H_2N_2O_2$  d. $NH_3$

## One or More than One Options Correct Type

Working Space

***Direction*** *(Q. Nos. 9-18) This section contains 10 multiple choice questions. Each question has four choices (a), (b), (c) and (d), out of which ONE or MORE THAN ONE are correct.*

**9** Which of the following is/are example(s) of disproportionation reaction(s)?

a. $3ClO^- + 2Cr(OH)_4^- \longrightarrow 2CrO_4^{2-} + 3Cl^-$
b. $Cu_2O + 2H^+ \longrightarrow Cu + Cu^{2+} + H_2O$
c. $CuS + O_2 \longrightarrow Cu + SO_2$
d. $2HCuCl_2 \longrightarrow Cu + Cu^{2+} + 4Cl^- + 2H^+$

**10** Select the feasible reactions out of the following.

a. $2S_2O_3^{2-}(aq) + I_2(s) \longrightarrow S_4O_6^{2-}(aq) + 2I^-(aq)$
b. $S_2O_3^{2-}(aq) + 2Br_2(l) + 5H_2O(l) \longrightarrow 2SO_4^{2-}(aq) + 4Br^-(aq) + 10H^+(aq)$
c. $2KBr(aq) + I_2(s) \longrightarrow 2KI(aq) + Br_2(l)$
d. $2KI(aq) + Cl_2(aq) \longrightarrow 2KCl(aq) + I_2(s)$

**11** Some or more substances can act as reducing as well as oxidising agents. These can be

a. $Cl_2$
b. $H_2O_2$
c. $HNO_2$
d. $H_2SO_4$

**12** Corrosion reaction in which copper is converted into its basic carbonate can be thought in terms of

a. oxidation
b. reduction
c. acid-base reaction
d. precipitation reaction

**13** Properties shown by $CN^-$ can be

a. as a reducing agent
b. as an oxidising agent
c. as a complexing agent
d. as a precipitating agent

**14** If $Br^-$ and $I^-$ occur together in aqueous solution, then $Cl_2(aq)$ (in excess) can

a. oxidise $Br^-$ + $Br_2$ and extracted with $CS_2$
b. oxidise $I^-$ to $I_2$ and extracted with $CS_2$
c. oxidise $Br^-$ to $BrO_3^-$
d. oxidise $I^-$ to $IO_3^-$

**15** In the following redox reaction.

$BrO_2 \longrightarrow Br_2O + A$

$Br_2O$ and $A$ are formed by the change of equal number of oxidation number. Thus, $A$ is

a. $Br_2O_3$
b. $Br_2O_5$
c. $Br_2O_7$
d. BrO

**16** For *in situ* formation of $Br_2$, following reaction is used.

$$5Br^-(aq) + BrO_3^-(aq) + 6H^+(aq) \longrightarrow 3Br_2(l) + 3H_2O(l)$$

In this reaction,

a. $BrO_3^-$ disproportionates to $Br_2$
b. $Br^-$ disproportionates to $Br_2$
c. $Br^-$ has been oxidised and $BrO_3^-$ has been reduced
d. equivalent mass of $Br_2$ is equal to its molar mass

**17** Consider the following scheme involving oxides and oxyacids of nitrogen.

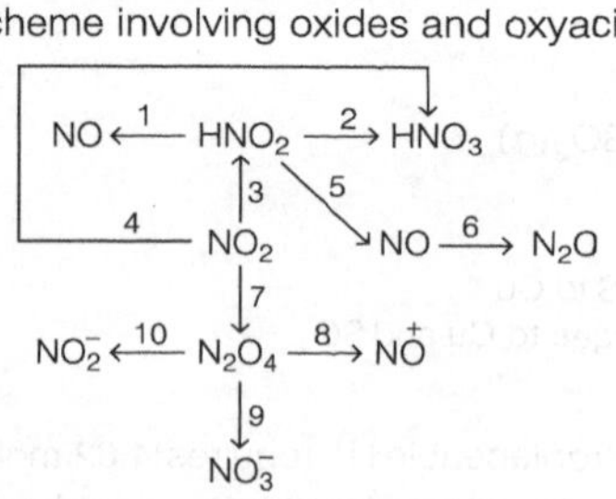


Disproportionation reaction-steps are

a. (1, 2)
b. (3, 4)
c. (8, 9)
d. (8,10)

**18** In the conversion of $Br^-$ to $BrO_3^-$

a. there is anodic oxidation in basic solution
b. there is cathodic reduction in basic solution
c. there is anodic oxidation in acidic mediums
d. equivalent mass of $Br^-$ is 20(Br = 80)

Working Space

## Comprehension Type

*Direction* *(Q. Nos. 19 and 20) This section contains a paragraph, each describing theory, experiments, data, etc. 2 questions related to the paragraph have been given. Each question has only one correct answer among the four given options (a), (b), (c) and (d).*

When Zn (*s*) is added to $K_2Cr_2O_7$ dissolved in HCl (*aq*), the colour of the solution changes from orange to green, then to blue and over a period of time, back to green.

**19** Formation of green colour on addition of zinc is due to

a. $Zn^{2+}$ ion
b. $CrO_4^{2-}$ ion
c. $Cr^{3+}$ ion
d. $Cr^{2+}$ ion

**20** Colour change of the solution from green → blue → green is due to

a. conversion of $Cr^{3+}$ (green) to $Cr^{2+}$ (blue) and then to green by atmospheric oxidation of $Cr^{2+}$

$$\underset{\text{Green}}{2Cr^{3+}} + Zn \longrightarrow Zn^{2+} + \underset{\text{Blue}}{2Cr^{2+}}$$

$$Cr^{2+} \xrightarrow{\text{Air}} \underset{\text{Green}}{Cr^{3+}} + e^-$$

b. conversion of $Cr^{2+}$ to $Cr^{3+}$ and then to $Cr^{2+}$
c. conversion of $CrO_4^{2-}$ to $Cr^{3+}$ and then to $Cr^{2+}$
d. None of the above

## Matching List Type

*Direction* *(Q. Nos. 21-23) Choices for the correct combination of elements from Column I and Column II are given as options (a), (b), (c) and (d), out of which one is correct.*

**21** Match the reactions given in Column I with their types in Column II.

| | Column I | | Column II |
|---|---|---|---|
| i. | $N_2 + O_2 \rightarrow 2NO$ | p. | Neutralison |
| ii. | $Pb(NO_3)_2 \rightarrow PbO + 2NO_2 + 0.5\,O_2$ | q. | Disproportionation redox |
| iii. | $NaH + H_2O \rightarrow NaOH + H_2$ | r. | Displacement redox |
| iv. | $2NO_2 + 2OH^- \rightarrow NO_2^- + NO_3^- + H_2O$ | s. | Decomposition redox |
| v. | $S_8 + 12OH^- \rightarrow 4S^{2-} + 2S_2O_3^{2-} + 6H_2O$ | t. | Combination redox |

**Codes**

| | i | ii | iii | iv | v |
|---|---|---|---|---|---|
| a. | p | q | r | s | t |
| b. | s | p | q | r | t |
| c. | r | s | p | q | t |
| d. | t | s | r | q | q |

**22** $KMnO_4$ is a good oxidising agent in acidic/basic/neutral medium. Match the reducing agent in Column I with its number of moles in Column II which can be oxidised by 1 mole of $KMnO_4$ in acidic medium.

| | Column I | | Column II |
|---|---|---|---|
| i. | Iron (II) oxalate | p. | 2.0 mol |
| ii. | Iron (II) sulphate (IV) | q. | 0.0 mol |
| iii. | Iron (II) oxalate | r. | 0.83 mol |
| iv. | Iron (II) sulphate (VI) | s. | 1.67 mol |
| v. | Iron (II) sulphate (IV) | t. | 2.50 mol |

**Codes**

| | i | ii | iii | iv | v |
|---|---|---|---|---|---|
| a. | r | s | s | q | s |
| b. | r | s | q | p | q |
| c. | p | q | r | s | t |
| d. | t | s | r | q | p |

Working Space

**23** Match the compounds in Column I with their related properties in Column II.

| | Column I | | Column II |
|---|---|---|---|
| i. | $H_2\underline{S}_2O_8$ | p. | Peroxy linkage |
| ii. | $\underline{Cr}O_5$ | q. | Oxidation number of the underline atom = + 6 per atom |
| iii. | $H_2\underline{S}O_5$ | r. | $H_2O_2$ is the hydrolysis product |
| iv. | $(NH_4)_2\underline{Cr}_2O_7$ | s. | Caro's acid |
| | | t. | Marshall's acid |

**Codes**

| | i | ii | iii | iv |
|---|---|---|---|---|
| a. | p,r,s | p,r | s,t | q,r |
| b. | p,q,r,s | p,q | p,q,r,s | q |
| c. | q,s | q,s | p,r,t | q,s |
| d. | r,t | p,q,r | t | r,s |

## One Integer Value Correct Type

***Direction** (Q. Nos. 24-28) This section contains 5 questions. When worked out will result in an integer from 0 to 9 (both inclusive).*

**24** The ion $A^{n+}$ is oxidised to $AO_3^-$ by $MnO_4^-$, changing to $Mn^{2+}$ in acidic medium. Given that $2.68 \times 10^{-3}$ mole of $A^{n+}$ requires $1.61 \times 10^{-3}$ mole of $MnO_4^-$. Thus, $n$ is

**25** $CrO_3$ changes to $CrO_5$ in presence of ether and $H_2O_2$. What is change in oxidation number of Cr atom?

**26** NaOCl a household beach can oxdise $[Cr(OH)_4]^-$ in basic solution to yield $CrO_4^{2-}$ and $Cl^-$ ion. How many moles of NaOCl are required by 4 moles of $[Cr(OH)_4]^-$ in basic solution?

**27** How many electrons are involved in the reduction of one mole of nitrate to hydrazine?

**28** $Cl_2$ disproportionates to $Cl^-$ and species $A$ such that equivalent mass of $Cl_2$ is 71. What is oxidation number of chlorine in species $A$?

# JEE Main & AIEEE Archive

## (Compilation of Last 13 Years Questions)

**1.** In which of the following reactions $H_2O_2$ acts as a reducing agent? **(JEE Main 2014)**

I. $H_2O_2 + 2H^+ + 2e^- \longrightarrow 2H_2O$
II. $H_2O_2 - 2e^- \longrightarrow O_2 + 2H^+$
III. $H_2O_2 + 2e^- \longrightarrow 2OH^-$
IV. $H_2O_2 + 2OH^- - 2e^- \longrightarrow O_2 + 2H_2O$

a. I, II  b. III, IV
c. I, III  d. II, IV

**2.** Consider the following reaction,

$$x\,MnO_4^- + y\,C_2O_4^{2-} + z\,H^+ \longrightarrow x\,Mn^{2+} + 2y\,CO_2 + \frac{z}{2}H_2O$$

The values of x, y and z in the reaction are, respectively **(AIEEE 2013)**

a. 5, 2 and 16  b. 2, 5 and 8
c. 2, 5 and 16  d. 5, 2 and 8

**3.** Which of the following chemical reactions depicts the oxidising behaviour of $H_2SO_4$?

a. $2HI + H_2SO_4 \longrightarrow I_2 + SO_2 + 2H_2O$
b. $Ca(OH)_2 + H_2SO_4 \longrightarrow CaSO_4 + 2H_2O$
c. $NaCl + H_2SO_4 \longrightarrow NaHSO_4 + HCl$
d. $2PCl_5 + H_2SO_4 \longrightarrow 2POCl_3 + 2HCl + SO_2Cl_2$

**(AIEEE 2006)**

**4.** The oxidation state of Cr in $[Cr(NH_3)_4Cl_2]^+$ is **(AIEEE 2005)**

a. 0  b. +1
c. +2  d. +3

**5.** The oxidation state of chromium in the final product formed by the reaction between KI and acidified potassium dichromate solution is **(AIEEE 2005)**

a. +3  b. +2
c. +6  d. +4

**6.** Oxidation number of Cl in $CaOCl_2$ (bleaching powder) is **(AIEEE 2002)**

a. zero, since it contains $Cl_2$
b. −1, since it contains $Cl^-$
c. +1, since it contains $ClO^-$
d. +1 and −1, since it contains $ClO^-$ and $Cl^-$

**7.** $MnO_4^-$ is a good oxidising agent in different medium changing to

$$MnO_4^- \longrightarrow Mn^{2+} \longrightarrow MnO_4^{2-} \longrightarrow MnO_2 \longrightarrow Mn_2O_3$$

Changes in oxidation number respectively, are **(AIEEE 2002)**

a. 1, 3, 4, 5
b. 5, 4, 3, 2
c. 5, 1, 3, 4
d. 2, 6, 4, 3

**8.** Which of the following is a redox reaction ? **(AIEEE 2002)**

a. $NaCl + KNO_3 \longrightarrow NaNO_3 + KCl$
b. $CaC_2O_4 + 2HCl \longrightarrow CaCl_2 + H_2C_2O_4$
c. $Ca(OH)_2 + 2NH_4Cl \longrightarrow CaCl_2 + 2NH_3 + 2H_2O$
d. $2K[Ag(CN)_2] + Zn \longrightarrow 2Ag + K_2[Zn(CN)_4]$

**9.** Which of the following reaction is possible at anode ? **(AIEEE 2002)**

a. $Fe_2 + 2e^- \longrightarrow 2F^-$
b. $2H^+ + \frac{1}{2}O_2 + 2e^- \longrightarrow H_2O$
c. $2Cr_2^{3+} + 7H_2O \longrightarrow Cr_2O_7^{2-} + 14H^+ + 6e^-$
d. None of the above

# 3. Chemical Bonding



DPP
DAILY PRACTICE PROBLEMS

# DPP-1 Ionic Bonds and Fajans' Rule

## Subjective Questions

Working Space

**Direction** (Q. Nos. 1-4) *These questions are subjective in nature, need to be solved completely on notebook.*

**1** Explain the formation of following ionic compounds by exchange of electrons.

i. NaF  ii. $MgCl_2$
iii. $Li_2O$  iv. KBr

**2** Name the following processes, with names of enthalpy change.

i. $M(s) \longrightarrow M(g)$
ii. $M(g) \longrightarrow M^+(g) + e^-$
iii. $X_2(g) \longrightarrow 2X(g)$
iv. $X(g) + e^- \longrightarrow X^-(g)$
v. $M^+(g) + X^-(g) \longrightarrow M^+X^-(s)$

**3** Atomic number of the four elements *A*, *B*, *C* and *D* are $Z - 1$, $Z$, $Z + 1$ and $Z + 2$. Assume $Z = 9$, then what is the type of bonding between

i. *A* and *D*  ii. *B* and *D*
iii. *A* and *A*  iv. *B* and *B*

**4** Compare the ionic bond strength of NaCl to KF; given

$$r_{Na^+} = 0.097 \text{ nm}$$
$$r_{K^+} = 0.133 \text{ nm}$$
$$r_{F^-} = 0.133 \text{ nm}$$
$$r_{Cl^-} = 0.181 \text{ nm}$$

## Only One Option Correct Type

**Direction** (Q. Nos. 5-20) *This section contains 16 multiple choice questions. Each question has four choices (a), (b), (c) and (d), out of which ONLY ONE option is correct.*

**5** The correct statement for the molecule $CsI_3$ is

a. it is a covalent compound
b. it contains $Cs^{3+}$ and $I^-$ ions
c. it contains $Cs^+$ and $I_3^-$ ions
d. it contains $Cs^+$, $I^-$ ions and lattice $I_2$ molecule

**6** $Mg^{2+}, O^{2-}$ is formed when

a. ionisation energy of Mg is low
b. electron affinity of O is low
c. ionisation energy of O is low
d. ionisation energy of Mg is high

Working Space

**7** Solubility of KCl is maximum in

a. $CH_3CH_2OH$
b. $CH_3OCH_3$
c. $H_2O$
d. $CH_3COOH$

**8** Lewis structure of $MgCl_2$ and $Al_2O_3$ are

a. 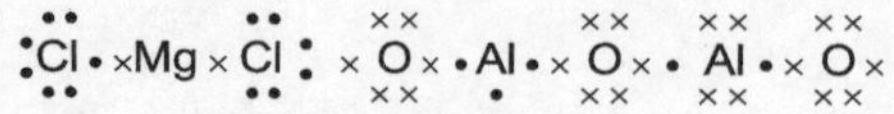

b. 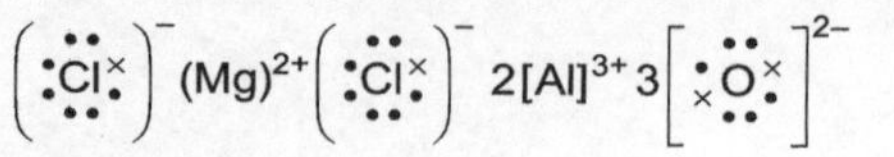

c. Both (a) and (b) are correct
d. None of the above

**9** Lewis structure of the elements *M* and *Z* are shown below.

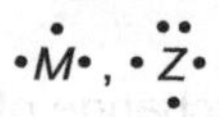

Compound formed is

a. $M_5Z_3$
b. $M_2Z_3$
c. $MZ$
d. $MZ_3$

**10** Different kinds of bonds and interaction present within $CuSO_4 \cdot 5H_2O$. They can be

I. $\sigma$-bond
II. $\pi$-bond
III. coordinate bond
IV. electrostatic force of attraction
V. H-bond due to dipole-dipole interaction
VI. H-bond due to ion-dipole interaction

Select the correct types of bonds/interactions.

a. I, II and III
b. II, III and IV
c. I, III, V and VI
d. All of these

**11** Consider the following compounds

I. $K_4[Fe(CN)_6]$
II. $NH_4Cl$
III. $H_2SO_4$
IV. $[Ni(CO)_4]$

Ionic, covalent and coordinate bonds are present in

a. all except II
b. all except IV
c. all except I
d. all except III

**12** Out of the following, maximum covalent nature is in

a. NaF
b. MgO
c. AlN
d. SiC

**13** Arrange NaCl, $MgCl_2$, $AlCl_3$, $SiCl_4$ in increasing solubility in ether (non-polar solvent).

a. $NaCl < MgCl_2 < AlCl_3 < SiCl_4$
b. $SiCl_4 < AlCl_3 < MgCl_2 < NaCl$
c. $SiCl_2 = AlCl_3 < MgCl_2 < NaCl$
d. Can't be decided

**14** Covalent nature of NaF, $Na_2O$ and $Na_3N$ in the increasing order is

a. $Na_3N < Na_2O < NaF$
b. $NaF < Na_2O < Na_3N$
c. $Na_2O < NaF < Na_3N$
d. $Na_2O < Na_3N < NaF$

Working Space

**15** Solubility of CuS (I), ZnS (II) and $Na_2S$ (III) in water is in the order

a. I < II < III
b. III < I < II
c. II < I < III
d. III < II < I

**16** Which pair is not in the correct order of lattice energy?

a. KCl < MgO
b. AlN > MgO
c. $BeCO_3 > MgCO_3$
d. $BeCO_3 = MgCO_3$

**17** Most stable compound is

a. LiF
b. LiCl
c. LiBr
d. LiI

**18** Out of the following ions, which pair will make the compound most covalent?

$$Na^+, Mg^{2+}, Al^{3+}, Si^{4+}, F^-, O^{2-}, N^{3-}, C^{4-}$$

a. $Na^+, F^-$
b. $Mg^{2+}, O^{2-}$
c. $Al^{3+}, N^{3-}$
d. $Si^{4+}, C^{4-}$

**19** Aqueous solution of a mixture contains LiCl, CuCl, NaCl and $AlCl_3$. This is shaken with ether. What is/are left in water?

a. LiCl, NaCl
b. NaCl, $AlCl_3$
c. Only CuCl
d. Only NaCl

**20** Select the correct statement(s).

a. CuCl is covalent and NaCl is ionic
b. Inner electrons have poor shielding effect on the nucleus and thus, electronegativity of the 18-electrons shell is increased
c. Cations with 18-electrons shell have greater polarising power than the cations with 8-electrons shell
d. All of the above are correct statements

## One or More than One Options Correct Type

**Direction** (Q. Nos. 21-24) *This section contains 4 multiple choice questions. Each question has four choices (a), (b), (c) and (d), out of which ONE or MORE THAN ONE are correct.*

**21** Which of the following can be calculated based on Born-Haber cycle of formation of a lattice $A^+B^-(s)$ from $A(s)$ and $B(g)$?

a. Lattice energy of $A^+B^-(s)$
b. Electron affinity of $B(g)$
c. Hydration energy of $A^+(g)$
d. Ionisation energy of $A(s)$.

**22** Which of the following Lewis acid-Lewis base interactions are associated with the expansion of octet?

a. $SbF_5 + HF \longrightarrow H[SbF_6]$
b. $NH_3 + BF_3 \longrightarrow [H_3N \longrightarrow BF_3]$
c. $AlCl_3 + Cl^- \longrightarrow [AlCl_4^-]$
d. $SF_4 + F^- \longrightarrow [SF_5^-]$

**23** An ionic compound $A^+B^-$ is most like to be formed from $A$ and $B$ when

a. ionisation energy of $A$ is low
b. electron affinity of $B$ is high
c. electron affinity of $A$ is high
d. ionisation energy of $A$ is high

**24** Consider following types of bonds.

I. Ionic bond
II. Covalent $\sigma$-bond
III. Coordinate bond
IV. Covalent $\pi$-bond

All these bonds are present in

a. $NH_4Cl$
b. $NaNO_3$
c. $H_2SO_4$
d. $NaN_3$

## Comprehension Type

Working Space

***Direction*** *(Q. Nos. 25 and 26) This section contains a paragraph, each describing theory, experiments, data, etc. Three questions related to the paragraphs have been given. Each question has only one correct answer among the four given options (a), (b), (c) and (d).*

$A^+B^-$ (ionic) is formed in the following steps from gaseous atoms:

$$A(g) \longrightarrow A^+(g) + e^-; \qquad \Delta H = 5.5 \text{ eV}$$

$$B(g) + e^- \longrightarrow B^-(g); \quad \Delta H = -1.5 \text{ eV}, \; r_{B^-} = 0.20 \text{ nm}$$

Ionic strength $= 1.44 \dfrac{z_1 z_2}{r^2}$ eV

**25** How much energy is required to transfer one electron from $A$ to $B$?

a. 4.0 eV
b. − 4.0 eV
c. 7.0 eV
d. − 7.0 eV

**26** What is the maximum size of $A^+$ that would lend itself for the formation of an energetically stable bond?

a. 0.40 nm
b. 0.36 nm
c. 0.60 nm
d. 0.80 nm

## One Integer Value Correct Type

***Direction*** *(Q.No. 27) This section contains 1 question. When worked out will result in an integer from 0 to 9 (Both inclusive).*

**27** An ionic bond is established between a positive ion $A^+$ and negative ion $B^-$. How many times strength of the ionic bond is affected by doubling the charge on $A^+$ and making the radius halved?

# DPP-2 Covalent and Coordinate Bonding

## Subjective Questions

Working Space

**Direction** *(Q. Nos. 1-5) These questions are subjective in nature, need to be solved completely on notebook.*

**1** Write structures of the following indicating coordinate bonds wherever exist :

i. $N_2O$ ii. $Cr_2O_7^{2-}$
iii. $CrO_5$ iv. $Al_2Cl_6$
v. $H_3PO_4$ vi. $H_3PO_3$
vii. $H_3PO_2$ viii. $NO_2$

**2** Valence electrons in the elements *X* and *Y* are 3 and 6. What is the formula of the compound formed from *X* and *Y*?

**3** Classify Lewis acid and Lewis base out of

i. $H_2O$ ii. $NH_3$
iii. $H^+$ iv. $BF_3$
v. $AlH_3$ vi. $CN^-$
vii. $Fe^{2+}$ viii. $^{\oplus}CH_3$

**4** B—F bond lengths in $BF_3$ is 130 pm and in $BF_4^-$ is 143 pm. Explain.

**5** Explain

i. The type of bonding in $B_2H_6$.
ii. $(CH_3)_3N$ is pyramidal but $N(SiH_3)_3$ is planar.
iii. $H_2O_3$, hydrogen trioxide, a close relative of hydrogen peroxide, has recently been synthesised. It is extremely unstable and can be isolated is very small quantities. Write a Lewis electron dot structure for $H_2O_3$.

## Only One Option Correct Type

**Direction** *(Q. Nos. 6-20) This section contains 15 multiple choice questions. Each question has four choices (a), (b), (c) and (d), out of which ONLY ONE option is correct.*

**6** By Lewis structure NO has following facts

a. Dot Lewis structure is $\overset{\times}{\underset{\times}{\times}}N\overset{\times}{\times}\;\overset{\cdot\cdot}{\underset{\cdot\cdot}{:O:}}$
b. Bond order is 2.5 since, five electrons are shared between N and O-atom
c. There is one unpaired electron hence, NO is paramagnetic
d. All are correct facts

Working Space

**7** Lewis structure of HCl, $H_2O$ and $NH_3$ are

a. 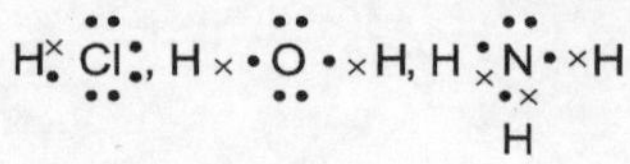

b. 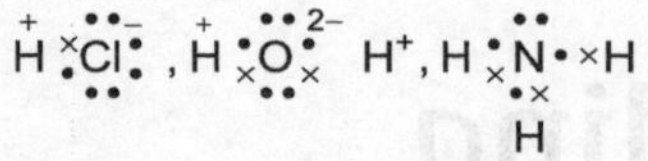

c. 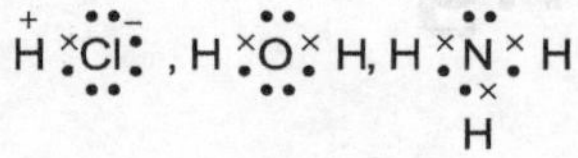

d. None of the above is correct

**8** Maximum covalency is equal to the number

a. paired *p*-electrons
b. unpaired *s*-electrons
c. unpaired *s* and *p*-electrons
d. *s*-and *p*-electrons in the valence shells

**9** Which of the following has incomplete octet?

a. $NH_3$ b. $BCl_3$
c. $CCl_4$ d. $PCl_3$

**10** Expansion of octet can not take place in

a. N b. S
c. Si d. P

**11** Which of the following is electron deficient?

a. $(CH_3)_2$ b. $(SiH_3)_2$
c. $(BH_3)_2$ d. $PH_3$

**12** Based on valence bond theory, types of bonding in $H_2O$ are

a. *s-p* b. $p_x$-$p_x$
c. $p_y$-$p_z$ d. $p_z$-$p_z$

**13** Bond order of (B—H) *x* and *y* bonds in diborane are respectively

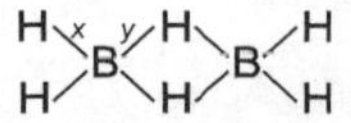

a. 1, 1 b. 0, 0.5
c. 0.5, 0.5 d. 1, 0.5

**14** Considering *x*-axis as the internuclear axis, which out of the following will not form sigma bond.

a. 1*s* and 1*s* b. 1*s* and $2p_x$
c. $2p_y$ and $2p_y$ d. 1*s* and 2*s*

**15** In which of the following molecules would you expect the N to N bond to be shortest?

a. $N_2H_4$ b. $N_2$
c. $N_2O_4$ d. $N_2O$

**16** Which of these molecules have non-bonding electron pairs on the central atom?

I. $SF_4$ II. $ICl_3$
III. $SO_2$ IV. $NH_3$
V. $SiF_4$.

a. All except V b. All except II and V
c. All except I and V d. All except III and IV

Working Space

**17** HCN and HNC molecules have equal number of

a. lone pairs and $\sigma$-bonds
b. $\sigma$-bonds and $\pi$-bonds
c. $\pi$-bonds and lone pair
d. lone pairs, $\sigma$-bonds and $\pi$-bonds

**18** When a chemical bond is formed, there is decrease in

a. kinetic energy
b. potential energy
c. repulsive force
d. attractive force

**19** Mean bond enthalpy of different bonds are given

$$N \equiv N \quad 946\ \text{kJ mol}^{-1}$$
$$N - N \quad 160\ \text{kJ mol}^{-1}$$
$$P \equiv P \quad 481\ \text{kJ mol}^{-1}$$
$$P - P \quad 215\ \text{kJ mol}^{-1}$$

Thus, possible structures of nitrogen and phosphorus in molecular state are

a. :N≡N:, 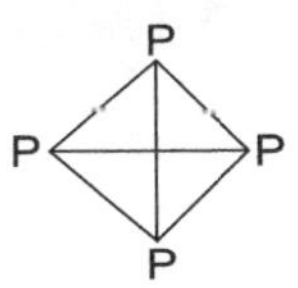

b. 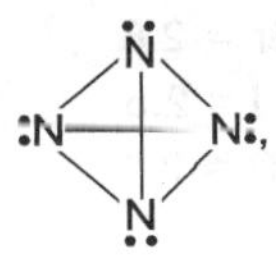 :P≡P

c.  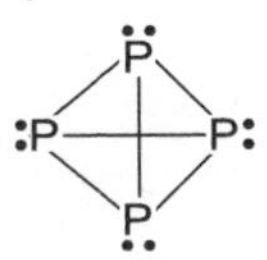

d. :N≡N:, :P≡P:

**20** Covalency of carbon in CO is three because

a. an unexcited carbon atom has two unpaired electrons
b. the C-atom can be an acceptor of an electron pair
c. the C-atom has four valence electrons
d. the maximum covalency of carbon is three

## One or More than One Options Correct Type

***Direction*** *(Q. Nos. 21-25) This section contains 5 multiple choice questions. Each question has four choices (a), (b), (c) and (d), out of which ONE or MORE THAN ONE are correct.*

**21** Which of the following species contain at least one atom that violates the octet rule?

a. O—Cl—O
b. F—Xe—F
c. $PCl_5$
d. $SF_6$

**22** Select the correct statement(s) about $NO_2$.

a. It is paramagnetic in nature
b. It forms dimer and paramagnetic is lost
c. $NO_2$ and dimer formed have $sp^2$-hybridised N-atom
d. Brown colour of $NO_2$ fades and (N—N) bond length is greater than normal (N—N) bond length

**23** Which of the following is/are correct statement(s)?

a. $s + p_y \longrightarrow$ two $sp$-hybrid orbitals lying in the $yz$ plane
b. $s + p_x \longrightarrow$ two $sp$-hybrid orbitals lying in $xz$ plane
c. $(s + p_x + p_z) \longrightarrow$ three $sp^2$-hybrid orbitals lying in $xz$ plane
d. $(s + p_y) \longrightarrow$ two $sp$-hybrid orbitals lying along $y$-axis

Working Space

**24** Electron-seeking species are called electrophiles. Which of the following sets consists of electrophiles only?

a. $BF_3, NH_3, H_2O$ b. $AlCl_3, SO_3, \overset{\oplus}{N}O_2$

c. $\overset{\oplus}{N}O_2, \overset{\oplus}{C}H_3, CH_3\overset{\oplus}{C}O$ d. $\overset{\ominus}{C_2H_5}, \overset{\bullet}{C_2H_5}, \overset{\oplus}{C_2H_5}$

**25** Ratio of $\sigma$ and $\pi$-bonds is equal to 1 in

a. benzene b. tetracyanomethane

c. allene ($C_3H_4$) d. $CO_2$

## Comprehension Type

***Direction*** *(Q. Nos. 26-30) This section contains 2 paragraphs, each describing theory, experiments, data, etc. Five questions related to the paragraphs have been given. Each question has only one correct answer among the four given options (a), (b), (c) and (d).*

### Passage I

Electronegativity values (EN) of elements have been given (in Pauling's scale)

H = 2.1 O = 3.5 Br = 2.8

C = 2.5 F = 4.0 I = 2.5

N = 3.0 Cl = 3.0 P = 2.1

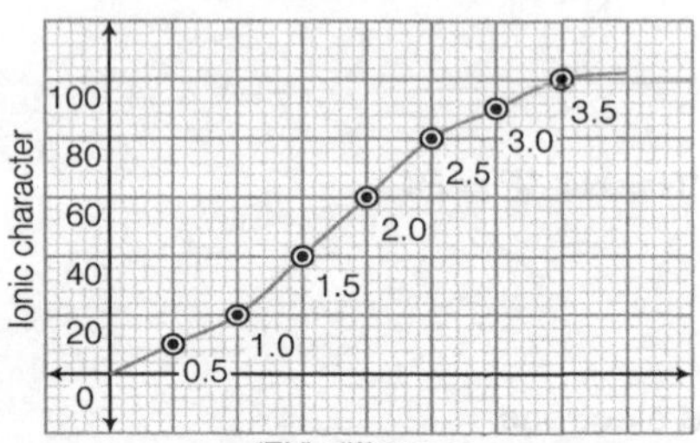


Graph shows variation of percentage ionic character with (EN) difference of two elements.

**26** Which of the following bonds is most polar?

a. H—Br b. N—H

c. N—O d. P—H

**27** A molecule of the type H—*X* is 50% ionic when EN difference is

a. 1.5 b. 2.0

c. 1.0 d. 1.7

### Passage II

Consider the following structure

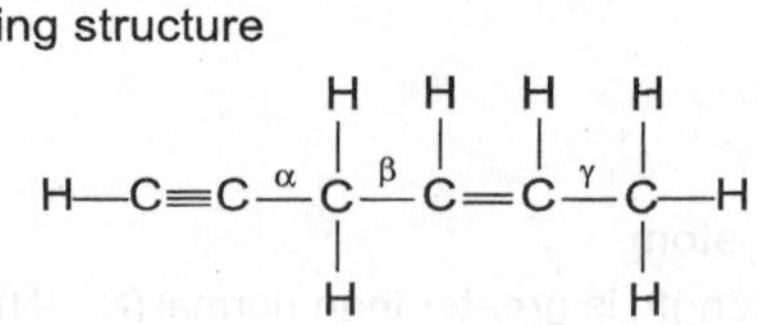

**28** Shortest (C—H) bond is

a. H—C≡C b. H—C=C

c. H—C—C d. all (C—H) bonds are of equal length.

Working Space

**29** Shortest (carbon-carbon) bond is

a. C—C
b. C=C
c. C≡C
d. equal bond length

**30** $(C \overset{\sigma}{—} C)$ bond is longest

a. α
b. β
c. γ
d. equal bond length

## Matching List Type

*Direction (Q. No. 31) Choices for the correct combination of elements from Column I and Column II are given as options (a), (b), (c) and (d), out of which one is correct.*

**31** Match the species in Column I with their properties in Column II.

| | Column I | | Column II |
|---|---|---|---|
| i. | $H_2N_2O_2$ | p. | σ and π-bonds are present. |
| ii. | $N_2O_4$ | q. | Exists as *cis, trans-* isomers. |
| iii. | $N_2O_5$ | r. | N—N bond length is greater than theoretical value. |
| iv. | $HNO_2$ | s. | Oxidising, reducing and complexing agent. |

**Codes**

| | i | ii | iii | iv |
|---|---|---|---|---|
| a. | p,q | p,r | p | p,s |
| b. | q | r,s | p | qs |
| c. | p | q | r | s |
| d. | s | r | q | p |

## One Integer Value Correct Type

*Direction (Q. Nos. 32 and 33) This section contains 2 questions. When worked out will result in an integer from 0 to 9 (both inclusive).*

**32** $CrO_5$ is the oxidation product of $CrO_3$ using $H_2O_2$. How many (O—Cr—O) bonds are there in it?

**33** $K_2Cr_2O_7$ is an oxidising agent in acidic medium. How many O-atoms are directly linked to each Cr-atom?

# DPP-3 Covalent Bonding

## Only One Option Correct Type

Working Space

***Direction** (Q. Nos. 1-16) This section contains 16 multiple choice questions. Each question has four choices (a), (b), (c) and (d), out of which ONLY ONE option is correct.*

**1** Octet rule is not followed in

a. $CCl_4$, $N_2O_4$, $N_2O_5$
b. $BF_3$, $BeCl_2$, $NO_2$
c. NaCl, $MgCl_2$, MgO
d. $PCl_3$, $NH_3$, $H_2O$

**2** Select the correct statement about valence bond approach.

a. Each bond is formed by maximum overlap for its maximum stability
b. It represents localised electron model of bonding
c. Most of the electrons retain the same orbital locations as in separated atoms
d. All of the above are correct statements

**3** Give $A—A$ bond length = $x$, C—C bond length = $y$ and $A—B$ bond length = $z$

Then B—C bond length is

a. $2(x-y)$
b. $x-y-z$
c. $(y-x)/2$
d. $z+\left(\frac{y-x}{2}\right)$

**4** Cyanide ion ($CN^-$), nitrosonium ion ($NO^+$) and nitrogen ($N_2$) are isoelectronic but $N_2$ is chemically inert because of

a. low bond energy
b. absence of bond polarity
c. unsymmetrical electron distribution
d. presence of more number of electrons in bonding

**5** The correct order of increasing bond length of

I. C—H
II. C—O
III. C—C
IV. C=C

a. I < II < III < IV
b. I < IV < II < III
c. III < IV < II < I
d. II < I < III < IV

**6** Ratio of $\sigma$ and $\pi$-bonds is maximum in

a. tetracyanomethane
b. benzene
c. 1-phenyl-prop-1-ene
d. enolic form of acetone

**7** In which of the following cases, covalent bonds are cleaved?

a. Boiling of $H_2O$
b. Melting of KCN
c. Boiling of $CF_4$
d. Melting of $SiO_2$

Working Space

**8** The compound ($NH_3$, $BF_3$) can be easily separated into its individual components because

a. $BF_3$ is highly reactive
b. $NH_3$ is highly reactive
c. $BF_3$ is a Lewis base
d. $BF_3$ and $NH_3$ are joined by weak coordinate bond and have independent existence

**9** Consider the following species

I. $S_4O_6^{2-}$      II. $S_3O_9$
III. $S_2O_5^{2-}$      IV. $P_3O_9^{3-}$

Which of the above species have same number of $X$—O—$X$ linkages? ($X$=P or S)

a. II and III      b. I and III
c. II and IV      d. I and IV

**10** Following types of overlapping can take place in the formation of bonds,

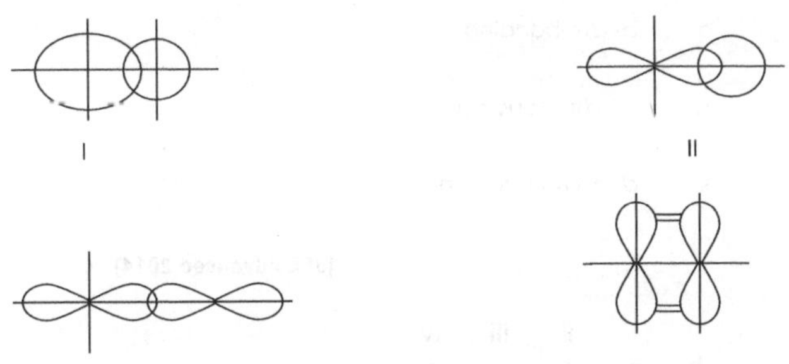


Strength of bonds formed by these types of overlapping is

a. I< III< II< IV      b. I< II< III< IV
c. IV < III< II< I      d. IV < I = III< II

**11** Following is the dot structure of $XY_4$

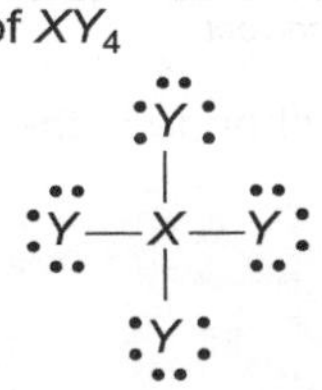


$X$ and $Y$ can be

a. Cl, F      b. I, Br      c. Xe, F      d. Si, Cl

**12** Hyperconjugation involve overlap of the following orbitals

a. $\sigma$-$\sigma$      b. $\sigma$-$\pi$      c. $p$-$p$      d. $\pi$-$\pi$

**13** $CH_4$, $NH_3$ and $H_2O$ are isoelectronic but not isostructured. Taking $A$ as the a central atom, $X$ as other bonding atoms and $E$ the lone pairs, then they are represented as

| | $CH_4$ | $NH_3$ | $H_2O$ | | $CH_4$ | $NH_3$ | $H_2O$ |
|---|---|---|---|---|---|---|---|
| a. | $AX_4$, | $AX_3$, | $AX_2$ | b. | $AX_4$, | $AX_3E$, | $AX_2E_2$ |
| c. | $AX_4$, | $AX_3E$, | $AX_2$ | d. | $AX_4$, | $AX_3E$, | $AX_2E$ |

**14** Which property could describe a covalent compound?

a. It is conducting in molten (fused) state
b. It is a gas at room temperature
c. It is composed of metal and non-metal
d. It is a solid with high melting point

Working Space

**15** Which of the following is non-polar?

a. $Cl_3CCH_3$ b. $PCl_5$ c. $CH_2Cl_2$ d. $NH_3$

**16** Which of these molecules would you expect to be polar? $Cl_2$, ICl, $BF_3$, NO, $SO_2$

a. $Cl_2$ and $BF_3$ b. $Cl_2$, ICl and $BF_3$ c. $BF_3$, NO and $SO_2$ d. ICl, NO and $SO_2$

## Matching List Type

***Direction** (Q. No. 17) Choices for the correct combination of elements from Column I and Column II are given as options (a), (b), (c) and (d), out of which one is correct.*

**17** Match the orbital overlap figures shown in Column I with the description given in Column II and select the correct answer using the code given below the columns.

| | Column I | | Column II |
|---|---|---|---|
| i. |  | p. | $p$-$d\pi$ anti-bonding |
| ii. |  | q. | $d$-$d\sigma$-bonding |
| iii. |  | r. | $p$-$d\pi$-bonding |
| iv. |  | s. | $d$-$d\sigma$ anti-bonding |

**[JEE Advanced 2014]**

**Codes**

| | i | ii | iii | iv | | i | ii | iii | iv |
|---|---|---|---|---|---|---|---|---|---|
| a. | q | p | r | s | b. | s | r | p | q |
| c. | q | r | p | s | d. | s | p | r | q |

## One or More than One Options Correct Type

***Direction** (Q. Nos. 18-19) This section contains 2 multiple choice questions. Each question has four choices (a), (b), (c) and (d), out of which ONE or MORE THAN ONE are correct.*

**18** As two H-atoms approach each other, potential energy change in the formation of a covalent compound $H_2$ is shown.

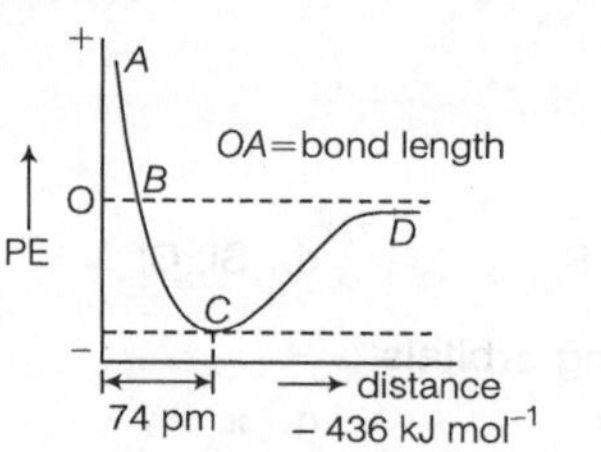


Thus, graph shows

a. a zero of energy when two H-atoms are separated by a great distance
b. a drop in potential energy (net attraction) as the two H-atoms approach each other
c. when stable $H_2$ molecule is formed, internuclear distance is about (74 pm) at a minimum potential energy (–436 kJ $mol^{-1}$)
d. an increase in potential energy as the atoms approach more closely

**19** Select the correct statement(s) about divalent anion of ortho-phosphorous acid.

a. It is amphiprotic
b. All bond length (P—O) are identical due to resonance
c. The bond order is 1.33 for (P—O) bonds
d. It is a base

# DPP-4 Chemical Bonding (VSEPR Model) and Hybridisation

## Subjective Question

***Direction*** *(Q. No. 1) This question subjective in nature, need to be solved completely on notebook.*

*Working Space*

**1** $XeF_2$ is represented as

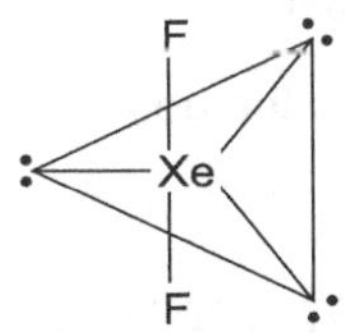

i. What is the type of structure?
ii. What is actual geometry?
iii. Write one species with identical arrangement of electron pair and geometry.

## Only One Option Correct Type

***Direction*** *(Q. Nos. 2-17) This section contains 16 multiple choice questions. Each question has four choices (a), (b), (c) and (d), out of which ONLY ONE option is correct.*

**2** The shape $[ClF_4]^-$ and $[ClF_2]^-$ ions is respectively

a. see-saw and linear
b. see-saw and bent
c. tetrahedral and linear
d. square planar and linear

**3** Which of the following sequences shows the correct bond angle order for isoelectronic species $O_3$, $NO_2^-$ and NOF?

a. $NO_2^- > NOF > O_3$
b. $O_3 > NO_2^- > NOF$
c. $NOF > NO_2^- > O_3$
d. Cannot be predicted

**4** $XeF_2$ is isostructural with

a. $TeF_2$
b. $ICl_2^-$
c. $SbCl_3$
d. $BaCl_2$

**5** The species having pyramidal shape is **[IIT JEE 2010]**

a. $SO_2$
b. $BrF_3$
c. $SiO_3^{2-}$
d. $OSF_2$

**6** Molecular shape of $CF_4$, $SF_4$ and $XeF_4$ are

a. the same, with 2, 0 and 1 lone pair of electrons respectively
b. the same with 1, 1 and 1 lone pair of electrons
c. different, with 0, 1 and 2 lone pair of electrons
d. different, with 1, 0 and 2 lone pair of electrons

*Working Space*

**7** Two of the following species have same shape

$NI_3$, $I_3^-$, $SO_3^{2-}$, $NO_3^-$. These are

a. $NI_3$, $I_3^-$ b. $NI_3$, $SO_3^{2-}$
c. $SO_3^{2-}$, $NO_3^-$ d. $I_3^-$, $NO_3^-$

**8** $ECl_3$ (where, $E$ = B, P, As, Bi) of these elements are known. Bond angles 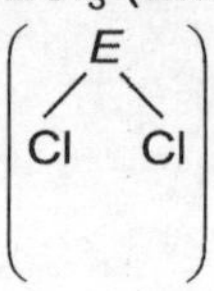 are in the following order

a. B > P > As > Bi b. B > P = As > Bi
c. Bi > As > P > B d. B > As > Bi > P

**9** Consider the two structures : $F_2Cl^-$ (I) $F_2Cl^+$ (II)

Select the correct statement(s).

a. I is linear, II is bent b. I is bent, II is linear
c. Both is bent d. Both are linear

**10** Which of the two ions from the list given below that have the geometry that is explained by the same hybridisation of orbitals?

$NO_2^{\ominus}$, $NO_3^{\ominus}$, $NH_2^{\ominus}$, $NH_4^{\oplus}$, $SCN^{\ominus}$

a. $NO_2^{\ominus}$ and $NH_2^{\ominus}$ b. $NO_2^{\ominus}$ and $NO_3^{\ominus}$
c. $NH_4^+$ and $NO_3^-$ d. $SCN^-$ and $NH_2^{\ominus}$

**11** The correct order of increasing bond angles in the following species is

a. $ClO_2^- < ClO_3^-$ $ClO_4^-$
b. $ClO_2^- < ClO_4^-$ $ClO_3^-$
c. $ClO_4^- < ClO_3^-$ $ClO_2^-$
d. equal since each Cl atom is $sp^3$-hybridised

**12** Which of the following molecule/species has the minimum number of lone pairs?

a. $ICl_3$ b. $BF_4^-$ c. $SnCl_2$ d. $XeF_4$

**13** $PCl_5$ has a shape of trigonal bipyramid whereas, $IF_5$ has a shape of square pyramid. It is due to

a. presence of unshared electron pair on I which is oriented so as to minimise repulsion while in $PCl_5$, P has no non-bonding pair
b. octet of P is complete that of I is incomplete
c. P and I are of different groups and periods
d. extent of different repulsion in F and I

**14** For $H_2O$ and $H_2S$, given

| | Hybridisation of | | Bond angle |
|---|---|---|---|
| $H_2O$ | O | $sp^3$ | 104.5° |
| $H_2S$ | S | $sp^3$ | 90° |

Difference in bond angle is due to

a. the larger size of S-atom as compared to O-atom which minimises repulsion and allows the bonds in $H_2S$ to be purely $p$-type
b. liquid state of $H_2O$ as compared to gaseous state of $H_2S$
c. Both of the above
d. None of the above

*Working Space*

**15** For which of the following sets of geometry, both axial and equatorial positions are present?

a. Octahedral, trigonal bipyramidal
b. Tetrahedral, pentagonal bipyramidal
c. Trigonal bipyramidal, pentagonal bipyramidal
d. Tetrahedral, octahedral

**16** Which of the following set of molecules have the same shape but different hybridisation?

a. $H_2O$, $H_2S$ b. $XeO_3$, $BrF_3$ c. $[Ni(CN)_4]^{2-}$, $[Ni(CO)_4]$ d. $BeCl_2$, $I_3^-$

**17** Which is the correct order of the bond angle?

a. $NH_3 < NF_3$ b. $H_2O > Cl_2O$ c. $PH_3 < SbH_3$ d. $H_2Te < H_2S$

## One or More than One Options Correct Type

**Direction** *(Q. Nos. 18-21) This section contains 4 multiple choice questions. Each question has four choices (a), (b), (c) and (d), out of which ONE or MORE THAN ONE are correct.*

**18** Which of the following statements are correct regarding $Cl_2O$ molecule?

a. The molecular planar
b. The shape of the molecule is bent
c. O-atom is $sp^3$-hybridised
d. The molecule is non-planar

**19** In $AX_4E_2$, atom $A$ is surrounded by four atoms $X$ and two lone pairs on it. This type of structure is predicted in

a. $ICl_4^-$ b. $XeF_4$ c. $POCl_3$ d. $XeO_2F_2$

**20** Which of the following sets of molecules have different shape but same hybridisation of the central atoms?

a. $NH_3$, $H_2O$ b. $H_2S$, $SO_2$ c. $Ni(CO)_4$, $[NiCl_4]^{2-}$ d. $SF_4$, $PCl_5$

**21** Select the correct statement(s) about $IF_7$.

a. I atom is $sp^3d^3$-hybridised
b. I atom is in highest oxidation state
c. There are five I—F longest and two I—F shortest bonds
d. It has pentagonal bipyramidal structure

## Comprehension Type

**Direction** *(Q. Nos. 22 and 23) This section contains a paragraph, each describing theory, experiments, data, etc. Two questions related to the paragraph have been given. Each question has only one correct answer among the four given options (a), (b), (c) and (d).*

Following reaction, $ClF_3 + AsF_5 \longrightarrow (ClF_2^+)(AsF_6^-)$

**22** Select the correct statements.

a. Hybridisation of Cl changes from $sp^2$ to $sp$
b. Hybridisation of As changes from $sp^3d$ to $sp^3d^2$.
c. Both (a) and (b) are correct
d. None of the above is correct

**23** Select the correct statement(s).

a. $ClF_3$ changes its structure from pyramidal to angular in $ClF_2^+$
b. $AsF_5$ changes to structure from angular bipyramidal to octahedral in $(AsF_6)^-$
c. Both (a) and (b) are correct
d. None of the above is correct

## Matching List Type

Working Space

**Direction** *(Q. Nos. 24 and 25) Choices for the correct combination of elements from Column I and Column II are given as options (a), (b), (c) and (d), out of which one is correct.*

**24** Match the species in Column I with the structure in Column II.

| | Column I | | Column II |
|---|---|---|---|
| i. | $ClF_3$ | p. | Linear |
| ii. | $ICl_4^-$ | q. | See-saw |
| iii. | $SF_4$ | r. | T-shaped |
| iv. | $I_3^-$ | s. | Square planar |

**Codes**

| | i | ii | iii | iv |
|---|---|---|---|---|
| a. | p | q | r | s |
| b. | s | q | p | r |
| c. | p | s | r | q |
| d. | r | s | q | p |

**25** Match the species in Column I with the shape in Column II.

| | Column I | | Column II |
|---|---|---|---|
| i. | $OSF_2$ | p. | Tetrahedral |
| ii. | $O_2SF_2$ | q. | Linear |
| iii. | $XeF_4$ | r. | Square planar |
| iv. | $ClO_4^-$ | s. | Trigonal pyramidal |
| v. | $I_3^-$ | t. | See-saw |

**Codes**

| | i | ii | iii | iv | v |
|---|---|---|---|---|---|
| a. | s | p | r | p | q |
| b. | s | p | r | q | t |
| c. | p | s | r | p | t |
| d. | q | r | s | t | p |

## One Integer Value Correct Type

**Direction** *(Q. No. 26 and 27) This section contains 2 questions. When worked out will result in an integer from 0 to 9 (both inclusive).*

**26** i. Based on VSEPR theory, the number of 90° $\widehat{F—Br—F}$ angles in $BrF_5$ is **[IIT JEE 2010]**

ii. Based on VSEPR theory, number of lone pairs in $XeF_2$ is......

iii. The total number of lone pair of electrons in melamine is ...... **[JEE Advanced 2013]**

**27** A list of species having the formula $XZ_4$ is given below :

$XeF_4$    $[FeCl_4]^{2-}$

$SF_4$    $[IF_4^-]$

$[BrF_4]^-$

$SiF_4$    $[CoCl_4]^{2-}$

$BF_4^-$    $[PtCl_4]^{2-}$

$[Cu(NH_3)_4]^{2+}$

Defining the shape on the basis of location $X$ and $Z$, the total number of species having a square planar shape is....... **[JEE Advanced 2014]**

# DPP-5 Hybridisation

## Subjective Questions

Working Space

**Direction** *(Q. Nos. 1-3) These questions are subjective in nature, need to be solved completely on notebook.*

**1** Benzene has following structure

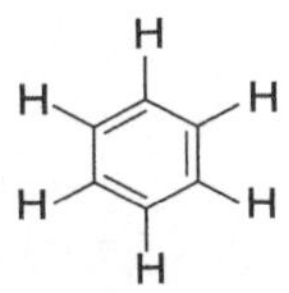

i. What is hybridisation of each carbon atom?

ii. There is delocalisation of $\pi$-electrons as shown

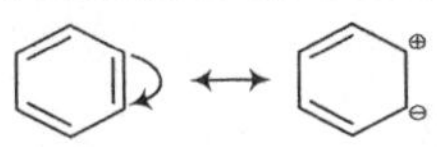

specify hybridisation on

a. $C^{\oplus}$ and  b. $C^{\ominus}$.

**2** Al (13) has following EC

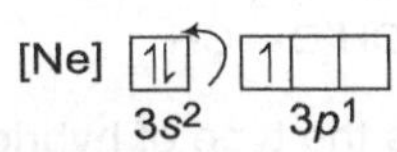

i. How will you explain trivalent nature of Al?

ii. What is type of hybridisation of Al in trivalent compounds?

iii. What is hybridisation of Al in $AlH_4^-$? Explain its formation as Lewis acid-Lewis base reaction.

iv. What is difference between $SiH_4$ and $AlH_4^-$?

**3** i. For the given hydrocarbon $\underset{A}{CH_3} \overset{p}{—} \underset{B}{CH} \overset{q}{=} CH \overset{r}{—} \underset{C}{C} \overset{s}{\equiv} CH$

a. Specify hybridisation of each carbon.

b. What is percentage of *s*-character in each?

c. What is order of electronegativity of each carbon?

d. Arrange bond length and bond energy of *p*, *q*, *r* and *s* (C—C) bond.

ii. What is hybridisation of each C-atom in allene?

## Only One Option Correct Type

**Direction** *(Q. Nos. 4-14) This section contains 11 multiple choice questions. Each question has four choices (a), (b), (c) and (d), out of which ONLY ONE option is correct.*

**4** Out of the following $NO_3^-$, $CO_3^{2-}$, $ClO_3^-$ and $SO_3$.

Isostructural and isoelectronic species are

a. $NO_3^-$, $CO_3^{2-}$  b. $SO_3$, $NO_3$

c. $ClO_3^-$, $CO_3^{2-}$  d. $CO_3^{2-}$, $SO_3$

Working Space

**5** Which of the following statement is incorrect regarding $NO_2$ molecule?

a. It has one unpaired electron is one of the $sp^2$-hybridised orbital
b. It is coloured due to unpaired electron
c. It is deeply coloured in dimeric form
d. Paramagnetic behaviour is lost in dimeric form

**6** Molecule $MX_3$ (atomic number $M < 21$) has zero dipole moment, the sigma bonding orbitals used by $M$ are

a. pure $p$
b. $sp$-hybridised
c. $sp^2$-hybridised
d. $sp^3$-hybridised

**7** $(C \overset{\sigma}{—} C)$ bond in between $(C_2—C_3)$ vinyl acetylene is formed by ...... overlapping

a. $sp$-$sp$
b. $sp^2$-$sp^2$
c. $sp^2$-$sp^3$
d. $sp^2$-$sp$

**8** Consider the following statements for diborane.

I. Boron atom is $sp^3$-hybridised.
II. B—H—B angle is 180°.
III. There are two terminal B—H bonds for each boron atom.
IV. There are only 12 bonding electrons available.

Of these, correct statements are

a. I, II and IV
b. I, II and III
c. II, III and IV
d. I, III and IV

**9** Considering the state of hybridisation of C-atoms, select the molecule among the following which is linear

a. $CH_3CH_2CH_2CH_3$
b. $CH_3CH{=}CHCH_3$
c. $CH_3C{\equiv}CCH_3$
d. $CH_2{=}CHCH_2C{\equiv}CH$

**10** In which of the following species, the central atom has the type of hybridisation which is not the same as that present in the other three?

a. $PCl_5$
b. $SF_4$
c. $I_3^-$
d. $SbCl_6^-$

**11** $BF_3$ is electron deficient (Lewis acid) and $NH_3$ is electron-rich (Lewis acid). In 1 : 1 complex of $BF_3$ and $NH_3$, then coordination geometry and hybridisation of N and B atoms are :

| | N | B |
|---|---|---|
| a. | tetrahedral, $sp^3$ | tetrahedral, $sp^3$ |
| b. | pyramidal, $sp^3$ | pyramidal, $sp^3$ |
| c. | pyramidal, $sp^3$ | planar, $sp^2$ |
| d. | pyramidal, $sp^3$ | tetrahedral, $sp^3$ |

**12** If there are five electron pairs in outer shell, then structure and bond angle as predicted by Sidgwick-Powell theory is

a. octahedral, (90°)
b. trigonal bipyramidal, (120° and 90°)
c. pentagonal bipyramidal (72° and 90°)
d. tetrahedral, (90°)

**13** By VSEPR model, correct matching for $BrF_5$ is

a. $AX_5$- Trigonal bipyramidal
b. $AX_5E$- Square pyramidal
c. $AX_5E_2$ - Square planar
d. $AX_2E_3$- Linear

**14** Based on VSEPR model, $AX_4E$ representation is for

a. $SF_4$
b. $XeF_4$
c. $SiF_4$
d. $BF_4^-$

Working Space

## Statement Type

*Direction (Q. Nos. 15-17) This section is based on Statement I and Statement II. Select the correct answers from the codes given below.*

a. Both Statement I and Statement II are correct and Statement II is the correct explanation of Statement I
b. Both Statement I and Statement II are correct and Statement II is not the correct explanation of Statement I
c. Statement I is correct but Statement II is incorrect
d. Statement II is correct but Statement I is incorrect

**15** **Statement I** $BF_3$ molecule is planar but $NF_3$ is pyramidal.

**Statement II** N atom is smaller than B.

**16** **Statement I** B atom is $sp^2$-hybridised in $B_2H_6$.

**Statement II** There is no lone pair or unpaired electron in $B_2H_6$.

**17** **Statement I** $BCl_3$ exists as monomer while $AlCl_3$ exists as dimer.

**Statement II** Boron atom is too small to form the stable 4-membered ring system.

## One or More than One Options Correct Type

*Direction (Q. Nos. 18-20) This section contains 3 multiple choice questions. Each question has four choices (a), (b), (c) and (d), out of which ONE or MORE THAN ONE are correct.*

**18** Select the correct order of bond lengths of the specified bonds.

a. $CO_3^{2-} < CO_2 < CO$ of (C—O) bond
b. $O_2 < O_2^- < O_2^{2-}$ of (O—O) bond
c. $NO < NO_2 < NO_3^-$ of (N—O) bond
d. $CH{\equiv}CH < CH_2{=}CH_2 < CH_3{-}CH_3$ of (C—H) bond

**19** $NH_4NO_3$ and $NH_4NO_2$ differ is

a. hybridisation of N of the anion
b. hybridisation of N of the cation and anion both
c. decomposition product
d. structure of the anion

**20** Select the correct statement(s) about ethene.

a. C-atom is $sp^2$-hybridised
b. On hydrogenation hybridisation of carbon and bond angle both changes
c. nodal plane of $\pi$-bond is located in molecular plane
d. C—H bond length is also affected on hydrogenation

## Comprehension Type

*Direction (Q. Nos. 21 and 22) This section contains a paragraph, each describing theory, experiments, data, etc. Two questions related to the paragraph have been given. Each question has only one correct answer among the four given options (a), (b), (c) and (d).*

In certain polar solvents, $PCl_5$ undergoes an ionisation reaction as shown

$$\underset{(I)}{2PCl_5} \rightleftharpoons \underset{(II)}{PCl_4^+} + \underset{(III)}{PCl_6^-}$$

**21** Select the incorrect statement (s).

a. Dissociation is a redox reaction
b. Hybridisation of I changes from $sp^3d$ to $sp^3$ in II and to $sp^3d^2$ in III.
c. I, II, and III have different structure
d. All of the above are correct statements

*Working Space*

**22** Number of lone pairs on P in I, II and III are

a. 0, 1, 2 b. 0, 0, 0
c. 1, 2, 3 d. 0, 2, 1

## Matching List Type

***Direction** (Q. Nos. 23 and 24) Choices for the correct combination of elements from Column I and Column II are given as options (a), (b), (c) and (d), out of which one is correct.*

**23** Match the following compounds given in Column I with their geometry given is Column II

| | Column I | | Column II |
|---|---|---|---|
| i. | $H_2S$ | p. | Linear |
| ii. | $N_2O_4$ | q. | Angular |
| iii. | HCN | r. | Planar |
| iv. | $SbCl_6^-$ | s. | Tetrahedral |
| v. | $BF_4^-$ | t. | Octahedral |

**Codes**

| | i | ii | iii | iv | v |
|---|---|---|---|---|---|
| a. | p | p | r | s | t |
| b. | p | q | r | t | s |
| c. | q | r | p | t | s |
| d. | r | p | q | s | t |

**24** Match the processes in Column I with type of changes in hybridisation in Column II.

| | Column I (Process) | | Column II (Change in hybridisation of underlined atoms) |
|---|---|---|---|
| i. | $\underline{Al}H_3 \longrightarrow \underline{Al}H_4^-$ | p. | $sp^2$ to $sp^3$ |
| ii. | $CH_3\overset{O}{\overset{\Vert}{\underline{C}}}OH \longrightarrow CH_3CH_2OH$ | q. | $sp^3$ to $sp^2$ |
| iii. | $CH_3\underset{OH}{\underline{C}}HCH_3 \longrightarrow CH_3CH{=}CH_2$ | r. | $sp^3$ to $sp$ |
| iv. | $CH_3\underset{Cl}{\underset{\vert}{CH}}-\underset{Cl}{\underset{\vert}{\underline{C}H_2}} \longrightarrow CH_3C\equiv CH$ | s. | $sp$ to $sp^2$ |
| v. | $CH_3\underline{C}\equiv\underline{C}H \longrightarrow CH_3\overset{O}{\overset{\Vert}{C}}CH_3$ | t. | $sp$ to $sp^3$ |

**Codes**

| | i | ii | iii | iv | v |
|---|---|---|---|---|---|
| a. | p | p | q | r | s |
| b. | p | p | q | r | s,t |
| c. | q | r | p | s | t |
| d. | p | p | q | r | s |

## One Integer Value Correct Type

***Direction** (Q. No. 25) This section contains a question. When worked out will result in an integer from 0 to 9 (both inclusive).*

**25** Most acidic H in the following compound is attached to C-atoms denoted by at S.N. ( ) ...

$$\underset{5}{CH_2}{=}\underset{4}{CH}-\underset{3}{CH_2}-\underset{2}{C}\equiv\underset{1}{CH}$$

# DPP-6 Molecular Orbital Theory

## Subjective Questions

Working Space

*Direction* (Q. Nos. 1-3) *These questions are subjective in nature, need to be solved completely on notebook.*

**1** i. Explain the formation of MO from atomic orbitals.
ii. Place different MO in increasing energy for the species with electrons $\leq 14$.
iii. Place different MO in increasing energy for the species with electrons $> 14$.

**2** For each of the following species
$C_2$, $C_2^+$, $C_2^{2-}$, CO, $CN^-$, CN, $CN^+$, $B_2$, $Li_2$.
i. Write the molecular orbital electronic configuration (EC).
ii. Determine bond order.
iii. Determine whether paramagnetic or diamagnetic.

**3** For each of the following species
$O_2$, $O_2^-$, $O_2^{2-}$, $O_2^+$ $F_2$, NO, OF and $Ne_2$.
i. Write the molecular orbital E.C.
ii. Determine bond order.
iii. Determine whether paramagnetic or diamagnetic.

## Only One Option Correct Type

*Direction* (Q. Nos. 4-17) *This section contains 14 multiple choice questions. Each question has four choices (a), (b), (c) and (d), out of which ONLY ONE option is correct.*

**4** Assuming that Hund's rule is violated, the bond order and magnetic nature of the diatomic molecule $B_2$ is **[IIT JEE 2010]**

a. 1 and diamagnetic
b. 0 and diamagnetic
c. 1 and paramagnetic
d. 0 and paramagnetic

**5** In which of the following pairs of molecules/ions both the species are not likely to exist? **[JEE Main 2013]**

a. $H_2^+$, $He_2^{2+}$
b. $H_2^-$, $He_2^{2-}$
c. $H_2^{2+}$, $He_2$
d. $H_2^-$, $He_2^{2+}$

**6** Stability of the species $Li_2$, $Li_2^-$ and $Li_2^+$ increases in the order **[JEE Main 2013]**

a. $Li_2 < Li_2^+ < Li_2^-$
b. $Li_2^- < Li_2^+ < Li_2$
c. $Li_2 < Li_2^- < Li_2^+$
d. $Li_2^- < Li_2 < Li_2^+$

**7** Assuming (2*s*-2*p*) mixing is not operative, the paramagnetic species among the following is **[JEE Advanced 2014]**

a. $Be_2$
b. $B_2$
c. $C_2$
d. $N_2$

Working Space

**8** Select the correct statement about $O_2^+$ and $O_2$.

a. $O_2^+$ and $O_2$ both are paramagnetic and bond order of $O_2^+$ is greater than that $O_2$
b. $O_2^+$ is paramagnetic and $O_2$ is diamagnetic and bond order of $O_2^+$ is greater than that of $O_2$
c. $O_2^+$ is paramagnetic and $O_2$ is diamagnetic and bond order of $O_2^+$ is less than that of $O_2$
d. None of the above is correct

**9** The common features among the species $CN^-$, CO, $NO^+$ and $N_2$ are

a. isoelectronic, paramagnetic, bond order three
b. isoelectronic, diamagnetic, bond order three
c. isoelectronic, paramagnetic, bond order two
d. isoelectronic, diamagnetic, bond order two

**10** According to MO theory which of the following lists ranks the nitrogen species in terms of increasing bond order?

a. $N_2^- < N_2 < N_2^{2-}$ b. $N_2^{2-} < N_2^- < N_2$
c. $N_2 < N_2^{2-} < N_2^-$ d. $N_2^- < N_2^{2-} < N_2$

**11** O—F and $F_2$ can be compared in terms of

a. OF is paramagnetic while $F_2$ is diamagnetic
b. OF is more stable towards dissociation into atoms
c. Both the above are correct
d. None of the above is correct

**12** Which of the following diatomic molecules would be stabilised by the removed of an electron?

a. $C_2$ b. CN
c. $N_2$ d. $O_2$

**13** $H_A + H_B \longrightarrow H_2$ ($\sigma$-bond)

Probability (electron charge density) of bonding and anti-bonding molecular orbitals are given.

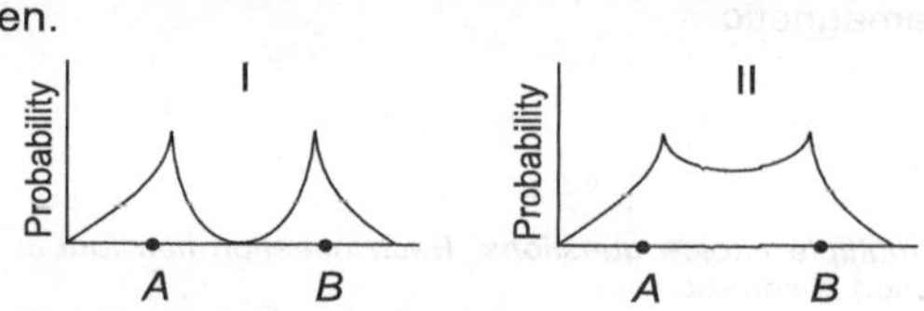


Select the correct probability,

a. I bonding, II anti-bonding
b. II bonding, I anti-bonding
c. Both I and II bonding
d. Both I and II anti-bonding

**14** If one of the electrons ($1s^2$) of helium is taken in excited state then bond order of $He_2$ is

a. 0 (zero) b. 1
c. 0.5 d. 2.0

**15** If one of the electrons in the $He_2$ molecule is taken to the next excited state, then bond order in $He_2$

a. increases by 1 unit
b. decreases by 1 unit
c. increases by 0.5 unit
d. no change

Working Space

**16** The bond energy of $H_2$ is 436 kJ mol$^{-1}$. Thus, bond energy of $H_2^+$ is

a. 436 kJ mol$^{-1}$
b. 218 kJ mol$^{-1}$
c. 327 kJ mol$^{-1}$
d. 872 kJ mol$^{-1}$

**17** Consider the following oxidation/reduction process,

I. $N_2 \xrightarrow{+e^-} N_2^-$

II. $N_2 \xrightarrow{-e^-} N_2^+$

III. $O_2 \xrightarrow{-e^-} O_2^+$

IV. $O_2^+ \xrightarrow{+2e^-} O_2^-$

Magnetic moment does not change in

a. I
b. II
c. III
d. IV

## Statement Type

***Direction*** *(Q. No. 18) This section is based on Statement I and Statement II. Select the correct answer from the codes given below.*

a. Both Statement I and Statement II are correct and Statement II is the correct explanation of Statement I
b. Both Statement I and Statement II are correct and Statement II is not the correct explanation of Statement I
c. Statement I is correct but Statement II is incorrect
d. Statement II is correct but Statement I is incorrect

**18** **Statement I** $N_2$ has a greater dissociation energy than $N_2^+$, whereas $O_2$ has lower dissociation energy than $O_2^+$.

**Statement II** $N_2$ has 14 electrons while $O_2$ has 16 electrons.

## One or More than One Options Correct Type

***Direction*** *(Q. Nos. 19-21) This section contains 3 multiple choice questions. Each question has four choices (a), (b), (c) and (d), out of which ONE or MORE THAN ONE are correct.*

**19** Set of species with identical bond order is/are

a. NO, $O_2^+$
b. CO, $NO^+$
c. $N_2$, $O_2^{2-}$
d. $F_2$, $O_2^2$

**20** One of the electrons of the highest energy level is taken to next excited state in the following diatomic species. Select the species which undergoes change in bond order?

a. $H_2$
b. $Li_2$
c. $B_2$
d. $C_2$

**21** In which of the following processes, does the value of magnetic moment change?

a. $CO \longrightarrow CO^+$
b. $N_2^- \longrightarrow N_2^+$
c. $Zn \longrightarrow Zn^{2+}$
d. $O_2 \longrightarrow O_2^-$

Working Space

## Comprehension Type

***Direction*** *(Q. Nos. 22 and 23) This section contains a paragraph, each describing theory, experiments, data, etc. Two questions related to the paragraph have been given. Each question has only one correct answer among the four given options (a), (b), (c) and (d).*

Valence shell MO electronic configuration of a diatomic species is shown

| $\sigma 2s$ | $\sigma^* 2s$ | $\sigma 2p$ | $\pi 2p$ | $\pi^* 2p$ | $\sigma^* 2p$ |
|---|---|---|---|---|---|
| ⇅ | ⇅ | ⇅ | ⇅ ⇅ | ↑ ↑ | |

* is for anti-bonding molecular orbital (MO).

**22** Bond order of this species is

a. 1 b. 2
c. 3 d. 2.5

**23** Divalent cation of this species

a. has same magnetic moment as divalent anion
b. has same bond order as divalent anion
c. has same bond energy as that of divalent anion
d. has all the above facts true

## Matching List Type

***Direction*** *(Q. No. 24) Choices for the correct combination of elements from Column I and Column II are given.*

**24** Match the conversion in Column I with the type of effect given in Column II.

| | Column I | | Column II |
|---|---|---|---|
| i. | $N_2 \longrightarrow N_2^+$ | p. | Bond order increases. |
| ii. | $O_2^+ \longrightarrow O_2^-$ | q. | Bond order decreases. |
| iii. | $N_2 \longrightarrow N_2^-$ | r. | Magnetic moment increases. |
| iv. | $O_2 \longrightarrow O_2^+$ | s. | Magnetic moment decreases. |
| v. | $O_2 \longrightarrow O_2^{2-}$ | t. | No change in magnetic moment. |

**Codes**

| | i | ii | iii | iv | v |
|---|---|---|---|---|---|
| a. | p, r | t, s | q, s | t, p | r |
| b. | p | q, r | s, t | r, p | q, r |
| c. | s, t | r, p | t, p | q, s | q, t |
| d. | q, r | q, t | q, r | p, s | q, s |

## One Integer Value Correct Type

***Direction*** *(Q. Nos. 25 and 26) This section contains 2 questions. When worked out will result in an integer from 0 to 9 (both inclusive).*

**25** Total number of electrons in anti-bonding MO in $O_2^-$ (superoxide ion) is .......

**26** How many bonding MO are used in the formation of NO?

# DPP-7 Dipole Moment

## Subjective Questions

Working Space

***Direction*** *(Q. Nos. 1-5) These questions are subjective in nature, need to be solved completely on notebook.*

**1** Arrange the following compounds in increasing order of dipole moment.

i. 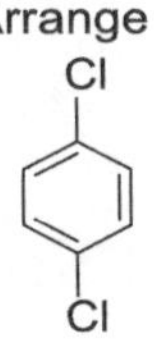

ii. 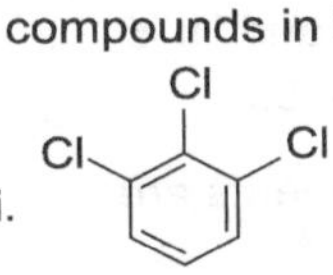

iii. 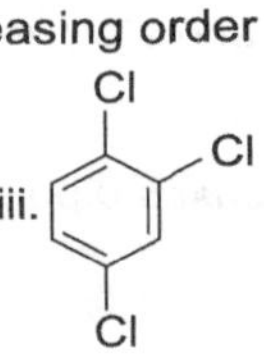

iv. 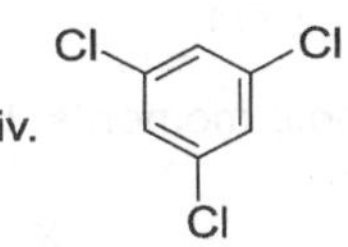

**2** Observed dipole moment of LiF is 6.32 D. Calculate percentage ionic character of LiF if bond length (Li—F) is 0.156 nm.

**3** Relate esu-cm with cm.

**4** $NH_3$ and $NF_3$ are isostructural with one lone pair on N-atom, but dipole moment of $NH_3$ is greater than that of $NF_3$. Explain.

**5** The polarity of B—$X$ bonds is in order B—F > B—Cl > B—Br but Lewis acidity shows the sequence $BF_3 < BCl_3 < BBr_3$. Explain.

## Only One Option Correct Type

***Direction*** *(Q. Nos. 6-17) This section contains 12 multiple choice questions. Each question has four choices (a), (b), (c) and (d), out of which ONLY ONE option is correct.*

**6** Consider the following molecule.

I. $CH_2Cl_2$ II. $BF_3$

III. 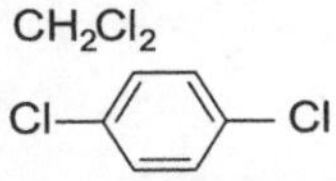 IV. $CO_2$

Species with zero dipole moment is/are

a. I, III and IV b. III and IV c. I, II, III and IV d. II, III and IV

**7** Correct decreasing order of dipole moment of $CH_3F$, $CH_3Cl$ and $CH_3Br$ is

a. $CH_3F > CH_3Cl > CH_3Br$ b. $CH_3F > CH_3Br > CH_3Cl$

c. $CH_3Cl > CH_3F > CH_3Br_2$ d. $CH_3Cl > CH_3Br > CH_3F$

**8** Consider the following compounds

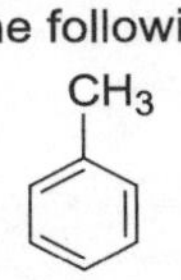

I

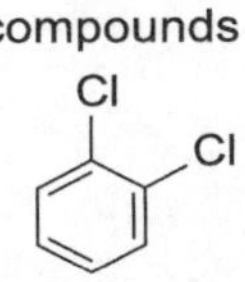

II

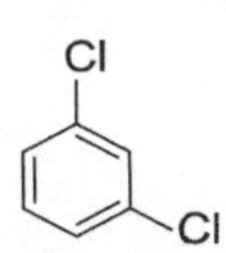

III

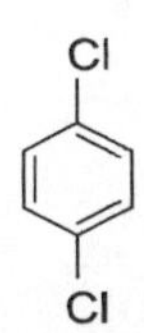

IV

Increasing order of dipole moment of these compounds is

a. I< II< III< IV b. IV < III< II< I c. IV < I< III< II d. IV < II< I< III

Working Space

**9** Dipole moment of 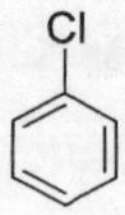 is 1.5 D. Thus, dipole moment of 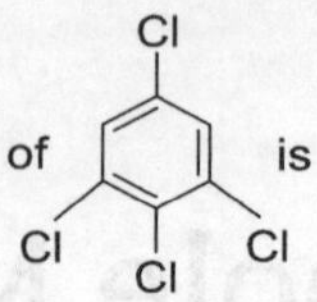 is

a. 0.00 D
b. 1.5 D
c. 2.0 D
d. 3.0 D

**10** Given for HCl and HI

| Species | Bond length | Dipole moment |
|---|---|---|
| HCl | 1.3 Å | 1.03 D |
| HI | 1.6 Å | 0.38 D |

Ratio of fraction of electric charge existing on each atom in HCl and HI is

a. 1.2:1 b. 2.7:1 c. 3.3:1 d. 1:3.3

**11** *Ortho*-isomers have dipole moments. In which cases dipole moments are maximum and minimum?

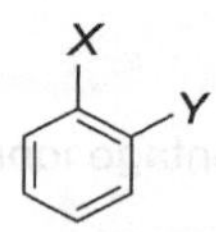

| | **Maximum** X | **Maximum** Y | **Minimum** X | **Minimum** Y |
|---|---|---|---|---|
| a. | Cl | Cl | $CH_3$ | $CH_3$ |
| b. | OH | $CH_3$ | Cl | Cl |
| c. | OH | $NO_2$ | $CH_3$ | $CH_3$ |
| d. | OH | $NO_2$ | Cl | Cl |

**12** Which of the following is non-polar but contains polar bonds?

a. HCl b. $CO_2$ c. $NH_3$ d. $NO_2$

**13** Bond length of H—Cl is 1.2476 Å. Thus, its dipole moment is

a. 6.0 D b. 1.02 D c. 1.25 D d. None of these

**14** $PCl_xF_{5-x}$ shows non-polar behaviour when $x$ is

a. 0, 3, 5 b. 0 c. 2, 1, 4 d. 0, 4, 1

**15** Which of the following bonds is the weakest?

a. Coordinate bond
b. Hydrogen bond
c. van der Waals' forces
d. Covalent bond

**16** In which of the following sets of compounds dipole moment of I is larger than that of II?

| | I | II |
|---|---|---|
| a. | 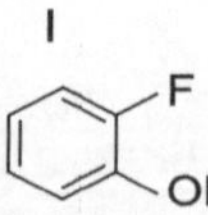 | 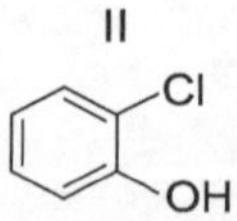 |
| b. | 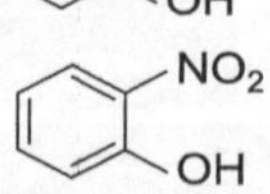 | 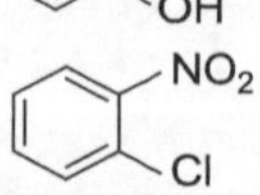 |
| c. | $SO_2$ | $CO_2$ |

d. In all cases

**17** Which of the following would have a permanent dipole?

a. $BF_3$ b. $SiF_4$ c. $SF_4$ d. $XeF_4$

Working Space

## One or More than One Options Correct Type

*Direction (Q. Nos. 18-20) This section contains 3 multiple choice questions. Each question has four choices (a), (b), (c) and (d), out of which ONE or MORE THAN ONE are correct.*

**18** Which of the following has/have zero dipole moments?

a. $SO_3$, $[Ni(CN)_4]^{2-}$, $CCl_4$
b. $SO_3$, CO, $SO_2$
c. $H_2O_2$, $NH_3$, HCl
d. $C_6H_6$, $CO_2$, $[Ni(CN)_4]^{2-}$

**19** Dipole moment is shown by

a. 1, 4-dichloro benzene
b. *cis*-1, 2-dichloro ethene
c. *trans*-1, 2-dichloro ethene
d. *trans*-1, 2-dichloro-2-pentene

**20** The molecules that will have dipole moment are

a. 2, 2-dimethyl propane
b. *trans*-2-pentene
c. *cis*-3-hexene
d. 2, 2, 3, 3-tetramethyl butane

## Comprehension Type

*Direction (Q. Nos. 21-22) This section contains a paragraph, describing theory, experiments, data, etc. Two questions related to the paragraph have been given. Each question has only one correct answer among the four given options (a), (b), (c) and (d).*

$CH_4$ is subjected to photochemical reaction and changes to $CH_3Cl$, $CH_2Cl_2$, $CHCl_3$ and $CCl_4$.

**21** Select the correct alternate.

a. All products have dipole moments
b. Only $CHCl_3$ and $CH_3Cl$ have dipole moments
c. Except $CCl_4$, all products have dipole moment
d. All products are non-polar

**22** One of the intermediates of the photochemical reaction is

a. methyl free radical with one unpaired electron, paramagnetic and $sp^2$-hybridised carbon
b. methyl carbocation with no lone pair $sp^2$- hybridised carbon
c. methyl carbanion with one lone pair $sp^3$- hybridised carbon
d. methyl free radical with one unpaired electron, paramagnetic $sp^3$-hybridised carbon

## One Integer Value Correct Type

*Direction (Q. No. 23 and 24) This section contains 2 questions. When worked out will result in an integer from 0 to 9 (both inclusive).*

**23** How many of the following molecules are polar?

$Cl_2$, ICl, $BF_3$, NO, $SO_2$
$XeF_4$, $CH_2Cl_2$, OCS, $CO_2$

**24** How many molecules out of the following have non-zero value of dipole moments?

$CH_2Cl_2$, $CH_4$, $CCl_4$, $H_2O$, $CHCl_3$, *p*-dichloro benzene, *o*-cresol, *p*-xylene, $SCl_2$, $BF_3$, IBr, HCHO

# DPP-8 Formal Charge and Resonance

## Only One Option Correct Type

*Direction* (Q. Nos. 1-15) *This section contains 15 multiple choice questions. Each question has four choices (a), (b), (c), and (d), out of which ONLY ONE option is correct.*

Working Space

**1** For the

I. benzene ($C_6H_6$) and

II. borazine ($B_3N_3H_6$)

Select the correct statement.

a. C—H bond length in I is identical with N—H and B—H bond lengths

b. The nature of double bond is perfectly identical in both

c. Both the molecules are planar

d. I is non-polar and II is polar

**2** Which is not the correct structure of $NO_3^-$ ion?

a. 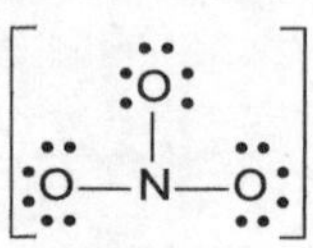

b. 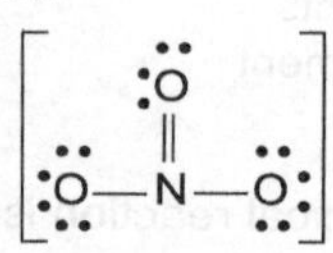

c. 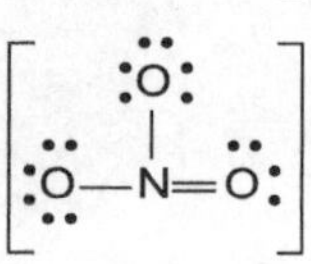

d. 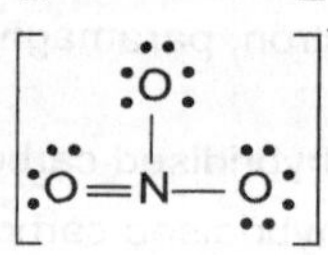

**3** Nitroethene can have following structure

a. 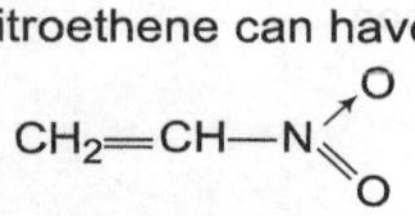

b. 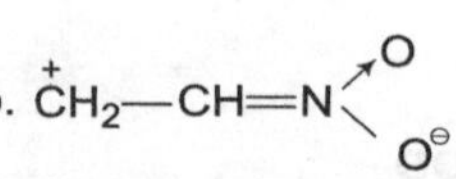

c. Both a. and b.

d. None of these

**4** Select the correct Lewis structures out of the following

a. 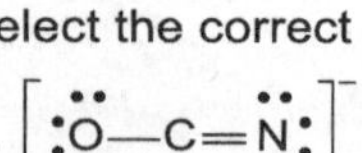 cyanate ion

b. $[C \equiv C:]^{2-}$ carbide ion

c. 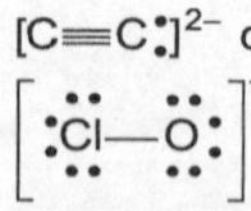 hypochlorite ion

d. $[:\dot{N}=O:]$ nitrogen (II) oxide

**5** Which of the following structures is more acceptable for $HNO_3$?

a. 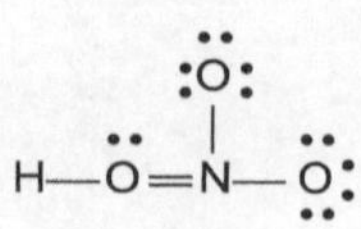

b. 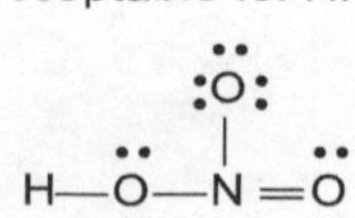

c. Both equally

d. None of these

Working Space

**6** Consider the following pairs

| | (X) | | (Y) |
|---|---|---|---|
| I. | $HNO_3$ | and | $NO_3^{\ominus}$ |
| II. | $H—C(=O)—O—H$ | and | $H—C(=O)—O^{\ominus}$ |
| III. | 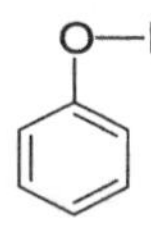 | and | 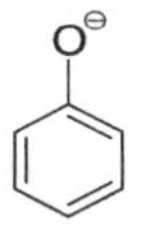 |

More stable species in each pair is

| | I | II | III |
|---|---|---|---|
| a. | *X* | *X* | *Y* |
| b. | *X* | *Y* | *X* |
| c. | *Y* | *X* | *X* |
| d. | *Y* | *Y* | *Y* |

**7** Azide ion $[N_3^-]$ exhibits an (N—N) bond order of 2 and may be represented by resonance structures I, II and III given below :

$$\overset{\ominus}{N}=\overset{\oplus}{N}=\overset{\ominus}{N} \leftrightarrow \overset{2-}{N}—\overset{\oplus}{N}\equiv N \leftrightarrow N\equiv\overset{\oplus}{N}—N^{2-}$$

I II III

Select the correct statement(s) about more contributions.

a. I and II
b. II and III
c. I and III
d. Equal of each

**8** C—Cl bond in $CH_2=CH—Cl$ (vinyl chloride) is stabilised in the same way as in

a. benzyl chloride
b. benzoyl chloride
c. chlorobenzene
d. allyl chloride

**9** $NO_2$ can be represented as $[O=N=O]^+$

Formal charge on each oxygen atom is

a. −2
b. −1
c. 0
d. +1

**10** Out of the following the best Lewis structure of nitrogen chloride is

| O=Cl—N | O—Cl=N | O=N—Cl | O—N=Cl |
|---|---|---|---|
| I | II | III | IV |

a. I
b. II
c. III
d. IV

**11** By $S_N1$ reaction allyl chloride changes to allyl alcohol as shown

$$\underset{I}{CH_2=CH—\overset{14}{C}H_2Cl} + H_2O \xrightarrow{S_N1} \underset{II}{CH_2=CHCH_2OH}$$

Select the correct product II. ($\overset{14}{C}$— an isotope of carbon)

a. $CH_2=CH\overset{14}{C}H_2OH$
b. $\overset{14}{C}H_2=CHCH_2OH$
c. Both (a) and (b) are correct
d. None of these is correct

Working Space

**12** Stability of (C—Cl) bond in chlorobenzene is similar to (C—Cl) bond in

a. benzyl chloride
b. vinyl chloride
c. allyl chloride
d. All of these

**13** Consider the following compounds

I. Vinyl chloride
II. $B_3N_3H_6$
III. $C_5H_5^-$

Delocalisation takes place

a. I, II and III
b. I and II
c. II and III
d. I and III

**14** Electron dot structure of $SO_2$ is given below

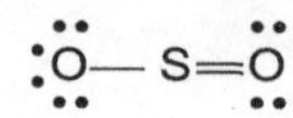

Formal charges on each atom are

| | O (left) | S | O (right) |
|---|---|---|---|
| a. | +1 | 0 | −1 |
| b. | +1 | 0 | 0 |
| c. | −1 | 0 | −1 |
| d. | 0 | 0 | 0 |

**15** Out of the following two structures of $N_2O$, which is more accurate?

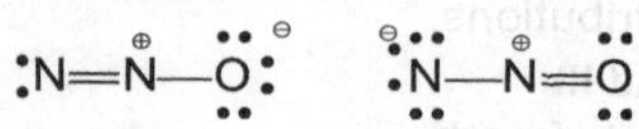

I II

a. I
b. II
c. Both equally
d. None of the above

## One or More than One Options Correct Type

***Direction*** *(Q. Nos. 16 and 17) This section contains 2 multiple choice questions. Each question has four choices (a), (b), (c) and (d), out of which ONE or MORE THAN ONE are correct.*

**16** Select the correct statements about $O_3$.

a. True structures is hybrid of I and II

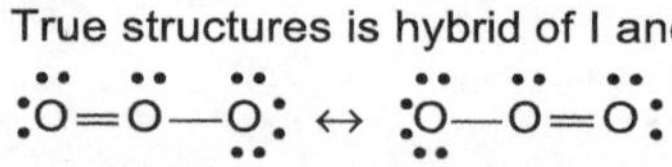

I II

b. Bond order is 1.5
c. (O—O) bond length is between (O—O) single bond and (O=O) double bond
d. It is paramagnetic in nature

**17** Select the correct statement(s) about benzene ($C_6H_6$).

a. (C=C) bond is shorter than (C—C) bond
b. All (C—H) bonds are equivalent
c. C-atom is $sp^2$-hybridised
d. All (C=C) and (C—C) bonds are of equal stability

Working Space

## Comprehension Type

*Direction (Q. Nos. 18-20) This section contains a paragraph, describing theory, experiments, data, etc. Three questions related to the paragraph have been given. Each question has only one correct answer among the four given options (a), (b), (c) and (d).*

Fluorine nitrate, $F\overset{*}{O}NO_2$ is an oxidising agent, used as a rocket propellant.

($\overset{*}{O}$ this oxygen is different from other two)

| Given, | Bond length |
|---|---|
| N—O | 129 pm |
| N—$\dot{O}$ | 139 pm |
| $\dot{O}$—F | 142 pm |
| | **Bond angles** |
| O—N—O | 125° |
| F—$\dot{O}$—N | 105° |

Also, $N\overset{*}{O}F$ plane is perpendicular to $O_2N\overset{*}{O}$ plane.

**18** Possible structure of $FONO_2$ is

a. 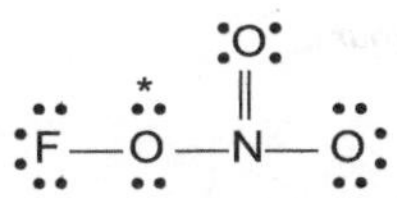

b. 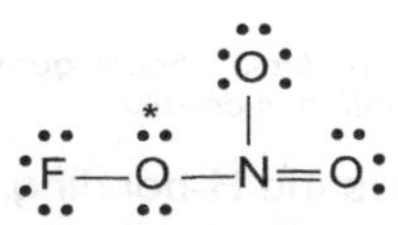

c. Both (a) and (b)

d. None of these

**19** N-atom is...... hybridised.

a. $sp^2$ b. $sp$ c. $sp^3$ d. $sp^3d$

**20** Which is/are correct σ -bonding scheme?

a. $F(p)$—$\overset{*}{O}(sp^3)$

b. $\overset{*}{O}(sp^3)$—$N(sp^2)$

c. $O(p)$—$N(sp^2)$

d. All are correct σ -bonding schemes

## One Integer Value Correct Type

*Direction (Q. Nos. 21 and 22) This section contains 2 questions. When worked out will result in an integer from 0 to 9 (both inclusive).*

**21** How many electrons are delocalised in pyridine?

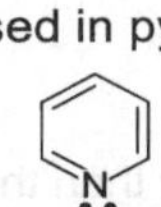

**22** Addition of HBr at position *A* and *B* in the following compound is called *x*, *y*-addition.

$$CH_3—\underset{A}{CH}=CH—CH=\underset{B}{CH}—CH_3$$

What is value of $(x + y)$?

# DPP-9 Hydrogen Bonding

## Subjective Questions

*Direction* (Q. Nos. 1-2) *These questions are subjective in nature, need to be solved completely on notebook.*

Working Space

**1** Explain the types of H-bonding in

i. $H_2O$ molecules
ii. HF molecules
iii. *o*-nitrophenol
iv. *p*-hydroxy benzaldehyde

**2** Out of $NH_3$, $PH_3$, $AsH_3$, boiling point of $NH_3$ is maximum and that of $PH_3$ is least. Explain.

## Only One Option Correct Type

*Direction* (Q. Nos. 3-15) *This section contains 13 multiple choice questions. Each question has four choices (a), (b), (c) and (d), out of which ONLY ONE option is correct.*

**3** Which of the following correctly describes the H-bonding present when a carboxylic acid dissolves in water?

a. 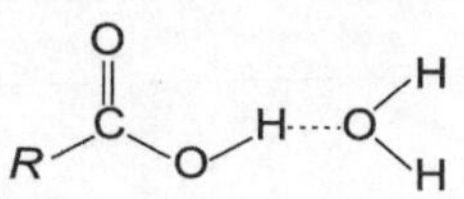

b. 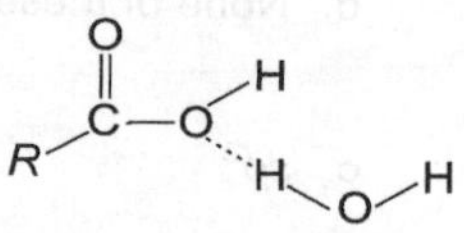

c. 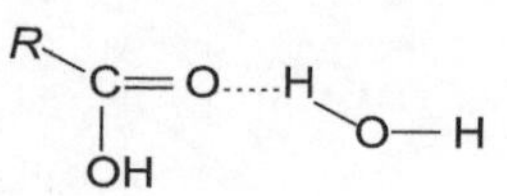

d. All of these

**4** Which of the following compounds are soluble in $H_2O$?

a. $CH_3CHO$
b. $CH_3COCH_3$
c. $CH_3CH_2OH$
d. All of these

**5** Which of the following has maximum boiling point?

a. $CH_2OH$
 |
 $CH_2OH$

b. $CH_3CH_2CH_2OH$

c. $H_2O$

d. $CH_3CH_2CH_2CH_3$

**6** In which of the following boiling point of Column I is not higher than that of Column II?

| | Column I | Column II |
|---|---|---|
| a. | $CH_3CH_2OH$ | $CH_3OH$ |
| b. | $CH_3CH_2OH$ | $CH_3CH_2CH_3$ |
| c. | $CH_3OH$ | $H_2O$ |
| d. | $CH_3CH_2CH_2CH_2OH$ | $H_2O$ |

**7** Which of the following correctly represents H-bonding?

a. $CH_3—\underset{\underset{H}{|}}{O}\cdots CH_3—O—H$

b. $CH_3—\underset{\underset{H}{|}}{O}\cdots H—O—CH_3$

c. $CH_3—O—H\cdots CH_3—O—H$

d. $CH_3—\underset{\underset{H}{|}}{O}\cdots H—CH_2—O—H$

Working Space

8 In dimer formation of *R*COOH, 2*R*COOH $\rightleftharpoons$ (*R*COOH)$_2$
which O-atom(s) is/are involved through H-bonding?

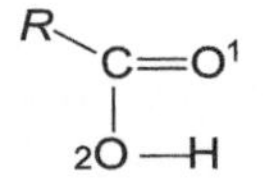

a. Only 1
b. Only 2
c. Both a. and b.
d. None of these

9 In the following compounds, select the points of intermolecular H-bonding (linear).

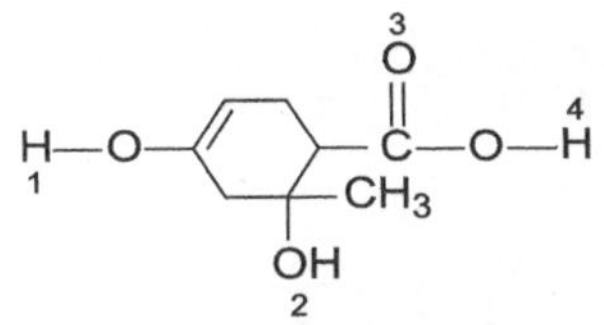

a. 1, 2, 3
b. 1, 3, 4
c. 1, 2, 4
d. 1, 4

10 In the following alcohol, which —OH group is involved to the maximum extent in H-bonding?

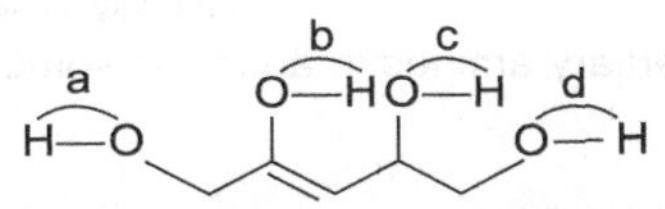

11 Among the following sets, highest boiling points are of the species.
I. HF, HCl, HBr, HI
II. $H_2O$, $H_2S$, $H_2Se$, $H_2Te$
III. $NH_3$, $PH_3$, $AsH_3$, $SbH_3$
a. HF, $H_2O$, $NH_3$
b. HCl, $H_2S$, $PH_3$
c. HF, $H_2Te$, $AsH_3$
d. HF, $H_2S$, $AsH_3$

12 The H-bond is shortest in
a. S—H···S
b. N—H···O
c. F—H···F
d. F—H···O

13 What is the dominant intermolecular force or bond that must be overcome in converting liquid $CH_3CH_2OH$ to vapours?
a. H-bonding
b. Dipole-dipole interactions
c. Covalent bonds
d. London dispersion force

14 Number of water molecules directly attached to one water molecule is
a. 1
b. 2
c. 3
d. 4

15 Consider the following compounds
I. HCl
II. HF
III. $CH_3COOH$
IV. $CH_4$
V. $CH_3OH$
VI. $CH_3COO^-$
H-bonding is not present in
a. I, III and IV
b. I, IVand VI
c. II, III and V
d. II, III and IV

## One or More than One Options Correct Type

Working Space

**Direction** *(Q. Nos. 16-19) This section contains 4 multiple choice questions. Each question has four choices (a), (b), (c) and (d), out of which ONE or MORE THAN ONE are correct.*

**16** In which of the following I is more volatile than II?

a. 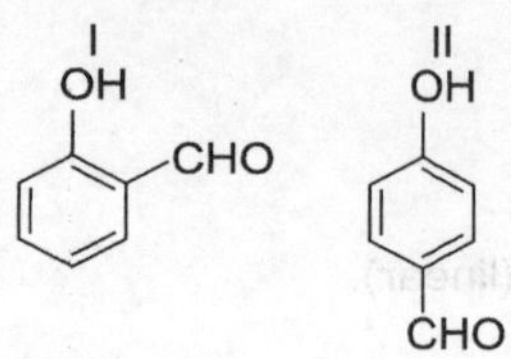

b. 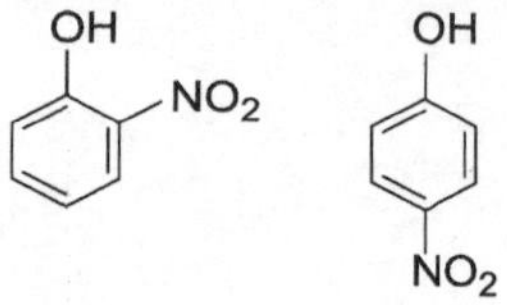

c. $H_2O$     $H_2S$

d. $CH_3\,CH_2\,OH$     $CH_3—O—CH_3$

**17** Hydrogen bonding plays a central role in the following phenomenon **[JEE Advanced 2014]**

a. ice floats in water

b. higher Lewis basicity of primary amines than tertiary amines in aqueous solution

c. formic acid is more acidic than acetic acid

d. dimerisation of acetic acid (in benzene)

**18** Which of the following molecules are stabilised by intramolecular H-bonding?

a. *o*-nitrophenol

b. *p*-nitrophenol

c. Acetyl acetone in enolic form

d. Benzoic acid

**19** Out of

I. $CCl_3CH(OH)_2$     II. $CH_3CH(OH)_2$     III. $CH_2OH$—$CH_2OH$ (III: two $CH_2OH$ groups bonded to each other)

unstable compound is/are

a. I and II

b. Only I

c. Only II

d. Only III

## Matching List Type

**Direction** *(Q. No. 20) Choices for the correct combination of elements from Column I and Column II are given as options (a), (b), (c), and (d), out of which one is correct.*

**20** Match the species in Column I with rank order of their boiling point in Column II. (Species with lowest boiling points is at $S_N1$).

| | Column I | | Column II |
|---|---|---|---|
| i. | $CH_3OH$ | p. | 1 |
| ii. | $CH_3CH_2OH$ | q. | 2 |
| iii. | $H_2O$ | r. | 3 |
| iv. | $CH_3CH_2CH_2CH_2OH$ | s. | 4 |
| v. | $CH_3CH_2CH_2CH_3$ | t. | 5 |

**Codes**

| | i | ii | iii | iv | v |
|---|---|---|---|---|---|
| a. | p | q | r | t | s |
| b. | q | r | s | t | p |
| c. | q | r | t | s | p |
| d. | r | q | s | t | p |

SET - 1

# Revisal Problems
## for JEE Main

### Only One Option Correct Type

Working Space

**Direction** *(Q. Nos. 1-20) This section contains 20 multiple choice questions. Each question has four choices (a), (b), (c), and (d), out of which ONLY ONE option is correct.*

**1** Some of the following compounds have more than one type of bonding.

$KCN, NH_4Cl, NaCl, H_2O, H_2SO_4, CaO, MgCl_2, CH_4, H_2$

These are

a. $KCN, NH_4Cl, H_2O, H_2SO_4$
b. $KCN, NaCl, CaO, MgCl_2$
c. $NaCl, H_2O, CH_4, H_2$
d. $H_2O, H_2SO_4, MgCl_2, CH_4$

**2** Out of the following paramagnetic species are $OH^-$, $OH$, $NO_3$, $SO_3$, $SO_3^{2-}$, $HO_2$

a. $OH^-, SO_3, SO_3^{2-}$
b. $OH, NO_3, HO_2$
c. $OH^-, OH, NO_3$
d. All are paramagnetic

**3** The Cl—C—Cl angles in

I. 1, 1, 2, 2-dichloroethene
II. tetrachloromethane and
III. 1, 1-dichloroethane

respectively will be

| | I | II | III |
|---|---|---|---|
| a. | 120° | 90° | 60° |
| b. | 120° | 109.5° | 109.5° |
| c. | 109.5° | 90° | 60° |
| d. | 120° | 120° | 120° |

**4** Metallic bonding is explained by

a. band model
b. electron-sea model
c. Both 'a' and 'b'
d. None of these

**5** The shape of $IF_7$ molecule is

a. pentagonal bipyramidal
b. trigonal pyramidal
c. tetrahedral
d. square planar

**6** $AsF_5$ molecule is trigonal bipyramidal. The hybrid orbitals used by As atoms for bonding are

a. $d_{x^2-y^2}, d_{z^2}, s, p_x, p_y$
b. $d_{xy}, s, p_x, p_y, p_z$
c. $s, p_x, p_y, p_z, d_{xy}$
d. $d_{x^2-y^2}, s, p_x, p_y, p_z$

**7** $P_4$ is stable form of phosphorus.The percentage of $p$-character in the orbital forming (P—P) bonds in $P_4$ is

a. 25
b. 33
c. 50
d. 75

**8** Central O-atom in $O_3$ is ······ hybridised.

a. $sp$
b. $sp^2$
c. $sp^3$
d. $dsp^2$

**9** $Mg_2C_3 + H_2O \longrightarrow C_3H_4; C_3^{4-}$ ion has

Working Space

a. two σ and two π-bonds
b. three σ and one π-bond
c. two σ and one π-bond
d. two σ and three π-bonds

**10** Among the following species, $NF_3$, $NO_3^-$, $BF_3$, $H_3O^+$, $N_3H$ isostructural pairs are

a. $[NF_3, NO_3^-]$; $[BF_3, H_3O^+]$
b. $[NF_3, N_3H]$; $[BF_3, NO_3^-]$
c. $[NF_3, H_3O^+]$; $[BF_3, NO_3^-]$
d. $[NF_3, H_3O^+]$; $[N_3H, BF_3]$

**11** If there are five electron pairs in outer shell of an atom, then structure and bond angle as predicted by Sidgwick–Powell theory is

a. octahedron; 90°
b. trigonal bipyramidal; 120° and 90°
c. pentagonal bipyramidal; 72° and 90°
d. tetrahedron; 109.5°

**12** Consider the following molecules

I. Acetic acid
II. *o*-nitrophenol
III. *m*-nitrophenol
IV. *o*-boric acid

Select the species showing intermolecular (*X*) and intermolecular the H-bonding (*Y*).

| | *X* | *Y* |
|---|---|---|
| a. | I, II | III, IV |
| b. | II | I, III, IV |
| c. | I, III | II, IV |
| d. | II, III | I, IV |

**13** In which of the following molecules would you expect the nitrogen-to-nitrogen bond to be the shortest?

a. $N_2H_4$
b. $N_2$
c. $N_2O_4$
d. $N_2O$

**14** Chelation can occur in

a. complexes with bidentate ligands
b. intramolecular H-bonding
c. Both 'a' and 'b'
d. None of the above

**15** The correct order of dipole moment of HF, $H_2S$ and $H_2O$ molecules is in order.

a. $HF < H_2S < H_2O$
b. $HF < H_2S > H_2O$
c. $HF > H_2S > H_2O$
d. $HF > H_2O < H_2S$

**16** Solubility of the salts of group II A (alkaline earth metals) is in order

a. $Be(OH)_2 < Mg(OH)_2 < Ca(OH)_2 < Sr(OH)_2 < Ba(OH)_2$
b. $BeCO_3 < MgSO_4 < CaSO_4 < SrSO_4 < BaSO_4$
c. $BeCO_3 < MgCO_3 < CaCO_3 < SrCO_3 < BaCO_3$
d. All of the above are correct variations

**17** Out of $KO_2$, $NO_2$, $\overset{+}{N}O_2$, NO, $K_2O_2$ and $K_2O$ unpaired electrons are present in

a. $KO_2$, $NO_2$, $K_2O$
b. $KO_2$, $NO_2$, NO, $K_2O_2$
c. $KO_2$, $NO_2$, NO
d. $NO_2$, $\overset{+}{N}O_2$, NO

**18** EN values of elements are given

| | | | |
|---|---|---|---|
| K = 0.8 | Ge = 1.8 | H = 2.1 | Cs = 0.7 |
| P = 2.1 | Cl = 3.0 | S = 2.5 | F = 4.0 |

Out of KCl, $PH_3$, GeCl, $H_2S$ and CsF ionic compounds are

a. KCl, GeCl, CsF
b. KCl, GeCl
c. $PH_3$, $H_2S$, KCl, CsF
d. KCl, CsF

Working Space

**19** Ionisation potential of sodium is (5.1 eV) and the electron affinity of chlorine is (3.6 eV). Energy required to transfer one electron from and an isolated sodium atom to an isolated gaseous atom is

a. − 1.5 eV
b. + 1.5 eV
c. 8.7 eV
d. − 8.7 eV

**20** Consider the following complexes.

$$\underset{\text{I}}{[Fe(H_2O)_6]^{3+}}, \underset{\text{II}}{[Fe(CN)_6]^{3-}}, \underset{\text{III}}{[Fe(CN)_6]^{4-}}, \underset{\text{IV}}{[FeF_6]^{3-}}$$

Fe is $sp^3d^2$ hybridisation in which of the above complexes

a. I and II
b. II and III
c. I, II and IV
d. I and IV

## Statement Type

**Direction** *(Q. Nos. 21-24)* ***This section is based on statement I and statement II. Select the correct answer from the codes given below.***

a. Both Statement I and Statement II are correct and Statement II is the correct explanation of Statement I
b. Both Statement I and Statement II are correct and Statement II is not the correct explanation of Statement I
c. Statement I is correct but Statement II is incorrect
d. Statement II is correct but Statement I is incorrect

**21** **Statement I** The chlorine to oxygen bond distance in $ClO_4^-$ is 1.44 Å.

**Statement II** There must be considerable double bond character.

**22** **Statement I** (N—N) bond length in $N_2O_4$ is larger than normal (N—N) bond length in $N_2O_4$.

**Statement II** Nitrogen completes its octet in $N_2O_4$.

**23** **Statement I** $I_3^-$ is linear in structure.

**Statement II** $I_2$ is a Lewis acid and $I^-$ is a Lewis base.

**24** **Statement I** $I_3^-$ is formed but $F_3^-$ is not formed.

**Statement II** $F^-$ is not a Lewis base.

SET - 2

# Revisal Problems for JEE Advanced

## Only One Option Correct Type

Working Space

***Direction*** *(Q. Nos. 1-5) This section contains 5 multiple choice questions. Each question has four choices (a), (b), (c), and (d), out of which ONLY ONE option is correct.*

**1** Select correct matching about $[Ni(CN)_4]^{2-}$, $Ni(CO)_4$, $[Ni(H_2O)_4)]^{2+}$

| | $[Ni(CN)_4]^{2-}$ | $Ni(CO)_4$ | $[Ni(H_2O)_4)]^{2+}$ |
|---|---|---|---|
| a. | $dsp^2$, square planar | $sp^3$, tetrahedral | $sp^3$, tetrahedral |
| b. | $sp^3$, tetrahedral | $sp^3$, tetrahedral | $sp^3$, tetrahedral |
| c. | $dsp^2$, square planar | $dsp^2$, square planar | $dsp^2$, square planar |
| d. | None of the above is correct | | |

**2** Energy gap between the highest filled bands and the lowest empty bands in elements are given :

| Element | Energy gap (kJ $mol^{-1}$) |
|---|---|
| *P* | 0 |
| *Q* | 70.0 |
| *R* | 530.0 |
| *S* | 111.0 |

Thus,

a. element *P* is a metal
b. *Q* and *S* are semiconductors
c. *R* is an insulator
d. All are correct conclusions about the elements *P*, *Q*, *R*, and *S*

**3** Match Column I with Column II and select the correct answer from the codes given below the table.

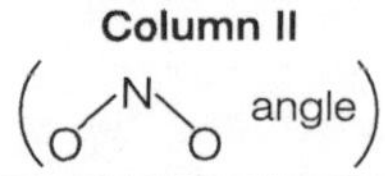

| | Column I (Species) | | Column II (O–N–O angle) |
|---|---|---|---|
| i. | $NO_2^+$ | p. | 180° |
| ii. | $NO_2$ | q. | 132° |
| iii. | $NO_2^-$ | r. | 120° |
| iv. | $NO_3^-$ | s. | 115° |
| | | t. | 109° |

**Codes**

| | i | ii | iii | iv | | | i | ii | iii | iv |
|---|---|---|---|---|---|---|---|---|---|---|
| a. | t | s | r | q | | b. | t | q | s | r |
| c. | p | s | t | r | | d. | p | s | r | q |

**4** Hybridisation of the underlined atoms is/are affected in the following transformations :

a. $LiH + \underline{Al}H_3 \longrightarrow Li\underline{Al}H_4$
b. $\underline{N}H_3 + HCl \longrightarrow NH_4Cl$
c. $3H_3\underline{P}O_2 \xrightarrow{\Delta} PH_3 + 2H_3PO_3$
d. $CH_3\underline{C}H_2CH_3 \xrightarrow{Cl_2} CH_3\underset{\displaystyle Cl}{\underset{|}{C}}HCH_3$

**5** In $BrF_3$ molecule, the lone pairs occupy equatorial positions to minimise

a. *lp* – *lp* repulsion
b. *bp* – *bp* repulsion
c. *lp* – *bp* and *lp* – *lp* repulsion
d. *lp* – *bp* repulsion

## Statement Type

Working Space

*Direction* (Q. Nos. 6 - 8) *This section is based on Statement I and Statement II. Select the correct answer from the codes given below.*

a. Both Statement I and Statement II are correct and Statement II is the correct explanation of Statement I
b. Both Statement I and Statement II are correct and Statement II is not the correct explanation of Statement I
c. Statement I is correct but Statement II is incorrect
d. Statement II is correct but Statement I is incorrect

**6** **Statement I** Two different bond lengths are observed in $PF_5$ molecule but only one bond length is observed in $SF_6$ molecule.

**Statement II** One lone pair exists in $PF_5$.

**7** **Statement I** The ion $ClF_2^-$ is linear but the ion $ClF_2^+$ is bent.

**Statement II** Cl-atom is $sp^2$-hybridised in each.

**8** **Statement I** Carbon and silicon are in group 14 in but (Si=Si) bond is not formed but (C=C) bond is formed.

**Statement II** Larger size of Si results in poor sideway overlap of $p$-orbitals to form $\pi$-bond.

## One or More than One Options Correct Type

*Direction* (Q. Nos. 9-21) *This section contains 13 multiple choice questions. Each question has four choices (a), (b), (c), and (d), out of which ONE or MORE THAN ONE are correct.*

**9** Select correct statement(s).

a. $K_{a_1}$ of maleic acid is higher than that of fumaric acid
b. $K_{a_2}$ of fumaric acid is higher than that of maleic acid
c. Dimethyl glyoxime forms stable complex with $Ni^{2+}$
d. Benzoic acid is dimerised in benzene solvent

**10** Select the correct variation of solubility of different salts in water.

a. $LiF < LiCl < LiBr < LiI$
b. $LiF < NaF < KF < RbF$
c. $CsI < CsBr < CsCl < CsF$
d. $CsNO_3 < KNO_3 < NaNO_3 < LiNO_3$

**11** Which of the following molecules is expected to exhibit diamagnetic behaviour?

a. $C_2$ b. $N_2$ c. $O_2$ d. $S_2$

**12** Which is/are correct variation(s) for

$O_2^+, O_2, O_2^-, O_2^{2-}$

a. $O_2^+ < O_2 < O_2^- < O_2^{2-}$ (O—O) bond length
b. $O_2^+ > O_2 > O_2^- > O_2^{2-}$ (bond order)
c. $O_2^+ > O_2 > O_2^- > O_2^{2-}$ (bond energy)
d. $O_2^{2-} < O_2^- = O_2^+ < O_2$ (magnetic moment)

**13** Which is/are not correct variation NI $I_2^-$, $NH_3$ and $NH_4^+$

a. $NH_2^- < NH_3 < NH_4^+$, (HNH) bond angle
b. $NH_2^- < NH_3 < NH_4^+$ lone pairs on N-atom
c. $NH_2^- < NH_3 < NH_4^+$ (N—H) bond length
d. $NH_2^- < NH_3 < NH_4^+$ number of $s$ and $p$-orbitals used in hybridisation

**14.** The hybridisation scheme for the central atom includes a $d$-orbital contribution in

a. $I_3^-$ b. $PCl_3$ c. $XeF_2$ d. $SF_4$

**15** Select the correct statement(s).

a. Larger the energy gap, larger the conduction in metals
b. The delocalised electrons in the $\pi$-orbitals of graphite cause this allotropic form to conduct electricity
c. Diatomic lithium molecules $Li_2$ are stable towards dissociation
d. Diamond and graphite behave similarly in conduction

**16** One or more species out of the following uses $sp^2$-hybrid orbitals in bonding is/are

a. $PH_3$ b. $\overset{+}{C}H_3$ c. $\overset{\bullet}{C}H_3$ d. $\overset{+}{N}O_2$

**Working Space**

**17** Consider the following species :

I. $\underline{C}O_2$ II. $\underline{S}O_2$ III. $\underline{N}O_2$ IV. $\underline{C}_2H_2$
V. $\underline{S}O_3$ VI. $\underline{N}H_3$

Sets of species with same type of hybridisation on underlined atom is/are

a. I, II and III b. IV, V and VI c. II, III and V d. I and IV

**18** In which of the following cases, driving force is to complete the octet of B-atom permanently

a. $2BH_3 \rightleftharpoons B_2H_6$

b. $B_2H_6 + 2$ 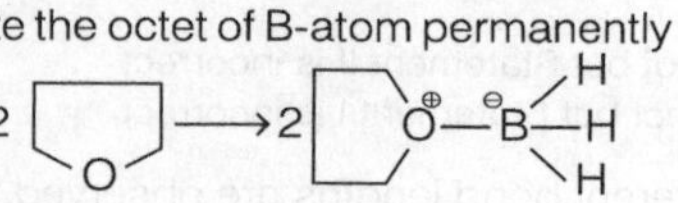

c. $NH_3 + BH_3 \xrightarrow{THF} (NH_3 \longrightarrow BH_3)$

d. $BH_3 \cdot NH_3 \xrightarrow{\Delta} B_3N_3H_6$

**19** Which of the following molecules has the maximum number of $A—X$ bonds of identical bond length when $A$ is the central atom and $X$ is the surrounding atom?

a. $SF_4$ b. $IF_7$ c. $PF_5$ d. $ClO_4^-$

**20** Select the correct statement(s) about $PCl_5$.

a. Three (P—Cl) equatorial bonds lie in one plane and make an angle of 120° with each other
b. Two (P—Cl) axial bonds one lying above and the other lying below the equatorial plane make an angle of 90° with the plane
c. Axial bonds are slightly longer and slightly weaker than equatorial bonds
d. Axial bonds are slightly shorter and slightly stronger than equatorial bonds

**21** Which of the following structures assume linear structure?

a. $Cl_2C{=}CCl_2$ b. $N{\equiv}C—C{\equiv}N$ c. $F_3C—C{\equiv}N$ d. $[S—C{\equiv}N]^-$

## Comprehension Type

***Direction*** (Q. Nos. 22-25) ***This section contains 2 paragraphs, each describing theory, experiments, data, etc. four questions related to the paragraphs have been given. Each question has only one correct answer among the four given options (a), (b), (c) and (d).***

### Passage I

Homolytic and heterolytic bond cleavage of organic compounds may result in short lived intermediates.

$$CH_3Cl \longrightarrow \overset{+}{C}H_3 + Cl^-$$

$$CH_3NO_2 \longrightarrow \overset{-}{C}H_3 + NO_2^+$$

$$CH_3—H \longrightarrow \overset{\bullet}{C}H_3 + \overset{\bullet}{H}$$

**22** Hybridisation of C-atom in the species formed is of the type

| | $\overset{+}{C}H_3$ | $\overset{-}{C}H_3$ | $\overset{\bullet}{C}H_3$ |
|---|---|---|---|
| a. | $sp^2$ | $sp^3$ | $sp^3$ |
| b. | $sp^2$ | $sp^3$ | $sp^2$ |
| c. | $sp^2$ | $sp^2$ | $sp^3$ |
| d. | $sp^3$ | $sp^3$ | $sp^3$ |

**23** Bond angle H–C–H (∠HCH) in the increasing order is

a. $\overset{\bullet}{C}H_3 < \overset{+}{C}H_3 < \overset{-}{C}H_3$
b. $\overset{-}{C}H_3 < \overset{+}{C}H_3 < \overset{\bullet}{C}H_3$
c. $\overset{-}{C}H_3 < \overset{\bullet}{C}H_3 = \overset{+}{C}H_3$
d. $\overset{+}{C}H_3 < \overset{-}{C}H_3 < \overset{\bullet}{C}H_3$

### Passage II

Consider the following compounds of nitrogen and answer the given questions :

$X: N(CH_3)_3$ $Y: N[Si(CH_3)_3]_3$ $Z: NH_3$

**24** Hybridisation of nitrogen atom can be

| | X | Y | Z |
|---|---|---|---|
| a. | $sp^2$ | $sp^2$ | $sp^2$ |
| b. | $sp^3$ | $sp^3$ | $sp^3$ |
| c. | $sp^2$ | $sp^3d$ | $sp^3$ |
| d. | $sp^2$ | $sp^3$ | $sp^2$ |

Working Space

**25** Structures can be

I. Trigonal planar
II. Trigonal pyramidal
III. Square planar

Select the correct structures

| | X | Y | Z | | X | Y | Z |
|---|---|---|---|---|---|---|---|
| a. | I | I | I | b. | II | II | I |
| c. | II | I | II | d. | I | II | III |

## Matching List Type

*Direction (Q. Nos. 26 and 27) Choices for the correct combination of elements from Column I and Column II are given as options.*

**26** Match the reactions in Column I with the nature of reactions/type of the products in Column II. **[IIT JEE 2007]**

| | Column I | | Column II |
|---|---|---|---|
| i. | $O_2^- \longrightarrow O_2 + O_2^{2-}$ | p. | Redox reaction |
| ii. | $CrO_4^{2-} + H^+ \longrightarrow$ | q. | One of the products has triangular planar structure. |
| iii. | $MnO_4^- + NO_2^- + H^+ \longrightarrow$ | r. | Dimeric bridged tetrahedral metal ion |
| iv. | $NO_3^- + H_2SO_4 + Fe^{2+} \longrightarrow$ | s. | Disproportionation |

**Codes**

| | i | ii | iii | iv | | i | ii | iii | iv |
|---|---|---|---|---|---|---|---|---|---|
| a. | p | r | s | q,p | b. | p,s | r | p,q | p |
| c. | q | r,s | q | p,r | d. | s,p | q,r | p | p,r |

**27** Match each of the diatomic molecules in Column I with the property/properties in Column II. **[IIT JEE 2009]**

| | Column I | | Column II |
|---|---|---|---|
| i. | $B_2$ | p. | Paramagnetic |
| ii. | $N_2$ | q. | Undergoes oxidation |
| iii. | $O_2^-$ | r. | Undergoes reduction |
| iv. | $O_2$ | s. | Bond order $\geq 2$ |
| | | t. | Mixing of *s* and *p* orbitals |

**Codes**

| | i | ii | iii | iv |
|---|---|---|---|---|
| a. | r,p,q,s | p,q,r,t | p,r,q | t,s,p,q |
| b. | t,r,s,p | p,r,s,q | p,r,s,t | p,q,r,s,t |
| c. | p,q,r,t | q,r,s,t | p,q,r,t | p,q,r,s,t |
| d. | p,q | r,s,t | p,q,r,s,t | q,r,s |

## One Integer Value Correct Type

*Direction (Q. Nos. 28-30) This section contains 3 questions. When worked out will result in an integer from 0 to 9 (both inclusive).*

**28** At 300 K and1.00 atm, density of gaseous HF is 3.17 $gL^{-1}$. How many HF molecules are associated by H-bonding?

**29** $AX_4$ forms square planar type structure. Distance between *A* and *X*-atoms is 4 units. What is distance between two *X*-atoms in *trans*-position?

**30** A total of $n \times 10^{20}$ energy levels are present in 3s conduction band of single crystal of sodium weighing 26.8 mg. What is the value of *n*?

# JEE Main & AIEEE Archive

## (Compilation of Last 13 Years Questions)

**1** The correct statement for the molecule $CsI_3$ is **(JEE Main 2014)**

a. It is a covalent molecule
b. It contains $Cs^+$ and $I_3^-$
c. It contains $Cs^{3+}$ and $I^-$ ions
d. It contains $Cs^+$, $I^-$ and lattice $I_2$ molecule

**2** For which of the following molecules significant $\mu \neq 0$? **(JEE Main 2014)**

(i) 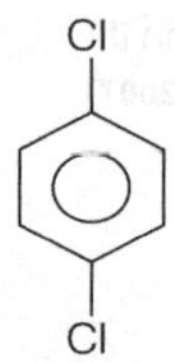


(ii) 

(iii) 

(iv) 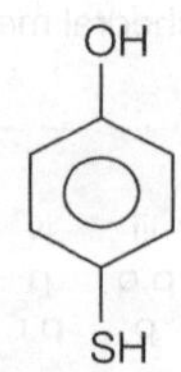


a. Only (i)
b. (i) and (ii)
c. Only (iii)
d. (iii) and (iv)

**3** Which one of the following molecules is expected to exhibit diamagnetic behaviour? **(JEE Main 2013)**

a. $C_2$
b. $N_2$
c. $O_2$
d. $S_2$

**4** In which of the following pairs of molecules/ions both the species are not likely to exist? **(JEE Main 2013)**

a. $H_2^+$, $He_2^{2-}$
b. $H_2^-$, $He_2^{2-}$
c. $H_2^{2+}$, $He_2$
d. $H_2^-$, $He_2^{2+}$

**5** Stability of the species $Li_2$, $Li_2^-$ and $Li_2^+$ increases in the order of **(JEE Main 2013)**

a. $Li_2 < Li_2^+ < Li_2^-$
b. $Li_2^- < Li_2^+ < Li_2$
c. $Li_2 < Li_2^- < Li_2^+$
d. $Li_2^- < Li_2 < Li_2^+$

**6** The molecule having smallest bond angle is **(AIEEE 2012)**

a. $NCl_3$
b. $AsCl_3$
c. $SbCl_3$
d. $PCl_3$

**7** In which of the following pairs, the two species are not isostructural? **(AIEEE 2012)**

a. $CO_3^{2-}$ and $NO_3^-$
b. $PCl_4^+$ and $SiCl_4$
c. $PF_5$ and $BrF_5$
d. $AlF_6^{3-}$ and $SF_6$

**8** Among the following, the maximum covalent character is shown by the compound **(AIEEE 2011)**

a. $FeCl_2$
b. $SnCl_2$
c. $AlCl_3$
d. $MgCl_2$

**9** The hybridisation of orbitals of N atom in $NO_3^-$, $NO_2^+$ and $NH_4^+$ are respectively

a. $sp$, $sp^2$, $sp^3$
b. $sp^2$, $sp$, $sp^3$
c. $sp$, $sp^3$, $sp^2$
d. $sp^2$, $sp^3$, $sp$ **(AIEEE 2011)**

**10** The structure of $IF_7$ is **(AIEEE 2011)**

a. square pyramid
b. trigonal bipyramid
c. octahedral
d. pentagonal bipyramid

**11** Which of the following has maximum number of lone pairs associated with Xe?

a. $XeO_3$
b. $XeF_4$ **(AIEEE 2011)**
c. $XeF_6$
d. $XeF_2$

**12** The number of types of bonds between two carbon atoms in calcium carbide is **(AIEEE 2011)**

a. one sigma, two pi
b. one sigma, one pi
c. two sigma, one pi
d. two sigma, two pi

**13** Using MO theory, predict which of the following species has the shortest bond length? **(AIEEE 2009)**

a. $O_2^{2+}$
b. $O_2^+$
c. $O_2^-$
d. $O_2^{2-}$

**14** Which one of the following constitutes a group of the isoelectronic species? **(AIEEE 2008)**

a. $C_2^{2-}$, $O_2^-$, CO, NO
b. $NO^+$, $C_2^{2-}$, $CN^-$, $N_2$
c. $CN^-$, $N_2$, $O_2^{2-}$, $CO_2^{2-}$
d. $N_2$, $O_2^-$, $NO^+$, CO

**15** Which one of the following pairs of species have the same bond order? **(AIEEE 2008)**

a. $CN^-$ and $NO^+$
b. $CN^-$ and $CN^+$
c. $O_2^-$ and $CN^-$
d. $NO^+$ and $CN^+$

**16** Which of the following species exhibits diamagnetic behaviour? **(AIEEE 2007)**

a. $O_2^{2-}$
b. $O_2^+$
c. $O_2$
d. NO

**17** In which of the following ionisation processes, the bond order has increased and the magnetic behaviour has changed?

a. $C_2 \rightarrow C_2^+$
b. $NO \rightarrow NO^+$
c. $O_2 \rightarrow O_2^+$
d. $N_2 \rightarrow N_2^+$ **(AIEEE 2007)**

**18** Which of the following hydrogen bond is the strongest? **(AIEEE 2007)**

a. O—H···N
b. F—H···F
c. O—H···O
d. O—H···F

**19** In which of the following molecules/ions, all the bonds are not equal? **(AIEEE 2006)**

a. $SF_4$ b. $SiF_4$
c. $XeF_4$ d. $BF_4^-$

**20** The decreasing values of bond angles from $NH_3$ (107°) to $SbH_3$ (91°) down group-15 of the periodic table is due to **(AIEEE 2006)**

a. increasing *bp*-*bp* repulsion
b. increasing *p*-orbital character in $sp^3$
c. decreasing *lp*-*bp* repulsion
d. decreasing electronegativity

**21** Which of the following molecules/ions does not contain unpaired electrons? **(AIEEE 2006)**

a. $O_2^{2-}$ b. $B_2$
c. $N_2^+$ d. $O_2$

**22** Lattice energy of an ionic compound depends upon **(AIEEE 2005)**

a. charge on the ion and size of the ion
b. packing of ions only
c. size of the ion only
d. charge on the ion only

**23** The molecular shapes of $SF_4$, $CF_4$ and $XeF_4$ are

a. different with 1, 0 and 2 lone pair of electrons on the central atom, respectively
b. different with 0, 1 and 2 lone pair of electrons on the central atom, respectively
c. the same with 1, 1 and 1 lone pair of electrons on the central atom, respectively
d. the same with 2, 0 and 1 lone pair of electrons on the central atom, respectively **(AIEEE 2005)**

**24** Which one of the following species is diamagnetic in nature? **(AIEEE 2005)**

a. $H_2^-$ b. $H_2^+$
c. $H_2$ d. $He_2^+$

**25** The maximum number of 90° angles between bond pair-bond pair of electrons is observed in **(AIEEE 2004)**

a. $dsp^3$-hybridisation b. $sp^3d$-hybridisation
c. $dsp^2$-hybridisation d. $sp^3d^2$-hybridisation

**26** Which one of the following has the regular tetrahedral structure ? **(AIEEE 2004)**

(Atomic number : B = 5, S = 16, Ni = 28, Xe = 54)

a. $XeF_4$ b. $SF_4$
c. $BF_4^-$ d. $[Ni(CN)_4]^{2-}$

**27** The correct order of bond angles (smallest first) in $H_2S$, $NH_3$, $BF_3$ and $SiH_4$ is **(AIEEE 2004)**

a. $H_2S < SiH_4 < NH_3 < BF_3$
b. $NH_3 < H_2S < SiH_4 < BF_3$
c. $H_2S < NH_3 < SiH_4 < BF_3$
d. $H_2S < NH_3 < BF_3 < SiH_4$

**28** The states of hybridisation of boron and oxygen atoms in boric acid ($H_3BO_3$) are respectively

a. $sp^2$ and $sp^2$ b. $sp^2$ and $sp^3$ **(AIEEE 2004)**
c. $sp^3$ and $sp^2$ d. $sp^3$ and $sp^3$

**29** The bond order in NO is 2.5 while that in $NO^+$ is 3. Which of the following statements is true for these two species? **(AIEEE 2004)**

a. Bond length in $NO^+$ is greater than in NO
b. Bond length in NO is greater than in $NO^+$
c. Bond length in $NO^+$ is equal to that in NO
d. Bond length is unpredictable

**30** The pair of species having identical shapes for molecules of both species, is **(AIEEE 2003)**

a. $CF_4$, $SF_4$ b. $XeF_2$, $CO_2$
c. $BF_3$, $PCl_3$ d. $PF_5$, $IF_5$

**31** Which one of the following compounds has the smallest bond angle? **(AIEEE 2003)**

a. $SO_2$ b. $OH_2$
c. $SH_2$ d. $NH_3$

**32.** Which one of the following pairs of molecules will have permanent dipole moments for both members? **(AIEEE 2003)**

a. $SiF_4$ and $NO_2$ b. $NO_2$ and $CO_2$
c. $NO_2$ and $O_3$ d. $SiF_4$ and $CO_2$

**33** Select the correct statement. **(AIEEE 2002)**

a. When a covalent bond is formed, transfer of electrons takes place
b. Pure $H_2O$ does not contain any ion
c. A bond is formed when attractive forces overcome repulsive forces
d. HF is less polar than HBr

**34** The number of lone pairs on Xe in $XeF_2$, $XeF_4$ and $XeF_6$ respectively, are **(AIEEE 2002)**

a. 3, 2, 1 b. 2, 4, 6
c. 1, 2, 3 d. 6, 4, 2

**35** Bond angle of 109°28′ is found in

a. $NH_3$ b. $H_2O$ **(AIEEE 2002)**
c. $\overset{\oplus}{C}H_5$ d. $\overset{\oplus}{N}H_4$

**36** Hybridisation of the underlined under lined atom changes in

a. <u>Al</u>$H_3$ changes to $AlH_4^-$ **(AIEEE 2002)**
b. $H_2$<u>O</u> changes to $H_3O^+$
c. <u>N</u>$H_3$ changes to $NH_4^+$
d. In all cases

**37** Increasing order of bond strength of $O_2$, $O_2^-$, $O_2^{2-}$ and $O_2^+$ is **(AIEEE 2002)**

a. $O_2^+ < O_2 < O_2^- < O_2^-$
b. $O_2 < O_2^+ < O_2^- < O_2^{2-}$
c. $O_2^- < O_2^{2-} < O_2^+ < O_2$
d. $O_2^{2-} < O_2^- < O_2 < O_2^+$

# JEE Advance and IIT Archive

## (Compilation of Last 10 Years Questions)

**1** Assuming 2*s*-2*p* mixing is not operative, the paramagnetic species among the following is **(2014 Adv., Single Option Correct Type)**

a. $Be_2$ b. $B_2$
c. $C_2$ d. $N_2$

**2** Hydrogen bonding plays a central role in the following phenomena
**(2014 Adv., One or More than Option Correct Type)**

a. ice floats in water
b. higher Lewis basicity of primary amines than tertiary amines in aqueous solutions
c. formic acid is more acidic than acetic acid
d. dimerisation of acetic acid in benzene

**3** Match the orbital overlap figures shown in Column I with the description given in Column II and select the correct answer using the codes given below the columns. **(2014 Adv., Matching Type)**

| | Column I | | Column II |
|---|---|---|---|
| i. | 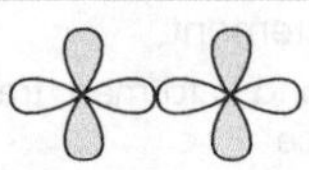 | p. | p-d $\pi$ antibonding |
| ii. | 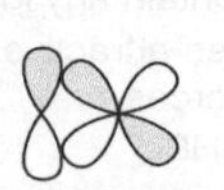 | q. | d-d $\sigma$ bonding |
| iii. | 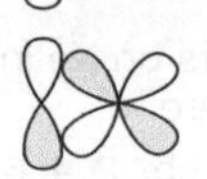 | r. | p-d $\pi$ bonding |
| iv. | 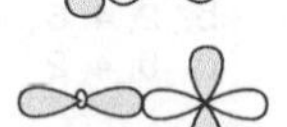 | s. | d-d $\sigma$ antibonding |

**Codes**

| | i | ii | iii | iv |
|---|---|---|---|---|
| a. | q | p | r | s |
| b. | s | r | p | q |
| c. | q | r | p | s |
| d. | s | p | r | q |

**4** Assuming that Hund's rule is violated, the bond order and magnetic nature of the diatomic molecule $B_2$ is
**(2010, Single Option Correct Type)**

a. 1 and diamagnetic
b. 0 and diamagnetic
c. 1 and paramagnetic
d. 0 and paramagnetic

**5** The species having pyramidal shape is
**(2010, Single Option Correct Type)**

a. $SO_3$ b. $BrF_3$
c. $SiO_3^{2-}$ d. $OSF_2$

**6** Match each of the diatomic molecules in Column I with its property/properties in Column II.
**(2009, Matching Type)**

| | Column I | | Column II |
|---|---|---|---|
| i. | $B_2$ | p. | Paramagnetic |
| ii. | $N_2$ | q. | Undergoes oxidation |
| iii. | $O_2^-$ | r. | Undergoes reduction |
| iv. | $O_2$ | s. | Bond order $\geq 2$ |
| | | t. | Mixing of '*s*' and '*p*' orbitals |

**Codes**

| | i | ii | iii | iv |
|---|---|---|---|---|
| a. | q,t | p | r,p | s |
| b. | s,p | r | p,s | q |
| c. | q | r,q | p | s,r |
| d. | q,r | q,r,t | p,q,r,t | p,r,s,t |

**7** Hyperconjugation involves overlap of the following orbitals **(2008, Single Option Correct Type)**

a. $\sigma - \sigma$ b. $\sigma - p$ c. $p - p$ d. $\pi - \pi$

**8** The species having bond order different from that in CO is **(2007, Single Option Correct Type)**

a. $NO^-$ b. $NO^+$
c. $CN^-$ d. $N_2$

**9** Among the following, the paramagnetic compound is **(2007, Single Option Correct Type)**

a. $Na_2O_2$ b. $O_3$
c. $N_2O$ d. $KO_2$

**10** Match the reactions in Column I with nature of the reactions/type of the products in Column II.
**(2007, Matching Type)**

| | Column I | | Column II |
|---|---|---|---|
| i. | $O_2^- \longrightarrow O_2 + O_2^{2-}$ | p. | Redox reaction |
| ii. | $CrO_4^{2-} + H^+ \longrightarrow$ | q. | One of the products has trigonal planar structure |
| iii. | $MnO_4^- + NO_2^- + H^+ \rightarrow$ | r. | Dimeric bridged tetrahedral metal ion |
| iv. | $NO_3^- + H_2SO_4 + Fe^{2+} \rightarrow$ | s. | Disproportionation |

**Codes**

| | i | ii | iii | iv |
|---|---|---|---|---|
| a. | q | p | r | s |
| b. | s | r | p | q |
| c. | s | r | q | p |
| d. | s | p | r | q |

**11.** Which of the following contains maximum number of lone pairs on the central atom?
**(2005, Single Option Correct Type)**

a. $ClO_3^-$ b. $XeF_4$
c. $SF_4$ d. $I_3^-$

# 4. Stoichiometry



DPP
DAILY PRACTICE PROBLEMS

# DPP-1 Mole Concept

## Subjective Questions

Working Space

*Direction* *(Q. Nos. 1-4) These questions are subjective in nature, need to be solved completely on notebook.*

**1** i. Analysis of an alloy of magnesium has 24.0% Mg. How many atoms of Mg are contained in 1.00 g of alloy?

ii. A diatomic gas occupies 0.8 L per gram at NTP. How many atoms are contained 0.8 L at NTP?

iii. 16 g of an ideal gas $SO_x$ occupied same volume as 20 g $SO_3$ at STP. How many atoms of oxygen are contained in given amount of $SO_x$?

**2** i. Mass of one atom of an element of $_6C$ is $2.324 \times 10^{-23}$ g.

a. Express this in atomic mass unit.

b. Determine number of protons, neutrons and electrons in one atom.

ii. Mass of one atom of an element of *A* is $4.0 \times 10^{-23}$g. How many atoms are contained in 1 g of it?

**3** i. How many moles of sucrose ($C_{12}H_{22}O_{11}$) are contained in a table spoon that contains 6.84 g?

ii. How many grams of urea ($NH_2CONH_2$) are there in 0.01 mole of it?

iii. In a bank, there are as many coins as number of electrons in 1.6 μg of $CH_4$. How many coins are there in the bank?

**4** i. By chemical analysis, an oxide of nitrogen is found to be composed of 30.43% N and 69.57% O. Its molecular mass is found to be 92. Derive molecular formula.

ii. Urea ($NH_2CONH_2$) and acetic acid ($CH_3COOH$) have same molar mass. Determine percentage of each element.

## Only One Option Correct Type

*Direction* *(Q. Nos. 5-23) This section contains 19 multiple choice questions. Each question has four choices (a), (b), (c) and (d) out of which ONLY ONE option is correct.*

**5** How many moles of $SO_2$ contain the same number of atoms as in 1 mole of $CH_4O$ ?

a. 1.0 b. 2.0 c. 3.0 d. 0.5

**6** The ratio of masses of oxygen and nitrogen of a particular gaseous mixture is 1 : 4. The ratio of number of their molecules is **[JEE Main 2014]**

a. 1 : 4 b. 7 : 32 c. 1 : 8 d. 3 : 16

**7** Total number of electrons in 1 mole of $CO_2$ is (in terms of $N_0$)

a. 3 $N_0$ b. 10 $N_0$

c. 16 $N_0$ d. 22 $N_0$

**8** Largest number of atoms is in

a. 1.6 g $CH_4$ b. 1.7 g $NH_3$

c. 1.8 g $H_2O$ d. Equal number of atoms

Working Space

**9** Number of moles in 1 kg electron is

a. $1.83 \times 10^6$ b. $9.1 \times 10^{-31}$
c. $9.1 \times 10^{-28}$ d. $1.1 \times 10^{28}$

**10** There are three isotopes of iron with their abundances given as

$^{54}_{26}Fe$ - 5.0%

$^{56}_{26}Fe$ - 90.0%

$^{57}_{26}Fe$ - 5.0%

Thus, atomic mass of Fe is

a. 55.85 b. 55.95 c. 55.75 d. 56.05

**11** How many moles are there in one atom?

a. $6.02 \times 10^{23}$ b. $1.66 \times 10^{-24}$ c. 1.00 d. None of these

**12** A flask of 50 L capacity and containing $H_2$ gas at 27°C exert a pressure of 10 atm. Thus, number of moles of $H_2$ taken is

a. 20.0 b. 10.2 c. 50.0 d. 20.3

**13** The number of lead atoms present in a small piece of lead with a volume of 0.105 $cm^3$ is (density of lead = 11.34 g/$cm^3$, lead = 206)

a. $3.48 \times 10^{21}$ b. $1.56 \times 10^{17}$
c. $5.57 \times 10^{21}$ d. $6.50 \times 10^{25}$

**14** A particular lead-cadmium alloy is 8% cadmium by mass. What mass of this alloy, in grams, must you weight out to obtain a sample containing $6.50 \times 10^{23}$ Cd-atoms ? (Cd = 112)

a. 0.121 kg b. 1.512 kg c. 0.1315 kg d. 0.0097 kg

**15** In rhombic sulphur, S-atoms are joined into $S_8$ molecules with density 2.07 g/$cm^3$.

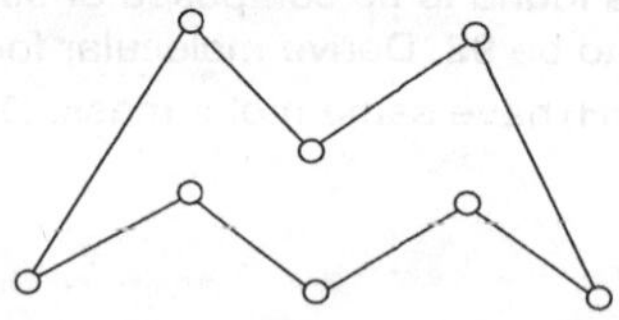

Number of $S_8$ molecules in a crystal of 0.568 $mm^3$ is

a. $1.336 \times 10^{21}$ b. $2.212 \times 10^{22}$
c. $2.765 \times 10^{21}$ d. $2.765 \times 10^{18}$

**16** A divalent cation is isoelectronic of $SO_2$ and has $(Z + 2)$ neutrons ($Z$ is atomic number of divalent cation). Thus, ionic mass of divalent cation is

a. 66 b. 68 c. 70 d. 72

**17** Ionic mass of $X^{3-}$ is 31 g $mol^{-1}$. It has 16 neutrons. Thus, number of electrons in $X^{3-}$ is

a. 15 b. 16 c. 17 d. 18

**18** Dot mark (.) made by lead pencil on the paper is 0.0001 g. How many atoms of the element of which pencil is made are there in one dot mark?

a. $2.92 \times 10^{17}$ b. $5.02 \times 10^{18}$
c. $6.02 \times 10^{16}$ d. $6.02 \times 10^{12}$

Working Space

**19** A drop of water is about 0.05 mL with density 1.0 g $mL^{-1}$. Number of water molecules present in 1 drop of water is

a. $1.67\times 10^{21}$ b. $6.02\times 10^{23}$
c. $3.01\times 10^{22}$ d. $0.05\times 10^{23}$

**20** Density of water is 1.0 g $mL^{-1}$. Assume water molecule spherical in nature, radius of the spherical molecule is

a. 0.96 Å b. 3.851 Å
c. 1.925 Å d. None of these

**21** The empirical formula of a gaseous fluorocarbon is $CF_2$. At a certain temperature and pressure, a 1 L volume hold 8.93 g of this fluorocarbon whereas under the same conditions, the 1 L volume holds only 1.70 g gaseous $F_2$. Thus, fluorocarbon is (F = 19)

a. $CF_2$ b. $C_2F_4$
c. $C_3F_6$ d. $C_4F_8$

**22** Pure nitrogen dioxide ($NO_2$) is formed when dinitrogen oxide ($N_2O$) and oxygen ($O_2$) are mixed in the presence of a certain catalyst. What volumes of $N_2O$ and $O_2$ are needed to produce 4.0 L $NO_2$ of all gases are held at the same condition of temperature and pressure?

| | $N_2O$ | $O_2$ | | $N_2O$ | $O_2$ |
|---|---|---|---|---|---|
| a. | 4 | 4 | b. | 2 | 3 |
| c. | 2 | 2 | d. | 3 | 2 |

**23** Nitrogen (N) and silicon (Si) form two binary compounds with the following compositions:

| Compound | Mass % (N) | Mass % (Si) |
|---|---|---|
| X | 33.28 | 66.72 |
| Y | 39.94 | 60.06 |

These two binary compounds follows

a. Law of multiple proportion
b. Law of reciprocal proportion
c. Law of combining volume
d. None of the above

## Matching List Type

***Direction*** *(Q. No. 24) Choices for the correct combination of elements from Column I and Column II are given as options (a), (b), (c) and (d), out of which one is correct.*

**24** Equal mass (1 g) of $CH_4$, $O_2$, $SO_2$ and $SO_3$ are taken separately. Match the number of moles of these species in Column II and select answer from the codes.

| | Column I | | Column II |
|---|---|---|---|
| i. | $CH_4$ | p. | 1.0/80 |
| ii. | $O_2$ | q. | 1.25/80 |
| iii. | $SO_2$ | r. | 2.5/80 |
| iv. | $SO_3$ | s. | 5/80 |

**Codes**

| | i | ii | iii | iv | | i | ii | iii | iv |
|---|---|---|---|---|---|---|---|---|---|
| a. | p | q | r | s | b. | s | r | q | p |
| c. | s | r | p | q | d. | r | s | q | p |

## Comprehension Type

Working Space

**Direction** *(Q. Nos. 25-28) This section contains 2 paragraphs, each describing theory, experiments, data etc. four questions related to the paragraphs have been given. Each question has only one correct answer among the four given options (a), (b), (c) and (d).*

### Passage I

Mass percentage of $Na_2SO_3$ and $H_2O$ are equal in $Na_2SO_3 \cdot xH_2O$.

**25** How many water molecules are contained in one formula unit of salt?

a. 5
b. 6
c. 7
d. 8

**26** Select the correct statement.

a. Number of atoms in anhydrous salt is equal to number of water molecules in hydrated salt
b. There are total 28 atoms in one formula units
c. There are total 10 oxygen atoms in one formula unit
d. All above are correct statements

### Passage II

Mass of one atom of $X$ is $6.642 \times 10^{-23}$ g.

**27** Atomic mass of $X$ is

a. $6.642 \times 10^{-23}$ g mol$^{-1}$
b. $1.103 \times 10^{-46}$ g mol$^{-1}$
c. 40.0 g mol$^{-1}$
d. 0.025 g mol$^{-1}$

**28** Number of moles of $X$ in 40 kg of it, is

a. $6.02 \times 10^{23}$
b. 1000
c. 0.04
d. 2.5

## One Integer Value Correct Type

**Direction** *(Q. Nos. 29-31) This section contains 3 questions. Each question, when worked out will result in an integer from 0 to 9 (both inclusive).*

**29** How many moles of $CH_4$ are contained in $2.408 \times 10^{25}$ atoms of $CH_4$?

**30** A coloured dye of molar mass 20000 g mol$^{-1}$ is found to contain 0.84% Fe. Thus, number of Fe-atoms in one formula unit of dye is .......... .

**31** An alkene occupies 0.8 L g$^{-1}$ at STP. Thus, total number of atoms of carbon and hydrogen in one formula unit are .......... .

# DPP-2 Chemical Calculations

## Subjective Questions

Working Space

*Direction* (Q. Nos. 1-5) *These questions are subjective in nature, need to be solved completely on notebook.*

**1** The sulphide ore of zinc (zinc sulphide) is reduced to elemental zinc in following steps

I. $ZnS + 3/2O_2 \longrightarrow ZnO + SO_2$ (called roasting)

II. $ZnO + CO \longrightarrow Zn + CO_2$ (called reduction)

On treating 5.32 kg ZnS in this way, 3.30 kg of pure zinc is formed. Calculate

i. theoretical yield

ii. percentage yield $Zn = 65$, $S = 32$, $O = 16$

**2** $H_2SO_4$ is formed in the following reaction

$$2SO_2 + O_2 + 2H_2O \longrightarrow 2H_2SO_4$$

Suppose, 400 g $SO_2$, 175 g $O_2$ and 125 g $H_2O$ are mixed and the reaction proceeds until one of the reactants is used up.

i. What is the limiting reactant?

ii. What mass of the $H_2SO_4$ is produced?

iii. What masses of the other reactants remain?

**3** An alloy used in aircraft structures consists of 93.7% Al and 6.3% Cu by mass. The alloy has density of 2.85 g/cm$^3$. A 0.691 cm$^3$ piece of alloy reacts with an excess of HCl(*aq*). If only Al reacts, what mass of $H_2$ is obtained?

**4** A hydrochloric acid solution consists of 28.0% HCl, by mass, and has a density of 1.14 g/mL. What volume of this solution is required to react with 1.35 g Al?

**5** Calcium hypochlorite, $Ca(OCl)_2$, is used as a bleaching agent. It is produced based on following reaction :

$$2NaOH + Ca(OH)_2 + 2Cl_2 \longrightarrow Ca(OCl)_2 + 2NaCl + 2H_2O$$

How many moles of $Cl_2$ and NaOH react with 740 g $Ca(OH)_2$ and how many moles of $Ca(OCl)_2$ are produced?

## Only One Option Correct Type

*Direction* (Q. Nos. 6-15) *This section contains 10 multiple choice questions. Each question has four choices (a), (b), (c) and (d), out of which ONLY ONE option is correct.*

**6** How many moles of ammonia will be formed from 0.50 mole of $N_2$ and 1.5 moles of $H_2$?

a. 0.5 mol
b. 1.5 mol
c. 2.0 mol
d. 1.0 mol

Working Space

7 17 g of $NH_3$ is mixed with the same mass of hydrogen chloride (HCl). Thus, $NH_4Cl$ formed is

$$NH_3 + HCl \longrightarrow NH_4Cl$$

a. 17.00 g
b. 34.00 g
c. 53.50 g
d. 24.92 g

8 The iron (III) oxide, $Fe_2O_3$ reacts with CO to give iron and $CO_2$

$$Fe_2O_3 + 3CO \longrightarrow 2Fe + 3CO_2$$

The reaction of 480.0 g $Fe_2O_3$ with excess of CO yields 252.0 g iron. Thus, percentage yield is (Fe = 56)

a. 100.0%
b. 91.2%
c. 75.0%
d. 50.0%

9 If 0.50 mole of $BaCl_2$ is mixed with 0.20 mole of $Na_3PO_4$, the maximum number of moles of $Ba_3(PO_4)_2$ that can be formed Is

a. 0.70
b. 0.50
c. 0.20
d. 0.10

10 100 mL of $BaCl_2$ solution (20.8% weight/volume) and 50 mL of $K_2SO_4$ solution (17.4% weight/volume) are mixed. Thus, white precipitate formed is

a. 23.3 g
b. 11.65 g
c. 7.45 g
d. 3.725 g

11 Silicon nitride, ($Si_3N_4$), is made by direct combination of silicon and nitrogen at high temperature. How much silicon must react with excess nitrogen to prepare 140 g $Si_3N_4$ if the yield of the reaction is 75.0%? (Si = 28)

a. 84.0 g
b. 63.0 g
c. 112.0 g
d. 105.0 g

12 1.84 g of a dibromide of a metal *M* on reaction with excess of $AgNO_3$ gave 3.76 g of yellow precipitate. Thus, molar mass of metal dibromide is (Br = 80, Ag = 108)

a. 24.0 g $mol^{-1}$
b. 184 g $mol^{-1}$
c. 188 g $mol^{-1}$
d. 75.2 g $mol^{-1}$

13 2.76 g of $Ag_2CO_3(s)$ is strongly heated in an open vessel. Mass of residue formed is

a. 1.24 g
b. 2.52 g
c. 2.40 g
d. 2.16 g

14 Amount of NO(*g*) formed in grams by oxidation of 1 mole of $NH_3(g)$ by 1 mole of $O_2(g)$ is

a. 30.0 g
b. 54.0 g
c. 60.0 g
d. 24.0 g

15 1 mole each of $CaH_2$ and $H_2O$ are allowed to react to form

a. 1 mole $H_2$
b. 2 moles $H_2$
c. 0.5 mole $H_2$
d. 0.25 mole $H_2$

Working Space

## Comprehension Type

*Direction (Q. Nos. 16-19) This section contains 2 paragraphs, each describing theory, experiments, data, etc. Four questions related to the paragraphs have been given. Each question has only one correct answer among the four given options (a), (b), (c) and (d).*

### Passage I

Sodium azide ($NaN_3$), the explosive chemical used in automobile airbags, is made by the following reaction :

$$NaNO_3 + 3NaNH_2 \longrightarrow NaN_3 + 3NaOH + NH_3$$

One takes 15.0 g of $NaNO_3$ and 15.0 g of $NaNH_2$

**16** $NaN_3$ formed in the reaction is

a. 15.0 g
b. 30.0 g
c. 9.61 g
d. 8.32 g

**17** If this amount is used in automobile bags, how much $N_2$ is formed due to collision at NTP

a. 4.30 L
b. 12.93 L
c. 8.62 L
d. 2.15 L

### Passage II

Gypsum is a hydrate calcium sulphate. A 1.0 g sample contains 0.791 g. $CaSO_4$; and rest water

**18** Moles of $CaSO_4$ and $H_2O$ present in the sample are

a. $5.82 \times 10^{-3}$, $11.61 \times 10^{-3}$
b. $11.61 \times 10^{-3}$, $5.82 \times 10^{-3}$
c. $5.82 \times 10^{-3}$, $5.82 \times 10^{-3}$
d. $11.61 \times 10^{-3}$, $11.61 \times 10^{-3}$

**19** Molecular formula of gypsum based on above analysis is

a. $CaSO_4 \cdot H_2O$
b. $CaSO_4 \cdot 1.5H_2O$
c. $CaSO_4 \cdot 2H_2O$
d. $CaSO_4 \cdot 3H_2O$

## One Integer Value Correct Type

*Direction (Q. Nos. 20-22) This section contains 3 questions. When worked out will result in an integer from 0 to 9 (both inclusive).*

**20** 1.25 g of gallium reacts with oxygen to give 1.68 g of $Ga_xO_y$. What is the value of $(x + y)$? (Ga = 70)

**21** If epsom salt $MgSO_4 \cdot xH_2O$ is heated to 250°C, all the water of hydration is lost. On heating a 1.687 g sample of the hydrate, 0.824 g of $MgSO_4$ remains. How many molecules of water occur per formula unit of $MgSO_4$?

**22** 3.36 g of a dibromide of element $X$, $XBr_2$ is completely converted into dichloride $XCl_2$ forming 1.58 g of it ($XCl_2$).

$$XBr_2 + Cl_2 \longrightarrow XCl_2 + Br_2$$

What is atomic mass of element $X$?

## Matching List Type

Working Space

***Direction*** *(Q. Nos. 23 and 24) Choices for the correct combination of elements from Column I and Column II are given as options (a), (b), (c) and (d), out of which one is correct.*

**23** Match the reaction in (Column I) with the maximum yield of the product in (Column II).

| | Column I | | Column II |
|---|---|---|---|
| i. | $\underset{1g}{C}+\underset{1g}{O_2} \longrightarrow CO_2$ | p. | 1.125 g |
| ii. | $\underset{1g}{2C}+\underset{1g}{2H_2} \longrightarrow C_2H_4$ | q. | 1.167 g |
| iii. | $\underset{1g}{H_2}+\underset{1g}{O_2} \longrightarrow H_2O_2$ | r. | 1.375 g |
| iv. | $\underset{1g}{2H_2}+\underset{1g}{O_2} \longrightarrow H_2O$ | s. | 1.0625 g |

Select the answer from following codes.

**Codes**

| | i | ii | iii | iv | | i | ii | iii | iv |
|---|---|---|---|---|---|---|---|---|---|
| a. | p | q | r | s | b. | s | r | q | p |
| c. | r | q | s | p | d. | p | r | s | q |

**24** 1 mole of the compounds in Column I is to be oxidised to the compound as indicated. Compare the reaction in Column I with the mass of $O_2$ required in Column II and select the answer from the codes given.

| | Column I | | Column II |
|---|---|---|---|
| i. | $CO \longrightarrow CO_2$ | p. | 0.5 mol |
| ii. | $N_2 \longrightarrow N_2O_4$ | q. | 1.5 mol |
| iii. | $P_4 \longrightarrow P_2O_5$ | r. | 2.0 mol |
| iv. | $Cl_2 \longrightarrow Cl_2O_3$ | s. | 5.0 mol |

**Codes**

| | i | ii | iii | iv | | i | ii | iii | iv |
|---|---|---|---|---|---|---|---|---|---|
| a. | s | q | p | r | b. | q | s | r | p |
| c. | p | r | s | q | d. | p | q | s | r |

# DPP-3 Concentration Factors

## Subjective Questions

Working Space

*Direction (Q. Nos. 1-3) These questions are subjective in nature, need to be solved completely on notebook.*

**1** i. Find the molarity of water. Given density = 1000 $kg\,m^{-3}$.

ii. Calculate the molality of 1.0 L solution of 93% $H_2SO_4$ (weight /volume). The density of solution is 1.84 $g\,mL^{-1}$.

**2** Calculate molarity, molality and normality of $H_2SO_4$ solution which is 98% by weight of solution (*W/W*) and density 1.8 $g\,mL^{-1}$.

**3** Ethanol-water mixture is 46% by weight.

i. What is the mole fraction of each component?

ii. What is molality as water solute and ethanol solute?

## Only One Option Correct Type

*Direction (Q. Nos. 4-25) This section contains 22 multiple choice questions. Each question has four choices (a), (b), (c) and (d), out of which ONLY ONE option is correct.*

**4** A mixture contains 9.2 g of ethanol ($C_2H_5OH$) and 18 g of water ($H_2O$). Thus, mole fraction of ethanol in the mixture is

a. 0.2 b. 0.1
c. 0.25 d. 0.167

**5** Dissolving 120 g urea (molar mass 60 $g\,mol^{-1}$) in 1000 g water gave a solution of density 1.15 $g\,mL^{-1}$. The molarity of the solution is **[IIT JEE 2011]**

a. 1.78 M b. 2.00 M
c. 2.05 M d. 2.22 M

**6** The molarity of a solution obtained by mixing 750 mL of 0.5 M HCl with 250 mL of 2 M HCl will be **[JEE Main 2013]**

a. 0.875 M b. 1.00 M
c. 1.75 M d. 0.0975 M

**7** Select the correct statement.

a. Molality and mole fraction are independent of temperature
b. Molality of a very dilute solution approaches molarity, if density of the solution approaches unity
c. Normality of 1 M $H_3PO_2$ solution is 1 N
d. All the above are correct statements

**8** The mass of 1 L of oxygen is 1.4 g. Volume of oxygen needed to react with 21 g of magnesium whose equivalent is 0.5 mole is

a. 14.0 L b. 28.0 L
c. 32. 0 L d. 10.0 L

Working Space

9 26.5 g of $Na_2CO_3$ is dissolved in enough water to make 250 mL of solution. Thus, molar concentration of $Na^+$ and $CO_3^{2-}$ ions are respectively

a. 1 M, 1 M
b. 1 M, 2 M
c. 2 M, 2 M
d. 2 M, 1 M

10 Water is evaporated from 135 mL of 0.188 M $K_2SO_4$ solution until the solution becomes 105 mL. Thus, molarity of the resultant solution is

a. 0.179 M
b. 0.242 M
c. 0.146 M
d. None of these

11 Which of the following represents a 1.00 M KCl solution?

a. 1.0 L of solution containing 100 g KCl
b. 250.0 mL of solution containing 74.5 g KCl
c. A solution containing 7.45 mg KCl/mL solution
d. 5.00 L of solution containing 372.5 g KCl

12 How many grams of sodium must react with 155 mL $H_2O$ to produce a solution that is 0.175 M NaOH?
(Assume final volume of solution as 155 mL)

a. 0.312 g
b. 0.543 g
c. 1.085 g
d. 0.624 g

13 Aqueous urea solution is 25% by mass of solvent. Its concentration by mass of solution is

a. 75%
b. 35%
c. 25%
d. 20%

14 In aqueous glucose solution mole fraction of glucose is 0.1. Thus, molality of glucose solution is

a. 10.1
b. 6.17
c. 2.92
d. 1.46

15 Equal moles of urea and $H_2O$ are taken in a flask. Thus, mass percentage of urea by mass of solution is

a. 76.92
b. 23.08
c. 50.00
d. 60.00

16 An aqueous solution is 17.4 ppm of $K_2SO_4$. Thus, molarity of $K^+$ ion is

a. 0.348 M
b. 0.174 M
c. $2 \times 10^{-4}$ M
d. $1 \times 10^{-4}$ M

17 An aqueous solution of NaOH is 40% by weight of solution. Thus, its molality is

a. 10.1 molal
b. 10.4 molal
c. 20.15 molal
d. 16.67 molal

18 Sodium acetate ($CH_3COONa$) is 0.50 molal in aqueous solution. Mass of this solution to obtain 0.150 mole of sodium acetate approximately is

a. 300 g
b. 312 g
c. 410 g
d. None of these

19 On evaporation of 1 L of urea solution of density 1.2 g $mL^{-1}$, 120 g of residue is formed. Thus, molality of solution is

a. 2.000 molal
b. 1.852 molal
c. 1.000 molal
d. 1.200 molal

Working Space

20 Mole fraction of ethanol in ethanol-water mixture is 0.040. If density of water is 1.0 g $mL^{-1}$ and there is no change in volume on addition of solute, molarity of ethanol solution is

a. 0.04 M b. 2.315 M c. 2.092 M d. 0.96 M

21 Volume strength of 1.5 N $H_2O_2$ solution is

a. 0.75 b. 3.00 c. 8.40 d. 16.8

22 A drop of 12.0 M HCl is spread over a sheet of thin aluminium foil. Assume that all the acid reacts with and thus, dissolves through the foil, (20 drops make 1 mL), density of Al foil = 2.7 g/$cm^3$ and thickness = 0.10 mm. Thus, area of the cylindrical hole (in $cm^2$) is

a. 2.0 $cm^2$ b. 0.2 $cm^2$ c. 4.0 $cm^2$ d. 0.1 $cm^2$

23 Following solutions have been given.

*P* : 0.01 M iron (II) sulphate (VI)
*Q* : 0.01 M iron (III) sulphate (IV)
*R* : 0.01 M iron (III) phosphate (V)

1 mL of each solution is diluted to 100 mL.

Molar concentration of iron ion in diluted solution is

| | *P* | *Q* | *R* |
|---|---|---|---|
| a. | $1\times10^{-4}$ M | $1\times10^{-4}$ M | $1\times10^{-4}$ M |
| b. | $1\times10^{-4}$ M | $2\times10^{-4}$ M | $2\times10^{-4}$ M |
| c. | $2\times10^{-4}$ M | $3\times10^{-4}$ M | $3\times10^{-4}$ M |
| d. | $1\times10^{-4}$ M | $2\times10^{-4}$ M | $1\times10^{-4}$ M |

24 How many millilitre of 0.715 M $NH_4NO_3$ solution must be diluted with water to produce 1.00 L of solution of 2.37 mg N per mL?

a. 118.4 mL b. 236.4 mL c. 59.2 mL d. 236.8 mL

25 A solution contains 6.0 micromoles of $Na_2SO_4$ in 250 mL solution. Assuming no change in volume on dissolution, $Na^+$ in ppm present in solution is

a. 0.552 b. 1.104 c. 55.2 d. 552

## Matching List Type

***Direction*** *(Q. No. 26) Choices for the correct combination of elements from Column I and Column II are given as options (a), (b), (c) and (d), out of which one is correct.*

26 A sugar syrup (density = 1.2 g $mL^{-1}$) of weight 214.2 g contains 34.2 g of sugar ($C_6H_{12}O_6$). Match the concentration parameters of Column I with their values in Column II and select answer from the codes given below the columns.

| | Column I | | Column II |
|---|---|---|---|
| i. | Kilomoles of water | p. | $5.55\times10^{-4}$ |
| ii. | Mole fraction of sugar | q. | $9.9\times10^{-3}$ |
| iii. | Molality of solution | r. | 0.01 |
| iv. | Molarity of solution | s. | 0.56 |

**Codes**

| | i | ii | iii | iv | | i | ii | iii | iv |
|---|---|---|---|---|---|---|---|---|---|
| a. | r | q | p | s | b. | p | q | s | p |
| c. | p | r | q | s | d. | r | p | s | q |

## Comprehension Type

Working Space

*Direction* (Q. Nos. 27-31) *This section contains 2 paragraphs, each describing theory, experiments, data, etc. Five questions related to the paragraphs have been given. Each question has only one correct answer among the four given options (a), (b), (c) and (d).*

### Passage I

An ionic compound used as a chemical fertiliser has the composition (by mass)

48.46% oxygen  23.45% phosphorus

21.21% nitrogen  6.87% hydrogen

**27** Ionic salt is

a. $NH_4H_2PO_4$  b. $NH_4H_2PO_3$

c. $(NH_4)_2HPO_4$  d. $NH_2(H_2PO_4)_2$

**28** Ionic salt is 1 M. What is its concentration as an acid?

a. 1 N  b. 2 N

c. 3 N  d. 0.5 N

**29** What is the percentage in terms of $NH_3$?

a. 28.08%  b. 25.76%

c. 21.21%  d. None of these

### Passage II

$8.0575 \times 10^{-2}$ kg of Glauber's salt is dissolved in water to obtain 1 $dm^3$ of solution of density 1077.2 kg $m^{-3}$.

**30** Molality of this solution is

a. 0.24 molal  b. 0.25 molal

c. 0.28 molal  d. 1.00 molal

**31** Molarity of this solution is

a. 0.24 M  b. 0.25 M

c. 0.26 M  d. 1.05 M

## One Integer Value Correct Type

*Direction* (Q. Nos. 32-34) *This section contains 3 questions. When worked out will result in an integer from 0 to 9 (both inclusive).*

**32** 29.2% (*W/W*) stock solution of HCl has density 1.25 g $mL^{-1}$. The molecular weight of HCl is 36.5 g $mol^{-1}$. The volume (mL) of stock solution required to prepare a 200 mL solution of 0.4 M HCl is .......... . **[IIT JEE 2012]**

**33** A compound $H_2X$ with molecular weight 80 g $mol^{-1}$ is dissolved in a solvent having density 0.4 g $mL^{-1}$. Assuming no change in volume upon dissolution, the molality of a 3.2 molar solution is .......... . **[JEE Advanced 2014]**

**34** Urea solution has molarity 4.0 mol $L^{-1}$ and molality 2.273 mol $kg^{-1}$ $H_2O$. Thus, density (g $mL^{-1}$) of the solution is .......... .

# DPP-4 Acid-Base Neutralisation Titration

## Subjective Question

Working Space

*Direction* (Q. No. 1) *This question is subjective in nature, need to be solved completely on notebook.*

**1** i. 0.24 g of metal on burning in oxygen formed 0.40 g of metal oxide. What is equivalent mass of metal?

ii. 6.0 g of a monobasic acid is dissolved in 100 mL solution to give normal solution. What is equivalent and molar mass of acid?

## Only One Option Correct Type

*Direction* (Q. Nos. 2-12) *This section contains 11 multiple choice questions. Each question has four choices (a), (b), (c) and (d), out of which ONLY ONE option is correct.*

**2** An aqueous solution of 6.3 g oxalic acid dihydrate is made up to 250 mL. The volume of 0.1 N NaOH required to completely neutralise 10 mL of this acid solution is **[IIT JEE 2001]**

a. 40 mL b. 20 mL
c. 10 mL d. 4 mL

**3** 0.265 g of pure $Na_2CO_3$ required 28.5 mL of aqueous HCl for titration to the equivalence point. Thus, HCl is

a. 0.350 M b. 0.175 M
c. 0.0875 M d. 0.700 M

**4** Oxalic acid ($H_2C_2O_4 \cdot 2H_2O$), a dibasic acid can be estimated by titration of NaOH solution. 1.575 g of this acid required *V* mL of 1 N-NaOH solution. Thus, *V* is

a. 25 mL b. 50 mL
c. 75 mL d. 100 mL

**5** 5.3 g of $M_2CO_3$ is dissolved in 150 mL of 1 N HCl. Unused acid required 100 mL of 0.5 N NaOH. Thus, equivalent weight of *M* is

a. 23 b. 12
c. 24 d. 13

**6** 2.4 g of Mg is treated with 200 mL of 1 M $H_2SO_4$ and solution diluted to 1 L. Molar concentration of $H_2SO_4$ in dilute solution is

a. 1.0 M b. 0.5 M
c. 0.2 M d. 0.1 M

**7** Volumes of 0.01 M HCl and 0.01 M $H_2SO_4$ required to decompose 0.10 g $CaCO_3$ separately are *x* and *y* mL. Thus, *x* and *y* are

| | Volume of 0.01 M HCl ($x$) | Volume of 0.01 M $H_2SO_4$ ($y$) |
|---|---|---|
| a. | 200 | 200 |
| b. | 100 | 100 |
| c. | 200 | 100 |
| d | 400 | 200 |

Working Space

**8** Sulphuric acid and orthophosphoric acid have the same molar mass. Ratio of the masses of these acids needed to neutralise the same amount of NaOH if the sulphate and dihydrogen orthophosphate were formed is

a. 1 : 2 b. 2 : 1
c. 1 : 3 d. 1 : 1

**9** Resultant $[H^+]$ in a mixture of 100 mL of 0.1 M $H_2SO_4$ and 200 mL of 0.1 M $H_3PO_3$ is

a. 0.1 M b. 0.2 M
c. 0.267 M d. 0.133 M

**10** 0.31 g of N containing compound on reaction with NaOH gave $NH_3$ which required 100 mL of 0.1 N HCl. Hence, % of nitrogen (N) in N-containing compound is

a. 45.16% b. 90.32%
c. 22.58% d. 11.29%

**11** 100 mL of 0.1 M KOH is mixed with 200 mL of 0.1 M $Ca(OH)_2$ and 300 mL of 0.1 M $H_2SO_4$. Thus, resultant mixture is

a. 0.0 M as $[OH^-]$ b. 0.0 M as $[H^+]$
c. 0.017 as $[H^+]$ d. 0.017 as $[OH^-]$

**12** When 11.2 g of iron combines with sulphur 17.6 g of iron sulphide is formed. Thus, number of equivalent present in 11.2 g of iron is

a. 0.1 b. 0.2
c. 0.3 d. 0.4

## One or More than One Options Correct Type

**Direction** *(Q. Nos. 13-17) This section contains 5 multiple choice questions. Each question has four choices (a), (b), (c) and (d), out of which ONE or MORE THAN ONE are correct.*

**13** 1 g equivalent of a substance is the weight of that amount of a substance which is equivalent to

a. 0.25 mol $O_2$ b. 8 g $O_2$
c. 16 g $O_2$ d. 0.50 mol $O_2$

**14** 2.84 g of $P_4O_{10}$ is dissolved in 1 L aqueous solution. Thus,

a. resulting solution is 0.01 M
b. it is neutralised to $Na_3PO_4$ by 400 mL of 0.3 M NaOH
c. it is neutralised to $CaHPO_4$ by 400 mL of 0.1 M $Ca(OH)_2$
d. it cannot be neutralised by basic solution

**15** Select the correct statement (s) about $H_3PO_2$.

a. It is tribasic acid
b. One mole of it is neutralised by 0.5 mole of $Ca(OH)_2$
c. $NaH_2PO_2$ is an acid salt
d. On heating it disproportionates to $PH_3$ and $H_3PO_3$

**16** 100 mL of 1 M $BaF_2$ solution is mixed with 100 mL of 2 M $H_2SO_4$. Resulting mixture contains

a. 0.1 mol $BaSO_4$ b. 0.2 M $H^+$
c. 2.0 M $H^+$ d. 0.4 M $H^+$

Working Space

**17** 1 mole of an aqueous solution of $SO_2Cl_2$ is neutralised by

a. 2 moles of NaOH
b. 2 moles of $Ca(OH)_2$
c. 1 mole of $Ca(OH)_2$
d. 4 moles of NaOH

## Comprehension Type

**Direction** (Q. Nos. 18 and 19) *This section contains one paragraph, each describing theory, experiments, data, etc. Two questions related to the paragraph have been given. Each question has only one correct answer among the four given options (a), (b), (c) and (d).*

11.2 millimoles of $CO_2$ were passed into 1.0 L of 0.15 N NaOH solution.

**18** Resultant molarity of NaOH solution is

a. 0.1276 M
b. 0.0112 M
c. 0.1500 M
d. 0.0224 M

**19** Molarity of $Na_2CO_3$ solution formed is

a. 0.0224 M
b. 0.0112 M
c. 0.1500 M
d. 0.1276 M

## One Integer Value Correct Type

**Direction** (Q. No. 20) *This section contains 1 question. When worked out will result in an integer from 0 to 9 (both inclusive).*

**20** Chromium (III) ion forms many compounds with ammonia. In a titration of 1.580 g of $Cr(NH_3)_xCl_3$, 24.26 mL of 1.500 M HCl is used.

$$Cr(NH_3)_xCl_3\,(aq) + xHCl(aq) \longrightarrow xNH_4^+(aq) + Cr^{3+}(aq) + (x+3)\,Cl^-(aq)$$

What is the value of $x$?

# DPP-5 Redox Based Titration

## Only One Option Correct Type

**Direction** *(Q. Nos. 1-15) This section contains 15 multiple choice questions. Each question has four choices (a), (b), (c) and (d), out of which ONLY ONE option is correct.*

*Working Space*

**1** How many millilitres of 0.200 M $MnO_4^-$ are required to oxidise 32.0 mL of 0.125 M $Cr^{3+}$ to $Cr_2O_7^{2-}$ solution in acidic medium?

a. 20.0 mL b. 5.0 mL c. 12.0 mL d. 24.0 mL

**2** 20 mL of 0.100 M $Sn^{2+}$ are required to titrate 0.250 M $Fe^{3+}$ solution upto equivalence point. Thus, volume of solution of 0.250 M $Fe^{2+}$ is

a. 16 mL b. 32 mL c. 8 mL d. 20 mL

**3** 0.2495 g of $CuSO_4 \cdot 5H_2O$ is dissolved in enough water and volume made upto 100 mL. 100 mL of this solution was treated with excess KI and liberated $I_2$ required 25 mL of $Na_2S_2O_3$ solution in iodometric titration. Thus, $Na_2S_2O_3$ is

a. 0.04 M b. 0.02 M c. 0.01 M d. 0.40 M

**4** Iron in ores can be estimated by titration using stand $KMnO_4$ in acidic medium. 1.0 g sample of iron containing ore required 25.0 mL of 0.02 M $KMnO_4$ in acidic medium. Thus, percentage of pure iron (in the form of Fe (II)) in the ore is

a. 7.0% b. 28.0% c. 14.0% d. 25.0%

**5** $MnO_4^-$ and $Cr_2O_7^{2-}$, both can be used to estimate Mohr's salt using suitable indicators ($MnO_4^-$ is self indicator and diphenylamine for $Cr_2O_7^{2-}$ and Mohr's salt). Ratio of moles of $MnO_4^-$ and $Cr_2O_7^{2-}$ required in acidic medium per mole of Mohr's salt is

a. 0.83 b. 1.00 c. 1.2 d. 2.0

**6** In the following reaction,

$$Cr_2O_7^{2-} + 3H_2SO_3 + 5H^+ \longrightarrow 2Cr^{3+} + 3HSO_4^- + 4H_2O$$

ratio of equivalents of $Cr_2O_7^{2-}$ and $H_2SO_3$ used in this redox reaction is

a. 1 : 3 b. 1 : 6 c. 6 : 1 d. 1 : 1

**7** $KMnO_4$ oxidises $C_2H_5OH$ to $CO_2$ and itself reduced to $Mn^{2+}$ in acidic medium

$$12MnO_4^- + 5C_2H_5OH + 36H^+ \longrightarrow 12Mn^{2+} + 10CO_2 + 33H_2O$$

Equivalents of $C_2H_5OH$ oxidised per mol of $KMnO_4$ is

a. 5 b. 5/12 c. 4/5 d. 1/5

**8** A mixture is 0.04 M is $Sn^{2+}$ and *x* M in $Fe^{2+}$. 15.0 mL of this mixture required 18.0 mL of 0.125 M $Cr_2O_7^{2-}$ solution to oxidise to $Sn^{4+}$ and $Fe^{3+}$ in acidic medium

$$Sn^{2+} + Fe^{2+} + Cr_2O_7^{2-} \longrightarrow Cr^{3+} + Sn^{4+} + Fe^{3+}$$

Thus, *x* is

a. 0.410 M b. 0.205 M c. 0.820 M d. 1.640 M

Working Space

9 The number of moles of $KMnO_4$ that will be needed to react with one mole of sulphite ion in acidic solution is

a. 0.4 b. 0.6
c. 0.8 d. 1.0

10 In a titration $H_2O_2$ is oxidised to $O_2$ by $MnO_4^-$. 24 mL of 0.1 M $H_2O_2$ require 16 mL of 0.1 M $MnO_4^-$ solution. Thus, $MnO_4^-$ changes to

a. $MnO_4^{2-}$ b. $MnO_2$
c. $MnO_4^{2-}$ d. $Mn_2O_7$

11 1 g of $X\%$ $H_2O_2$ required $X$ mL of $KMnO_4$ in acidic solution. Thus, molarity of $KMnO_4$ solution is

a. 0.120 M b. 0.60 M
c. 0.17 M d. None of these

12 100 mL of $NaHC_2O_4$ required 50 mL of 0.1 M $KMnO_4$ solution in acidic medium. Volume of 0.1 M NaOH required by 100 mL of $NaHC_2O_4$ is

a. 50 mL b. 100 mL
c. 125 mL d. 150 mL

13 Three moles of a mixture of iron (II) sulphate (VI) and iron (III) sulphate (VI) required 100 mL of 2 M $KMnO_4$ solution in acidic medium. Thus, mole fraction of iron (II) sulphate (VI) in the mixture is

a. 1/3 b. 2/3
c. 2/5 d. 3/5

14 100 mL of 1 M $KMnO_4$ oxidised 100 mL of $H_2O_2$ in acidic medium. Volume of this $KMnO_4$ required to oxidised 100 mL of $H_2O_2$ in basic medium is

a. $\frac{100}{3}$ mL b. $\frac{500}{3}$ mL
c. $\frac{300}{5}$ mL d. $\frac{500}{5}$ mL

15 Amount of Mohr's salt $(FeSO_4 \cdot (NH_4)_2SO_4 \cdot 6H_2O)$ having molar mass 392.0 g $mol^{-1}$ that must be dissolved in 250 mL solution to prepare an aqueous solution of density 1.00 g $mL^{-1}$ to have $Fe^{2+}$ 1 ppm by weight

a. $3.50 \times 10^{-3}$ g b. $1.75 \times 10^{-3}$ g
c. $7.00 \times 10^{-3}$ g d. $0.35 \times 10^{-3}$ g

## One or More than One Options Correct Type

*Direction* *(Q. Nos. 16 and 17) This section contains 2 multiple choice questions. Each question has four choices (a), (b), (c) and (d), out of which ONE or MORE THAN ONE are correct.*

16 Equal moles of $K_2Cr_2O_7$ in acidic medium are required for oxidation of by 1 mole of

a. $FeSO_3$ b. $FeS_2O_4$
c. $Fe_2(SO_4)_3$ d. $Al_2(SO_3)_3$

17 A double salt $KHC_2O_4 \cdot H_2C_2O_4 \cdot 2H_2O$ is 0.2 N as an acid. Thus, it is

a. 0.267 N as reducing agent
b. 0.6 M as an acid
c. 0.067 M as a reducing agent
d. 0.067 M as an acid

## Comprehension Type

*Working Space*

**Direction** (Q. Nos. 18 and 19) *This section contains a paragraph, each describing theory, experiments, data, etc. Two questions related to the paragraph have been given. Each question has only one correct answer among the four given options (a), (b), (c) and (d).*

A mixture containing $Na_2C_2O_4$ and $KHC_2O_4$ in 1:1 molar ratio is neutralised by 100 mL of 0.01 M KOH.

**18** Volume (in mL) of 0.01 M $KMnO_4$ required to oxidise above mixture is

a. 200 mL
b. 100 mL
c. 90 mL
d. 80 mL

**19** The number of milliequivalents of $C_2O_4^{2-}$ in the above mixture is

a. 0.80
b. 2.0
c. 4.0
d. 40.0

## Matching List Type

**Direction** (Q. No. 20) *Choices for the correct combination of elements from Column I and Column II are given as options (a), (b), (c) and (d), out of which one is correct.*

**20** $MnO_4^-$ is an oxidising agent in acidic, basic and neutral medium. Match the species in Column I with the number of moles of $KMnO_4$ in Column II required in acidic medium to oxidise (1 mole each of species in Column I). Select the answer from the codes given.

| | Column I | | Column II |
|---|---|---|---|
| i. | Iron (II) oxalate | p. | 0.0 mol |
| ii. | Iron (II) sulphate (IV) | q. | 1.0 mol |
| iii. | Iron (III) sulphate (VI) | r. | 0.6 mol |
| iv. | Iron (III) sulphate (IV) | s. | 1.2 mol |

**Codes**

| | i | ii | iii | iv |
|---|---|---|---|---|
| a. | p | r | s | q |
| b. | p | q | s | r |
| c. | q | r | p | s |
| d. | r | r | p | s |

## One Integer Value Correct Type

**Direction** (Q. Nos. 21-24) *This section contains 4 questions. When worked out will result in an integer from 0 to 9 (both inclusive).*

**21** 100 mL of sample solution containing 2.14 g of $KIO_3$ oxidise 60 mL of 0.5 M $Na_2SO_3$ solution of $Na_2SO_4$. Change in oxidation state of iodine is .......... .

**22** For a titration of a reducing agent, 20 mL of 1 M $KMnO_4$ was required in acidic medium and 33.4 mL was in alkali medium. To what state $MnO_4^-$ is reduced in alkali medium?

**23** The $SO_2$ present in air is mainly responsible for acid-rain phenomenon and can be estimated by $MnO_4^- / H^+$. A sample of air having $SO_2$ required 100 mL of 0.008M $KMnO_4$ based on

$$5SO_2 + MnO_4^- + 2H_2O \longrightarrow 5SO_4^{2-} + 2Mn^{2+} + 4H^+$$

Determine number of millimoles of $SO_2$ in air.

**24** 25 mL of a solution containing 6.1 g $L^{-1}$ of an oxalate $K_xH_y(C_2O_4)_z \cdot nH_2O$ required 18 mL of 0.1 N NaOH in titration I and 24 mL of 0.1 N $KMnO_4$ in acidic medium in titration II. What is the value of $(x + y + z)$?

# Revisal Problems
## for JEE Main

### Only One Option Correct Type

Working Space

*Direction* *(Q. Nos. 1-25) This section contains 25 multiple choice questions. Each question has four choices (a), (b), (c) and (d), out of which ONLY ONE option is correct.*

**1** Mole fraction of $H_2$ in a 2 g mixture of $H_2$ and He is 0.5. Thus, helium present in the mixture is

a. $\frac{2}{3}$ g  b. $\frac{4}{3}$ g
c. $\frac{1}{3}$ g  d. $\frac{3}{2}$ g

**2** Mixture of $CO_2$ and CO is passed over red hot graphite when 1 mole of mixture changes to 33.6 L (converted to STP). Hence, mole fraction of $CO_2$ in the mixture is

a. 0.25  b. 0.33
c. 0.50  d. 0.66

**3** 1 g each of $H_2$, He and $CH_4$ are placed in the closed vessel. Which is not placed at the correct position in terms of mole fraction of gases?

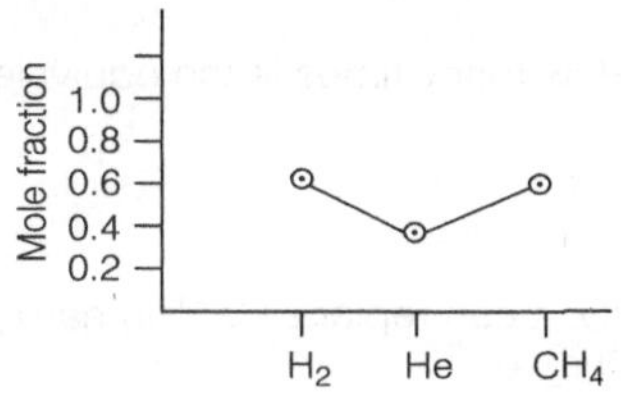


a. $CH_4$  b. He
c. $H_2$  d. All are at correct position

**4** 1 g methanol ($d$ = 0.75 g mL$^{-1}$) is mixed with 1 g water ($d$ = 1.00 g mL$^{-1}$) at 298 K. Thus, volume % of methanol ($V/V$) is

a. 42.86%  b. 57.14%
c. 50.00%  d. 25.00%

**5** 100 mL of concentrated $H_2SO_4$ ($d$ = 1.8 g mL$^{-1}$) is mixed with 100 mL of $H_2O$ ($d$ = 1.0 g mL$^{-1}$) thus, percentage of $H_2SO_4$ by mass of solution in the mixture is

a. 50.0%  b. 60.0%
c. 64.3%  d. 35.7%

**6** Hardness of water is expressed in terms of parts per million (ppm) `of $CaCO_3$. A sample of hard water is 200 ppm $CaCO_3$. Thus, mass percentage of $CaCO_3$ is

a. 0.20%  b. 0.02%
c. 0.002%  d. 0.0002%

**7** A sucrose solution 40% by mass of solution changes to 50% by mass of solution on heating. Thus, water lost due to heating of 100 g of solution is

a. 25 g  b. 20 g
c. 15 g  d. 10 g

**8** To prepare a solution that is 0.50 KCl starting with 100 mL of 0.40 M KCl

a. add 0.75 g KCl  b. add 20 mL $H_2O$
c. add 0.10 mol KCl  d. evaporate 10 mL $H_2O$

Working Space

**9** 800 g of a 40% solution by weight of solution was cooled. 100 g of the solute is precipitated. The percentage composition of the solute in the filtrate (after removing precipitate) is

a. 31.4% b. 55.6%
c. 67.8% d. 69.6%

**10** A mixture containing 1 mole each of $KHCO_3, Li_2CO_3$ and $K_2CO_3$ is strongly heated. Thus, number of moles of $CO_2$ obtained is

a. 2.0 b. 1.5
c. 1.0 d. 0.5

**11** Volume of 0.750 M $Pb(NO_3)_2$ solution required to react completely with 1.00 L of 2.25 M NaCl solution is

a. 1.125 L b. 1.500 L
c. 3.000 L d. 0.750 L

**12** Total volume occupied by a mixture containing 1 equivalent each of $N_2, H_2$ and $NH_3$ based on the following equation at STP is

$$N_2(g) + 3H_2(g) \longrightarrow 2NH_3(g)$$

a. 22.4 L b. 134.4 L
c. 67.2 L d. 11.2 L

**13** The same amount of metal combines with 0.4 g of oxygen and 4 g of halogen. Thus, equivalent mass of halogen is

a. 40.0 g $equiv^{-1}$
b. 80.0 g equiv
c. 160.0 g $equiv^{-1}$
d. None of these

**14** Equivalent mass of a metal is double that of oxygen. How many times is the equivalent mass of its oxide greater than that of the metal?

a. 4.0 times b. 2.0 times
c. 1.0 times d. 1.5 times

**15** $RH_2$ is an ion exchange resin used to purify water in RO. It can replace $Ca^{2+}$ in hard water.

$$RH_2 + Ca^{2+} \longrightarrow RCa + 2H^+$$

Water coming out of ion exchange resin has $[H^+] = 0.01\,M$. Thus, hardness of water in ppm $Ca^{2+}$ ion is

a. 100 b. 200
c. 50 d. 125

**16** 100 mL of $X$ molar $H_2O_2$ gives 3 L $O_2$ gas under the condition when 1 mole $O_2$ occupies 24 L. Thus, $X$ is if

$$2H_2O_2(l) \longrightarrow 2H_2O\,(l) + O_2(g)$$

a. 2.5 M b. 1.0 M
c. 0.5 M d. 0.25 M

**17** 0.6 g $NH_2CONH_2$ (urea) is treated with NaOH and $NH_3$ formed is passed into 300 mL of 0.1 N $H_2SO_4$. Volume of unreacted $H_2SO_4$ is neutralised by $V$ mL of 1 N NaOH. Thus, $V$ is

a. 5 mL b. 2.5 mL
c. 10 mL d. 15.0 mL

**18** Under the given conditions of temperature and pressure, 1 mole of $O_2$ gas occupies 30 L. Thus, volume occupied by 1 equivalent $O_2$ gas under the identical conditions is

a. 7.5 L b. 15.0 L
c. 30.0 L d. 22.4 L

**19** 20 mL of $H_2O_2$ oxidises 40 mL of 0.05 M $Cr^{3+}$ to $CrO_4^{2-}$ in acidic medium. Thus, $H_2O_2$ is

a. 0.15 M b. 0.30 M
c. 0.10 M d. 0.20 M

Working Space

20 $KMnO_4$ is used as an oxidising agent and is reduced to $K_2MnO_4$. Normality of $KMnO_4$ solution containing 12.64 g of it in 500 mL solution is

a. 0.08 N
b. 0.16 N
c. 0.32 N
d. 0.80 N

21 An aqueous solution of $NaHC_2O_4$ is 0.1 M when neutralised by NaOH. Its normality when oxidised to $CO_2$ by $MnO_4^- / H^+$ is

a. 0.5 N
b. 0.1 N
c. 0.2 N
d. 0.6 N

22 *X* and *Y* are two compounds of elements *A* and Cl. Molar mass % of Cl

| Compound | Molar mass | % of Cl |
|---|---|---|
| *X* | 137 | 77.5 |
| *Y* | 208 | 85.1 |

Thus, *X* and *Y* are

a. $ACl_3$, $ACl_4$
b. $A_2Cl_3$, $ACl_3$
c. $ACl$, $ACl_2$
d. $ACl_3$, $ACl_5$

23 The vapour density of a mixture of $NO_2$ and $N_2O_4$ is 38.3 at 27°C w.r.t. H taking as 1. Thus, number of moles of $NO_2$ in 100 g of mixture is

a. 1.309 mol
b. 0.2618 mol
c. 0.7372 mol
d. 0.6544 mol

24 1.947 g of iodine $I_2$ (I = 127) was allowed to react with 1.056 g of Sn (119) to form $Sn_xI_y$ along with 0.601 g of unreacted Sn. Thus, $Sn_xI_y$ is

a. $SnI_2$
b. $Sn_2I_3$
c. $SnI_4$
d. $SnI_3$

25 A piece of aluminium foil measuring 10.25 cm × 5.50 cm × 0.600 mm is dissolved in excess of HCl (*aq*). Density of foil is 2.70 g/cm$^3$. Thus, gaseous product formed is

a. 0.51 mol
b. 1.52 mol
c. 1.01 mol
d. 3.00 mol

SET - 2

# Revisal Problems for JEE Advanced

## Only One Option Correct Type

Working Space

**Direction** *(Q. Nos. 1-10) This section contains 10 multiple choice questions. Each question has four choices (a), (b), (c) and (d), out of which ONLY ONE option is correct.*

**1** Consider the following experiments being carried out separately

I. $\underset{1\text{g}}{Zn} + \underset{\text{Excess}}{NaOH(aq)} \longrightarrow$

II. $\underset{1\text{g}}{Zn} + \underset{\text{Excess}}{\text{dil. } H_2SO_4(aq)} \longrightarrow$

The ratio of volumes of $H_2$ gas evolved is

a. 1 : 1 b. 1 : 2 c. 2 : 1 d. 9 : 4

**2** A mixture of $X$ and $Y$ was treated with $AgNO_3$ and $BaCl_2$ separately

$X$ : 0.02 mole $[Co(NH_3)_5SO_4]Br + AgNO_3$ (excess) $\longrightarrow P$

$Y$ : 0.02 mole $[Co(NH_3)_5Br]SO_4 + BaCl_2$ (excess) $\longrightarrow Q$

Number of moles of $P$ and $Q$ obtained are

a. 0.01, 0.01 b. 0.01, 0.02 c. 0.02, 0.01 d. 0.02, 0.02

**3** In the Solvay process for producing sodium carbonate ($Na_2CO_3$), the following reactions occur in sequence

$$NH_3 + CO_2 + H_2O \longrightarrow NH_4HCO_3$$
$$NH_4HCO_3 + NaCl \longrightarrow NH_4Cl + NaHCO_3$$
$$2NaHCO_3 \xrightarrow{\Delta} Na_2CO_3 + H_2O + CO_2$$

How much of $Na_2CO_3$ would be produced per kg of $NH_3$ used if the process were 100% efficient?

a. 3.12 kg b. 1.00 kg c. 2.00 kg d. 0.56 kg

**4** 56 mL of 0.1 M solution of $A^{3+}(aq)$ required exactly 25 mL of 0.1 M sodium sulphate (IV) to oxidise it to sodium sulphate (VI). Thus, oxidation number of the reduced ion is

a. 0 b. 1 c. 2 d. 4

**5** In the photographic developing process, AgBr (silver bromide) is dissolved by adding $Na_2S_2O_3$ (sodium thiosulphate)

$$AgBr(s) + Na_2S_2O_3(aq) \longrightarrow Na_3[Ag(S_2O_3)_2] + NaBr$$

What volume of 0.015 M $Na_2S_2O_3$ is required to dissolve 0.3760 g of AgBr($s$)?

a. 150 mL b. 376 mL c. 134 mL d. 267 mL

**6** How many millilitre of 0.650 M $K_2CrO_4$ are needed to precipitate all the silver ion in 415 mL of 0.186 M $AgNO_3$ as $Ag_2CrO_4(s)$?

a. 100.0 mL b. 41.5 mL c. 59.4 mL d. 206.0 mL

**7** A mixture of $Na_2C_2O_4(A)$ and $KHC_2O_4 \cdot H_2C_2O_4(B)$ required equal volumes of 0.1 M $KMnO_4$ and 0.1 M NaOH separately. Molar ratio of $A$ and $B$ in the mixture is

a. 1 : 1 b. 1 : 5.5 c. 5.5 : 1 d. 3.1 : 1

**8** 10 L of hard water required 0.56 g of lime (CaO) for removing hardness

$$Ca(HCO_3)_2 + CaO \longrightarrow 2CaCO_3 + H_2O$$

Thus, temporary hardness in terms of ppm of $CaCO_3$ is

a. 10 b. 20 c. 100 d. 200

Working Space

**9** Atoms of the element $X$ (atomic mass = 23) are spherical. Atoms are placed at the corners of the cube as shown. Density of the element is 6.2 g/cm$^3$. Thus, $AB$ is

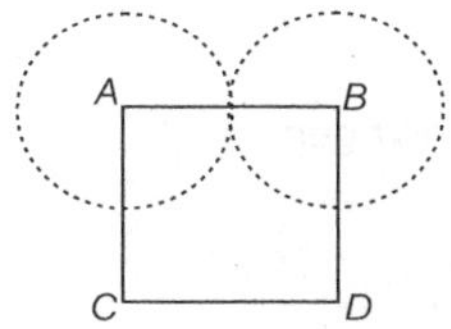


a. 2.274 Å  b. 1.137 Å  c. 2.274 nm  d. 1.137 nm

**10** 11.2 g of a mixture of $M$Cl (volatile) and NaCl gave 28.7 g of white precipitate with excess of $AgNO_3$ solution. 11.2 g of same mixture on heating gave a gas that on passing into $AgNO_3$ solution gave 14.35 g of white precipitate. Thus, molar ratio of $M$Cl and NaCl is

a. 1 : 1  b. 1 : 2  c. 2 : 1  d. 1 : 3

## One or More than One Options Correct Type

***Direction** (Q. Nos. 11-14) This section contains 4 multiple choice questions. Each question has four choices (a), (b), (c) and (d), out of which ONE or MORE THAN ONE are correct.*

**11** 0.6 g urea ($NH_2CONH_2$) on reaction with NaOH gave $NH_3$ which can be neutralised by

a. 100 mL of 0.1 M HCl
b. 200 mL of 0.2 M HCl
c. 100 mL of 0.2 M HCl
d. 100 mL of 0.1 M $H_2SO_4$

**12** In which of the following cases equivalent weight is half of the molecular weight?

a. $H_3PO_3$ is completely neutralised by NaOH
b. $H_3PO_3$ is oxidised to $H_3PO_4$
c. $MnSO_4$ is oxidised to $MnO_2$
d. $NaHC_2O_4$ is oxidised to $CO_2$

**13** Which of the following statements is/ are correct concerning glucose ($C_6H_{12}O_6$)?

a. The percentage, by mass of C and O are the same as in CO
b. The ratio of C to H to O atoms is the same as in dihydroxy acetone $(CH_2OH)_2CO$
c. The proportions, by mass of C and O are equal
d. The highest percentage, by mass is that of H

**14** 20 g of $CaCO_3$ impure sample gave 5.6 g of CaO and 4.4 g of $CO_2$. Thus, there are

a. 0.2 $N_0$ $CaCO_3$ molecules  b. 0.1 $N_0$ $CaCO_3$ molecules
c. $N_0$ atoms  d. 0.5 $N_0$ atoms

## Comprehension Type

***Direction** (Q. Nos. 15-23) This section contains 4 paragraphs, each describing theory, experiments, data, etc. Nine questions related to the paragraphs have been given. Each question has only one correct answer among the four given options (a), (b), (c) and (d).*

### Passage I

In one reaction with a 95% yield, 225 mL of 1.50 M $Na_2CO_3(aq)$, 22.1 g NO and a large excess $O_2$ are allowed to react

$$2Na_2CO_3(aq) + 4NO(g) + O_2(g) \longrightarrow 4NaNO_2(aq) + 2CO_2(g)$$

**15** What mass of $NaNO_2$ is obtained based on experimental yield?

a. 44.25 g  b. 46.58 g
c. 23.29 g  d. None of these

**16** Excess reagent is and ......... moles.

a. $Na_2CO_3$, 0.169 mol  b. NO, 0.1143 mol
c. NO, 0.2286 mol  d. Both are in exact stoichiometric ratio

Working Space

### Passage II

A compound contains

28% nitrogen by weight

72% metal by weight

3 atoms of metal combine with 2 atoms of nitrogen.

**17** Molar mass of the compound is

a. 72 g $mol^{-1}$
b. 28 g $mol^{-1}$
c. 100 g $mol^{-1}$
d. 124 g $mol^{-1}$

**18** Atomic weight of the metal is

a. 28 g $mol^{-1}$
b. 24 g $mol^{-1}$
c. 12 g $mol^{-1}$
d. 72 g $mol^{-1}$

### Passage III

The following reaction of potassium superoxide, $KO_2$ is used in life support systems to replace $CO_2(g)$ in expired air by $O_2(g)$

$$4KO_2(s) + 2CO_2(g) \longrightarrow 2K_2CO_3(s) + 3O_2(g)$$

**19** How many moles of $KO_2(g)$ are consumed per 100.0 g $CO_2(g)$ removed from expired air?

a. 2.27 mol  b. 3.41 mol  c. 4.54 mol  d. 1.13 mol

**20** How many moles of $O_2(g)$ are produced by the reaction of 176 g $CO_2(g)$ with excess $KO_2(s)$?

a. 3.0 mol  b. 4.0 mol  c. 5.33 mol  d. 6.00 mol

**21** How many micromoles of $O_2(g)$ are produced per milligram of $KO_2$ consumed?

a. 1.70  b. 5.35  c. 10.71  d. 10.56

### Passage IV

Copper metal can be prepared by roasting copper ore which contains cuprite ($Cu_2S$) and copper (II) sulphide.

I. $Cu_2S(s) + O_2(g) \longrightarrow 2Cu(s) + SO_2(g)$

II. $CuS(s) + O_2(g) \longrightarrow Cu(s) + SO_2(g)$

Suppose an ore sample contains 11.0% impurity in addition to a mixture of CuS and $Cu_2S$. Heating 100.0 g of the mixture produces 75.4 g of copper metal with a purity of 89.5%.

**22** The weight per cent of CuS in the ore is

a. 89.5%  b. 62.1%  c. 26.8%  d. 44.75%

**23** The weight per cent of $Cu_2S$ in the ore is

a. 44.5%
b. 62.2%
c. 26.9%
d. 25.2%

## One Integer Value Correct Type

***Direction** (Q. Nos. 24-26) This section contains 3 questions. When worked out will result in an integer from 0 to 9 (both inclusive).*

**24** Consider two sets as shown, containing $H_2$ and $N_2$.

| Gas | Moles | $V$ | $T$ | $p$ |
|---|---|---|---|---|
| $H_2$ | $x$ | 2 L | 400 K | 5 atm |
| $N_2$ | $y$ | 5 L | 300 K | 3 atm |

What is the ratio $\left(\frac{y}{x}\right)$?

**25** Automobile batteries are filled with an aqueous solution of $H_2SO_4$. How many moles of sulphuric acid in 500 mL of the battery solution are contained if the density of the solution is 1.303 g / $cm^3$ and if the solution is 30.08% sulphuric acid by mass?

Working Space

**26** A 60.0 g sample of a dry-cleaning fluid was analysed and found to contain

10.80 g carbon
1.36 g hydrogen
and 47.86 g chlorine

How many total atoms of C, H and Cl are present in one empirical formula unit?

## Matching List Type

***Direction** (Q. Nos. 27-29) Choices for the correct combination of elements from Column I and Column II are given as options (a), (b), (c) and (d), out of which one is correct.*

**27** Match the reaction in Column I with titration method in Column II and select answer from the answer code.

| | Column I | | Column II |
|---|---|---|---|
| i. | $2S_2O_3^{2-} + I_2 \longrightarrow S_4O_6^{2-} + 2I^-$ | p. | Redox with self indicator |
| ii. | $MnO_4^- + Fe^{2+} + H^+ \longrightarrow Mn^{2+} + Fe^{3+} + H_2O$ | q. | Iodometric |
| iii. | $NH_4^+ + OH^- \longrightarrow 2NH_3 + H_2SO_4 \longrightarrow (NH_4)_2SO_4$ | r. | Iodimetric |
| iv. | $Cr_2O_7^{2-} + I^- + H^+ \longrightarrow Cr^{3+} + I_2 + H_2O$ | s. | Kjeldahl |

**Codes**

| | i | ii | iii | iv | | i | ii | iii | iv |
|---|---|---|---|---|---|---|---|---|---|
| a. | p | s | q | p | b. | r | q | p | s |
| c. | p | r | q | s | d. | r | p | s | q |

**28** 1 mole each of $CuSO_4$, $K_2Cr_2O_7$, $H_2O_2$, $O_3$ and $HNO_2$ are allowed to react with KI in acidic medium. Match these compounds in Column I with number of moles of $I_2$ formed from them in Column II.

| | Column I | | Column II |
|---|---|---|---|
| i. | $CuSO_4$ | p. | 0.5 |
| ii. | $K_2Cr_2O_7$ | q. | 1.0 |
| iii. | $H_2O_2$ | r. | 1.5 |
| iv. | $O_3$ | s. | 2.0 |
| v. | $HNO_2$ | t. | 3.0 |

**Codes**

| | i | ii | iii | iv | v | | i | ii | iii | iv | v |
|---|---|---|---|---|---|---|---|---|---|---|---|
| a. | p | q | s | r | q | b. | p | t | s | r | t |
| c. | p | t | q | t | p | d. | t | p | s | r | q |

**29** Match the properties of the species in Column I with the calculated number in Column II.

| | Column I | | Column II |
|---|---|---|---|
| i. | Number of protons in the element with mass number 81 and having 31.7% more neutrons than protons. | p. | 50 |
| ii. | Ionic mass of $M^{2+}$ which is isoelectronic of $CO_2$ and has $(Z+2)$ neutrons. | q. | 26 |
| iii. | Ionic mass of $X^-$ with 17 protons and having 11.1% more neutrons than electrons. | r. | 35 |
| iv. | Atomic number of $M^{3+}$ with mass number 56 and having 30.4% more neutrons than electrons. | s. | 37 |

**Codes**

| | i | ii | iii | iv | | i | ii | iii | iv |
|---|---|---|---|---|---|---|---|---|---|
| a. | p | q | r | s | b. | s | r | p | q |
| c. | p | q | r | s | d. | r | p | s | q |

# JEE Main & AIEEE Archive

## (Compilation of Last 13 Years Questions)

**1** The ratio of masses of oxygen and nitrogen in a particular gaseous mixture is 1: 4. The ratio of number of their molecule is **(JEE Main 2014)**

a. 1 : 4 b. 7 : 32
c. 1 : 8 d. 3 : 16

**2** The molarity of a solution obtained by mixing 750 mL of 0.5 M HCl with 250 mL of 2M HCl will be **(JEE Main 2013)**

a. 0.875 M b. 1.00 M
c. 1.75 M d. 0.0975 M

**3** A gaseous hydrocarbon gives upon combustion, 0.72 g of water and 3.08 g of $CO_2$. The empirical formula of the hydrocarbon is **(JEE Main 2013)**

a. $C_2H_4$ b. $C_3H_4$
c. $C_6H_5$ d. $C_7H_8$

**4** The density of a solution prepared by dissolving 120 g of urea (mol. mass = 60 u) in 1000 g of water is 1.15 g/mL. The molarity of this solution is **(AIEEE 2012)**

a. 0.50 M b. 1.78 M
c. 1.02 M d. 2.05 M

**5** The mass of potassium dichromate crystals required to oxidise 750 $cm^3$ of 0.6 M Mohr's salt solution is (molar mass = 392) **(AIEEE 2011)**

a. 0.49 g b. 0.45 g
c. 22.05 g d. 2.2 g

**6** The molality of a urea solution in which 0.0100 g of urea, $[(NH_2)_2CO]$ is added to 0.3000 $dm^3$ of water at STP is **(AIEEE 2011)**

a. $5.55 \times 10^{-4}$ M
b. 33.3M
c. $3.33 \times 10^{-2}$ M
d. 0.555 M

**7** Amount of oxalic acid present in a solution can be determined by its titration with $KMnO_4$ solution in the presence of $H_2SO_4$. The titration gives unsatisfactory result when carried out in the presence of HCl because HCl **(AIEEE 2008)**

a. gets oxidised by oxalic acid to chlorine
b. furnishes $H^+$ ions in addition to those from oxalic acid
c. reduces permanganate to $Mn^{2+}$
d. oxidises oxalic acid to carbon dioxide and water

**8** In the reaction,

$$2Al\,(s) + 6HCl\,(aq) \longrightarrow 2Al^{3+}\,(aq) + 6Cl^-\,(aq) + 3H_2(g)$$

**(AIEEE 2007)**

a. 6L HCl (*aq*) is consumed for every 3L $H_2(g)$ produced
b. 33.6 L $H_2(g)$ is produced regardless of temperature and pressure for every mole Al that reacts
c. 67.2 L $H_2(g)$ at STP is produced for every mole Al that reacts
d. 11.2 L $H_2(g)$ at STP is produced for every mole HCl (*aq*) consumed

**9** How many moles of magnesium phosphate, $Mg_3(PO_4)_2$ will contain 0.25 mole of oxygen atoms? **(AIEEE 2006)**

a. 0.02 b. $3.125 \times 10^{-2}$
c. $1.25 \times 10^{-2}$ d. $2.5 \times 10^{-2}$

**10** If we consider that 1/6, in place of 1/12, mass of carbon atom is taken to be the relative atomic mass unit, the mass of one mole of a substance will **(AIEEE 2005)**

a. be a function of the molecular mass of the substance
b. remain unchanged
c. increase two fold
d. decrease twice

**11** What volume of hydrogen gas, at 273 K and 1 atm pressure will be consumed in obtaining 21.6 g of elemental boron (atomic mass = 10.8) from the reduction of boron trichloride by hydrogen? **(AIEEE 2003)**

a. 89.6 L b. 67.2 L
c. 44.8 L d. 22.4 L

**12** Number of atoms in 560 g of Fe (atomic mass = 56 g $mol^{-1}$) is **(AIEEE 2002)**

a. twice that of 70 g N b. half that of 20 g H
c. Both (a) and (b) d. None of these

**13** In an organic compound of molar mass 108 g $mol^{-1}$ C, H and N atoms are present in 9 : 1 : 3.5 by weight. Molecular formula can be **(AIEEE 2002)**

a. $C_6H_8N_2$ b. $C_7H_{10}N$
c. $C_5H_6N_3$ d. $C_4H_{18}N_3$

# JEE Advanced & IIT JEE Archive

(Compilation of Last 10 Years Questions)

1 A compound $H_2X$ with molar weight of 80 g is dissolved in a solvent having density of 0.4 g $mL^{-1}$. Assuming no change in volume upon dissolution, the molality of a 3.2 molar solution is **(2014 Adv, Integer Type)**

2 29.2% (*w/W*) HCl stock solution has density of 1.25 g $mL^{-1}$. The molecular weight of HCl is 36.5 g $mol^{-1}$. The volume (mL) of stock solution required to prepare a 200 mL solution 0.4 M HCl is **(2012, Integer Type)**

3. Dissolving 120 g of urea (mol. wt. 60) in 1000 g of water gave a solution of density 1.15 g/mL. The molarity of the solution is **(2011, Single Option Type)**

a. 1.78 M b. 2.00 M
c. 2.05 M d. 2.22 M

4. Given that the abundances of isotopes $_{54}Fe$, $_{56}Fe$ and $_{57}Fe$ are 5%, 90% and 5%, respectively, the atomic mass of Fe is **(2009, Single Option Correct Type )**

a. 55.85 b. 55.95
c. 55.75 d. 56.05

5. Consider a titration of potassium dichromate solution with acidified Mohr's salt solution using diphenylamine as indicator. The number of moles of Mohr's salt required per mole of dichromate is **(2007, Single Option Correct Type)**

a. 3 b. 4
c. 5 d. 6

6. Calculate the amount of calcium oxide required when it reacts with 852 g of $P_4O_{10}$. **(2005, Subjective Type )**

JEE Archive

# Answers *with* Explanations

# Answers

## 1. Structure of Atom

### DPP-1 Electromagnetic Radiations

**1.** $6 \times 10^{-2}$ m, $5 \times 10^{9}$ s$^{-1}$, 16.67 m$^{-1}$ **2.** $2.014 \times 10^{16}$ **3.** $4.9404 \times 10^{2}$ kJ mol$^{-1}$

**4.** (i) $1.986 \times 10^{-23}$ J per photon, (ii) $11.956 \times 10^{-3}$ kJ mol$^{-1}$, (iii) $2.857 \times 10^{-3}$ kcal mol$^{-1}$

**5.** *a* **6.** *d* **7.** *a* **8.** *c* **9.** *b* **10.** *d* **11.** *d* **12.** *b* **13.** *a, b, c, d*
**14.** *a, c, d* **15.** *a, b, c, d* **16.** *a* **17.** *b* **18.** *b* **19.** 8 **20.** 4

### DPP-2 Dual Nature of Electron (de-Broglie's Equation) Heisenberg's Uncertainty Principle and Photo Electric Effect

**1.** $6.13 \times 10^{-33}$ kg **2.** (i) 896 nm, (ii) $3.3 \times 10^{6}$ watt **3.** (i) $4.4118 \times 10^{14}$ s$^{-1}$, (ii) $2.921 \times 10^{-19}$ J
**4.** $7.20 \times 10^{-19}$ J **5.** *d* **6.** *d* **7.** *d* **8.** *c* **9.** *b* **10.** *c* **11.** *b* **12.** *d*
**13.** *a* **14.** *b* **15.** *a* **16.** *b* **17.** *b* **18.** 4 **19.** 6 **20.** 9

### DPP-3 Bohr's Theory of H-Atom

**1.** $6.62 \times 10^{-5}$ cm **2.** 122 nm **3.** (i) $5.25 \times 10^{6}$ J mol$^{-1}$, (ii) 656.5 nm
**4.** (i) $n_1 = 1, n_2 = 2$, (ii) $3.2881 \times 10^{15}$ s$^{-1}$ and $2.4661 \times 10^{15}$ s$^{-1}$ **5.** $-3.4$ eV **6.** *d* **7.** *b*
**8.** *b* **9.** *a* **10.** *a* **11.** *b* **12.** *c* **13.** *a* **14.** *b* **15.** *a* **16.** *b* **17.** *a*
**18.** *b* **19.** *a* **20.** *b* **21.** 2 **22.** 4 **23.** 2 **24.** 3

### DPP-4 Quantum Numbers and Electronic Configuration

**1.** (i) 3, (ii) [Ar] $3d^3$ **2.** $n = 5$, suborbital = 1 and orbitals = 9 **4.** 12 elements **5.** *d* **6.** *c*
**7.** *c* **8.** *a* **9.** *a* **10.** *c* **11.** *c* **12.** *a* **13.** *b* **14.** *c* **15.** *a, d*
**16.** *a,b* **17.** *c, d* **18.** *a, b, c, d* **19.** *b, c, d* **20.** *c* **21.** *b* **22.** *b*
**23.** 6 **24.** 9

### DPP-5 Quantum Mechanical Model of an Atom

**1.** *d* **2.** *b* **3.** *d* **4.** *d* **5.** *d* **6.** *c* **7.** *c* **8.** *c* **9.** *a* **10.** *a*
**11.** *c* **12.** *b* **13.** *a, b, c, d* **14.** *a, b, c, d* **15.** *a, c*

### Revisal Problems (JEE Main)

**1.** *a* **2.** *a* **3.** *b* **4.** *d* **5.** *c* **6.** *a* **7.** *d* **8.** *a* **9.** *b* **10.** *c*
**11.** *b* **12.** *b* **13.** *b* **14.** *b* **15.** *a* **16.** *b* **17.** *a* **18.** *d* **19.** *d* **20.** *a*
**21.** *a* **22.** *d* **23.** *d* **24.** *c*

### Revisal Problems (JEE Advanced)

**1.** *a* **2.** *d* **3.** *c* **4.** *a* **5.** *b* **6.** *b* **7.** *a* **8.** *a* **9.** *b* **10.** *d*
**11.** *a* **12.** *b* **13.** *a* **14.** *a, b, c, d* **15.** *a,b* **16.** *a, b, c* **17.** *d* **18.** *c* **19.** *c*
**20.** *a* **21.** *a* **22.** *b* **23.** 6 **24.** 5 **25.** 5 **26.** (i) 5, (ii) 8 **27.** *a* **28.** *c*

### JEE Main & AIEEE Archive

**1.** *a* **2.** *a* **3.** *b* **4.** *c* **5.** *b* **6.** *d* **7.** *b* **8.** *c* **9.** *b* **10.** *d*
**11.** *b* **12.** *a* **13.** *c* **14.** *d* **15.** *a* **16.** *a* **17.** *b* **18.** *a* **19.** *c* **20.** *b*
**21.** *a* **22.** *a* **23.** *b* **24.** *d* **25.** *b* **26.** *b* **27.** *a* **28.** *c*

### JEE Advanced & IIT JEE Archive

**1.** 6 **2.** 5 **3.** *c* **4.** 9 **5.** 4 **6.** *b* **7.** *c* **8.** *b* **9.** *b* **10.** *a*
**11.** *a* **12.** (i) $2.18 \times 10^{6}$ ms$^{-1}$, (ii) $0.33 \times 10^{-9}$ m, (iii) $\sqrt{2}\,(h/2\pi)$

## 2. Redox Reactions

### DPP-1 Oxidation Number

**1.** (i) +5, (ii) –1, (iii) +6, (iv) +6, (v) 8/3 per atom, (vi) –2, – 2, (vii) +2, –2 **2.** +2 **3.** (i) +3, (ii) 0

**4.** 2.5 **6.** *d* **7.** *b* **8.** *a* **9.** *d* **10.** *a* **11.** *d* **12.** *c* **13.** *a* **14.** *c*

**15.** *c* **16.** *c* **17.** *b* **18.** *a, b, c,* **19.** *a, b, d* **20.** *a* **21.** *c* **22.** *b*

**23.** 8 **24.** 7

### DPP-2 Oxidation-Reduction and Disproportionation Reactions

**3.** *b* **4.** *a* **5.** *d* **6.** *b* **7.** *a* **8.** *c* **9.** *d* **10.** *a* **11.** *b* **12.** *a*

**13.** *c* **14.** *d* **15.** *d* **16.** *d* **17.** *b* **18.** *c* **19.** *b* **20.** *d* **21.** *a, c*

**22.** *a, b, c, d* **23.** *a, b* **24.** *b, d* **25.** *a, b, c, d*

### DPP-3 Equivalent in Terms of Redox Reactions (Simple Calculations)

**1.** *c* **2.** *c* **3.** *d* **4.** *b* **5.** *c* **6.** *d* **7.** *d* **8.** *c* **9.** *b* **10.** *b*

**11.** *b* **12.** *d* **13.** *a* **14.** *b* **15.** *c* **16.** *a, b, c, d* **17.** *a, b* **18.** *b* **19.** *b*

**20.** *d* **21.** *c* **22.** 5 **23.** 6

### Revisal Problems (JEE Main)

**1.** *c* **2.** *c* **3.** *a* **4.** *d* **5.** *c* **6.** *c* **7.** *b* **8.** *b* **9.** *c* **10.** *d*

**11.** *c* **12.** *b* **13.** *c* **14.** *a* **15.** *a* **16.** *a* **17.** *d* **18.** *b* **19.** *a* **20.** *a*

### Revisal Problems (JEE Advanced)

**1.** *a* **2.** *b* **3.** *c* **4.** *a* **5.** *c* **6.** *c* **7.** *d* **8.** *a, d* **9.** *b, d*

**10.** *a, b, d* **11.** *a, b, c* **12.** *a,b, c, d* **13.** *a, c* **14.** *a, d* **15.** *c* **16.** *c*

**17.** *a, b, c* **18.** *a, d* **19.** *c* **20.** *a* **21.** *d* **22.** *a* **23.** *b* **24.** 2 **25.** 0

**26.** 6 **27.** 7 **28.** 1

### JEE Main & AIEEE Archive

**1.** *d* **2.** *c* **3.** *a* **4.** *d* **5.** *a* **6.** *d* **7.** *c* **8.** *d* **9.** *c*

## 3. Chemical Bonding

### DPP-1 Ionic Bonds and Fajans' Rule

**2.** (i) Sublimation, heat of sublimation; (ii) Ionisation, ionisation energy; (iii) Dissociation, bond energy; (iv) Electron gain, electron gain energy; (v) Lattice formation, lattice energy

**3.** (i) ionic, (ii) ionic, (iii) covalent, (iv) covalent **4.** 91.55% **5.** *c* **6.** *a* **7.** *c*

**8.** *b* **9.** *c* **10.** *d* **11.** *b* **12.** *d* **13.** *a* **14.** *b* **15.** *a* **16.** *d* **17.** *a*

**18.** *d* **19.** *d* **20.** *d* **21.** *a,b, d* **22.** *a,d* **23.** *a,b* **24.** *b* **25.** *a* **26.** *a* **27.** 8

### DPP-2 Covalent Coordinate Bonding

**2.** $X_2Y_3$ **3.** Lewis bases -(i), (ii), (vi) and Lewis acid - (iii), (iv), (v), (vii), (viii) **6.** *d* **7.** *a*

**8.** *d* **9.** *b* **10.** *a* **11.** *c* **12.** *a* **13.** *d* **14.** *c* **15.** *b* **16.** *a* **17.** *d*

**18.** *b* **19.** *a* **20.** *b* **21.** *a,b,c,d* **22.** *a,b,c,d* **23.** *a,c,d* **24.** *b,c* **25.** *b,d*

**26.** *b* **27.** *d* **28.** *a* **29.** *c* **30.** *c* **31.** *a* **32.** 5 **33.** 4

### DPP-3 Covalent Bonding

**1.** *b* **2.** *d* **3.** *d* **4.** *b* **5.** *b* **6.** *d* **7.** *d* **8.** *d* **9.** *d* **10.** *c*

**11.** *d* **12.** *b* **13.** *b* **14.** *b* **15.** *b* **16.** *d* **17.** *c* **18.** *a,b,c,d* **19.** *c,d*

## DPP-4 Chemical Bonding (VSEPR Model) and Hybridisation

2. *d* 3. *b* 4. *b* 5. *d* 6. *c* 7. *b* 8. *a* 9. *a* 10. *b* 11. *a*
12. *c* 13. *a* 14. *a* 15. *c* 16. *d* 17. *d* 18. *a,b,c* 19. *a,b* 20. *a,d*
21. *a,b,c,d* 22. *b* 23. *b* 24. *d* 25. *a* 26. (i) 0, (ii) 3, (iii) 6 27. 5

## DPP-5 Hybridisation

1. $sp^2$-hydridised; (ii) (a) no change (b) one lone pair extra 4. *a* 5. *c* 6. *c* 7. *d*
8. *d* 9. *c* 10. *d* 11. *a* 12. *b* 13. *b* 14. *a* 15. *b* 16. *d* 17. *a*
18. *b,c,d* 19. *b,c,d* 20. *a,b,c,d* 21. *a* 22. *b* 23. *c* 24. *b* 25. 1

## DPP-6 Molecular Orbital Theory

4. *a* 5. *c* 6. *b* 7. *c* 8. *a* 9. *b* 10. *b* 11. *c* 12. *d* 13. *b*
14. *d* 15. *a* 16. *b* 17. *d* 18. *b* 19. *a,b,d* 20. *a,b* 21. *a,d* 22. *b*
23. *a* 24. *d* 25. 7 26. 5

## DPP-7 Dipole Moment

1. (i) = (iv) < (iii) < (ii) 2. 84.4% 3. 1 esu cm = $3.336 \times 10^{-12}$ Cm 6. *d* 7. *c* 8. *c*
9. *b*
10. *c* 11. *d* 12. *a* 13. *a* 14. *a* 15. *c* 16. *d* 17. *c* 18. *a,d* 19. *b,d*
20. *b,c* 21. *c* 22. *a* 23. 5 24. 7

## DPP-8 Formal Charge and Resonance

1. *c* 2. *a* 3. *c* 4. *c* 5. *b* 6. *d* 7. *a* 8. *c* 9. *c* 10. *d*
11. *c* 12. *b* 13. *a* 14. *a* 15. *a* 16. *a,b,c* 17. *b, c, d* 18. *c* 19. *a*
20. *d* 21. 6 22. 5

## DPP-9 Hydrogen Bonding

1. (i) Intermolecular H-bonding, (ii) Intermolecular H-bonding, (iii) Intramolecular H-bonding (iv) Intermolecular H-bonding 3. *a* 4. *d* 5. *a* 6. *c* 7. *b* 8. *a*
9. *d* 10. *b* 11. *a* 12. *c* 13. *a* 14. *d* 15. *b* 16. *a,b* 17. *a,b,d*
18. *a,c* 19. *c* 20. *b*

## Revisal Problems (JEE Main)

1. *a* 2. *b* 3. *b* 4. *c* 5. *a* 6. *c* 7. *d* 8. *b* 9. *a* 10. *c*
11. *b* 12. *b* 13. *b* 14. *c* 15. *c* 16. *ad* 17. *c* 18. *d* 19. *b* 20. *d*
21. *a* 22. *b* 23. *b* 24. *c*

## Revisal Problems (JEE Advanced)

1. *b* 2. *d* 3. *c* 4. *a* 5. *c* 6. *c* 7. *c* 8. *a* 9. *a, b, c, d*
10. *a, b, c , d* 11. *a, b* 12. *a, b, c, d* 13. *b, c, d* 14. *a, c, d* 15. *b, c*
16. *b, c* 17. *c, d* 18. *b,c* 19. *b* 20. *a, b, c* 21. *b, d* 22. *b* 23. *c* 24. *b*
25. *c* 26. *b* 27. *c* 28. 4 29. 8 30. 7

## JEE Main & AIEEE Archive

1. *b* 2. *d* 3. *a, b* 4. *c* 5. *b* 6. *c* 7. *c* 8. *c* 9. *b* 10. *d*
11. *d* 12. *a* 13. *a* 14. *b* 15. *a* 16. *a* 17. *b* 18. *b* 19. *a* 20. *d*
21. *a* 22. *a* 23. *a* 24. *c* 25. *d* 26. *c* 27. *c* 28. *b* 29. *b* 30. *b*
31. *c* 32. *c* 33. *c* 34. *a* 35. *d* 36. *a* 37. *d*

## JEE Advanced & IIT JEE Archive

1. *c* 2. *a,b,d* 3. *c* 4. *a* 5. *d* 6. *d* 7. *b* 8. *a* 9. *d* 10. *c*
11. *d*

## 4. Stoichiometry

### DPP-1 Mole Concept

**1.** (i) $6.02 \times 10^{21}$ Mg atoms, (ii) $4.3 \times 10^{22}$ atoms, (iii) $3.01 \times 10^{23}$ atoms
**2.** (i) (a) 14 amu (b) electrons = 6, protons = 6, neutrons = 8; (ii) $2.50 \times 10^{22}$ atoms
**3.** (i) 0.02 mol, (ii) 0.60 g, (iii) $6.02 \times 10^{17}$ **5.** *d* **6.** *b* **7.** *d* **8.** *a* **9.** *a* **10.** *b*
**11.** *b* **12.** *d* **13.** *a* **14.** *b* **15.** *d* **16.** *c* **17.** *d* **18.** *b* **19.** *a* **20.** *c*
**21.** *d* **22.** *b* **23.** *a* **24.** *b* **25.** *c* **26.** *c* **27.** *c* **28.** *b* **29.** 8 **30.** 3
**31.** 6

### DPP-2 Chemical Calculations

**1.** (i) 3.565 kg, (ii) 92.57% **2.** (i) $SO_2$, (ii) 612.5 g, (iii) $O_2$- 75 g, $H_2O$ -12.5 g **3.** 0.2050 g
**4.** 22.3 mL **5.** 20 moles, 20 moles, 10 moles **6.** *d* **7.** *d* **8.** *c* **9.** *d*
**10.** *b* **11.** *c* **12.** *b* **13.** *d* **14.** *d* **15.** *a* **16.** *d* **17.** *a* **18.** *a* **19.** *c*
**20.** 5 **21.** 7 **22.** 8 **23.** *c* **24.** *c*

### DPP-3 Concentration Factors

**1.** (i) 55.55 mol $L^{-1}$, (ii) 10.43 kg $mol^{-1}$ **2.** 18 mol $L^{-1}$, 500 mol $kg^{-1}$, 36 N
**3.** (i) 0.75, 0.25; (ii) 65.22 mol $kg^{-1}$, 18.52 mol $kg^{-1}$ **4.** *d* **5.** *c* **6.** *a* **7.** *d* **8.** *d*
**9.** *d* **10.** *b* **11.** *d* **12.** *d* **13.** *d* **14.** *b* **15.** *a* **16.** *c* **17.** *d* **18.** *b*
**19.** *b* **20.** *b* **21.** *c* **22.** *b* **23.** *d* **24.** *a* **25.** *b* **26.** *a* **27.** *c* **28.** *a*
**29.** *b* **30.** *b* **31.** *b* **32.** 8 **33.** 8 **34.** 2

### DPP-4 Acid-Base Neutralisation Titration

**1.** (i) 12 g $equiv^{-1}$; (ii) 60 g $equiv^{-1}$, 60 g $mol^{-1}$ **2.** *a* **3.** *b* **4.** *a* **5.** *a* **6.** *d*
**7.** *c* **8.** *a* **9.** *b* **10.** *a* **11.** *c* **12.** *d* **13.** *a, b,* **14.** *b,c* **15.** *b,d* **16.** *a, c*
**17.** *b,d* **18.** *a* **19.** *b* **20.** 6

### DPP-5 Redox-Based Titration

**1.** *c* **2.** *a* **3.** *a* **4.** *c* **5.** *c* **6.** *d* **7.** *a* **8.** *c* **9.** *a* **10.** *b*
**11.** *a* **12.** *c* **13.** *a* **14.** *b* **15.** *b* **16.** *a,b* **17.** *a,d* **18.** *d* **19.** *c* **20.** *d*
**21.** 6 **22.** 4 **23.** 4 **24.** 6

### Revisal Problems (JEE Main)

**1.** *b* **2.** *c* **3.** *a* **4.** *b* **5.** *c* **6.** *b* **7.** *b* **8.** *a* **9.** *a* **10.** *b*
**11.** *b* **12.** *a* **13.** *b* **14.** *d* **15.** *b* **16.** *a* **17.** *c* **18.** *a* **19.** *a* **20.** *b*
**21.** *c* **22.** *d* **23.** *c* **24.** *c* **25.** *a*

### Revisal Problems (JEE Advanced)

**1.** *a* **2.** *d* **3.** *a* **4.** *c* **5.** *d* **6.** *c* **7.** *c* **8.** *d* **9.** *a* **10.** *a*
**11.** *c,d* **12.** *a,b,c,d* **13.** *b* **14.** *b,d* **15.** *a* **16.** *b* **17.** *c* **18.** *b* **19.** *c*
**20.** *d* **21.** *d* **22.** *b* **23.** *c* **24.** 2 **25.** 2 **26.** 8 **27.** *d* **28.** *c* **29.** *d*

### JEE Main & AIEEE Archive

**1.** *b* **2.** *a* **3.** *d* **4.** *d* **5.** *c* **6.** *a* **7.** *c* **8.** *d* **9.** *b* **10.** *b*
**11.** *b* **12.** *c* **13.** *a*

### JEE Advanced & IIT JEE Archive

**1.** 8 **2.** 8 **3.** *c* **4.** *b* **5.** *d* **6.** 1008

# 1. Structure of Atom

## DPP-1 Electromagnetic Radiations

**1.** **Plan** Period = $T$ (time)

Then, frequency, $\nu = \frac{1}{T}$

$$\nu = \frac{c}{\lambda} = c\bar{\nu}$$

Thus, frequency ($\nu$), wavelength ($\lambda$) and wave number ($\bar{\nu}$) can be calculated

Period = $2.0 \times 10^{-10}$ s

Thus, frequency $= \frac{1}{2.0 \times 10^{-10}} = 5 \times 10^{9}\ s^{-1}$

$$\nu = \frac{c}{\lambda}$$

$\therefore \quad \lambda = \frac{c}{\nu} = \frac{3 \times 10^{8}\ ms^{-1}}{5 \times 10^{9}\ s^{-1}} = 6.0 \times 10^{-2}$ m

Wave number, $\bar{\nu} = \frac{1}{\lambda} = \frac{1}{6.0 \times 10^{-2}\ m}$

$= 16.67\ m^{-1}$

**2.** **Plan** $E = N_0 h\nu = \frac{N_0 hc}{\lambda}$

$E$ and $\lambda$ are known, thus, $N_0$ (number of photons) can be calculated

$E = 1$ J and $\lambda = 4000$ pm $= 4000 \times 10^{-12}$ m

Thus, $\quad N_0 = \frac{E\lambda}{hc} = \frac{1\ J \times 4000 \times 10^{-12}\ m}{6.62 \times 10^{-34}\ Js \times 3 \times 10^{8}\ ms^{-1}}$

$= 2.014 \times 10^{16}$ photons

**3.** **Plan** $Na(g) \longrightarrow Na^{+}(g) + e^{-}; \Delta H = (IE)$

$E = h\nu = \frac{hc}{\lambda}$ per atom $= \frac{N_0 hc}{\lambda}$ per mol

$\lambda = 242$ nm $= 242 \times 10^{-9}$ m.

Thus, $\quad E = \frac{6.02 \times 10^{23}\ mol^{-1} \times 6.62 \times 10^{-34}\ Js \times 3 \times 10^{8}\ ms^{-1}}{242 \times 10^{-9}\ m}$

$= 4.9404 \times 10^{5}\ J\ mol^{-1} = 4.9404 \times 10^{2}\ kJ\ mol^{-1}$

**4.** Wave number $(\bar{\nu}) = 1.00\ cm^{-1}$

$\bar{\nu} = 100.00\ m^{-1}$

i. $E = h\nu = \frac{hc}{\lambda} = hc\bar{\nu}$

$= 6.62 \times 10^{-34}\ Js \times 3 \times 10^{8}\ ms^{-1} \times 100\ m^{-1}$

$= 1.986 \times 10^{-23}$ J per photon

ii. $E = 1.986 \times 10^{-23}$ J per photon

$= 1.986 \times 10^{-23} \times 6.0 \times 10^{23}$ J per mol

$= 11.956\ J\ mol^{-1}$

$= 11.956 \times 10^{-3}\ kJ\ mol^{-1}$

iii. $E = 11.956 \times 10^{-3}\ kJ\ mol^{-1}$

$= \frac{11.956 \times 10^{-3}}{4.184}\ kcal\ mol^{-1}$

$= 2.857 \times 10^{-3}\ kcal\ mol^{-1}$

**5.** (a) $\nu = \frac{c}{\lambda}$

$\therefore \quad \lambda = \frac{c}{\nu} = \frac{3.0 \times 10^{8}\ ms^{-1}}{900 \times 10^{3}\ s^{-1}}$

$= 333$ m

**6.** (d) Wavelength, $\lambda = 400$ nm $= 400 \times 10^{-9}$ m

(a) Frequency, $\nu = \frac{c}{\lambda} = \frac{3 \times 10^{8}\ ms^{-1}}{400 \times 10^{-9}\ m} = 7.5 \times 10^{14}\ s^{-1}$

$= 7.5 \times 10^{14}$ Hz (correct)

(b) $\frac{1}{\lambda} = \bar{\nu}$ (wave number)

$= \frac{1}{400 \times 10^{-9}} = 2.5 \times 10^{6}\ m^{-1}$ (correct)

(c) $\lambda = \frac{h}{p} = \frac{h}{mv}$

$\therefore$ Momentum, $mv = \frac{h}{\lambda} = \frac{6.62 \times 10^{-34}\ Js}{400 \times 10^{-9}\ m}$

$= 1.66 \times 10^{-27}\ J\ sm^{-1}$

$= 1.66 \times 10^{-27}\ kg\ m^2 s^{-2} m^{-1} s\ [1\ J = 1\ kg\ m^2 s^{-2}]$

$= 1.66 \times 10^{-27}\ kg\ ms^{-1}$ (correct)

**7.** (a) $\sqrt{\nu} = a(Z - b) = aZ - ab$

Thus, graph is a straight line

At $\sqrt{\nu} = 0, \quad aZ = ab$

$Z = b = OA = 1$

Thus, $b = 1$ and at $Z = 0, \sqrt{\nu} = -ab$

Slope $= a = \tan 45^{\circ} = 1$

Thus, $\quad \sqrt{\nu} = (Z - 1)$

If $\nu = 400$ Hz, then

$\sqrt{400} = Z - 1$

$20 = Z - 1$

$\therefore \quad Z = 21$

**8.** (c) Energy values are additive

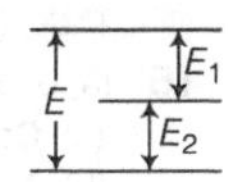


$E = E_1 + E_2$

$h\nu = h\nu_1 + h\nu_2$

$\frac{hc}{\lambda} = \frac{hc}{\lambda_1} + \frac{hc}{\lambda_2}$

$\therefore \quad \frac{1}{\lambda} = \frac{1}{\lambda_1} + \frac{1}{\lambda_2}$

$\frac{1}{355} = \frac{1}{680} + \frac{1}{\lambda_2}$

$\therefore \quad \frac{1}{\lambda_2} = \left(\frac{1}{355} - \frac{1}{680}\right) = 1.3463 \times 10^{-3}$ nm

$\lambda_2 = 743$ nm

**9.** (b) $\Delta H = 242\ kJ\ mol^{-1}$

$= 242 \times 10^{3}\ J\ mol^{-1} = \frac{242 \times 10^{3}\ J}{N_0}\ J\ molecule^{-1}$

$= \frac{242 \times 10^{3}\ J}{6.02 \times 10^{23}}$ J per (Cl—Cl) bond

Also $\Delta H = h\nu = \frac{hc}{\lambda}$

$\therefore \quad \lambda = \frac{hc}{\Delta H} = \frac{6.62 \times 10^{-34}\ Js \times 3 \times 10^{8}\ ms^{-1} \times 6.02 \times 10^{23}}{242 \times 10^{3}\ J}$

$= 4.94 \times 10^{-7}$ m $\quad 494 \times 10^{-9}$ m $= 494$ nm

**10.** (d)

(a) Electrons (−) and nucleus (+) are held together by electrostatic forces of attractions. Thus, **true**.

(b) Radius of the atom ($= 10^{-10}$ m) > radius of the nucleus ($10^{-15}$ m)

Thus, **true**.

(c) Most of the space is empty as given by radius of the nucleus and atom.

Thus, **true**.

(d) As electron moves in a circular path, it emits energy and ultimately falls into nucleus.

Thus, **false**.

**11.** (d)

(a) Cathode rays travel from cathode (source) to anode. **True**

(b) In the absence of these fields, they are not deflected but in the presence of electrical field, they are deflected indicating that they are charged. Thus, **True**

(c) Their nature is independent of the material of cathode. **True**

(d) Their nature is independent of the nature of the gas.

Thus, given statement is **false**.

**12.** (b) $E = N_0 h\nu$

$$= 6.02 \times 10^{23} \times 6.62 \times 10^{-34}\ \text{Js} \times 909 \times 10^3\ \text{s}^{-1}$$

$$= 3.62 \times 10^{-4}\ \text{J}$$

**13.** (a, b, c, d)

(a) Oscillations travel through space at a velocity of light, thus electromagnetic radiation is in the form of energy $E = h\nu$ -**true**.

(b) Radiowaves, microwaves, X-rays are part of visible light -thus, **true**.

(c) Some of the radiations are absorbed in the atmosphere, hence spectrum differ - **true**.

(d) Ozone layer absorbs ultraviolet radiation - **true**.

**14.** (a, c, d) Energy values are additives thus

(a) $\Delta E = \Delta E_1 + \Delta E_2$ (correct)

(c) $h\nu = h\nu_1 + h\nu_2$

$\nu = \nu_1 + \nu_2$ (correct)

(d) $\frac{hc}{\lambda} = \frac{hc}{\lambda_1} + \frac{hc}{\lambda_2}$

$\frac{1}{\lambda} = \frac{1}{\lambda_1} + \frac{1}{\lambda_2}$ (correct)

Thus, (b) is incorrect.

**15.** (a, b, c, d)

(a) Correct

(b) As temperature increases, distribution of frequency (and energy) increases. Thus, correct.

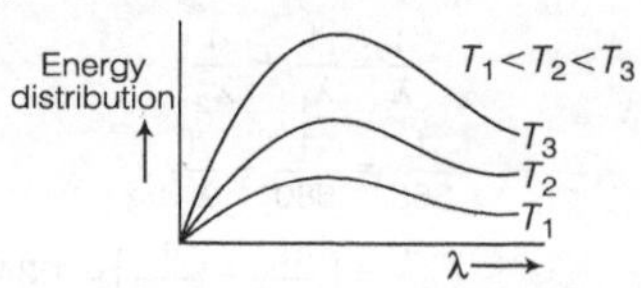


(c) As temperature increases, radiation emitted is shifted to lower wavelength, i.e. higher frequency. Correct

(d) All types of radiations have same speed $3 \times 10^8\ \text{ms}^{-1}$. Correct

**Explanations for Q. Nos. (16-17)**

*A* to *E* makes one complete wave.

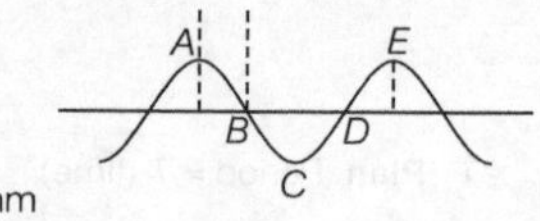


Thus, $AB = \frac{1}{4}\lambda$

$1.17\ \text{nm} = \frac{\lambda}{4}$

$\therefore \quad \lambda = 4.68\ \text{nm}$

$$E = h\nu = \frac{hc}{\lambda}$$

$$= \frac{6.62 \times 10^{-34}\ \text{Js} \times 3 \times 10^8\ \text{ms}^{-1}}{4.68 \times 10^{-9}\ \text{m}}$$

$$= 4.24 \times 10^{-19}\ \text{J}$$

If there are *n* nodes including those at the ends, in a string of length *l*, then

$$l = \frac{(n-1)\lambda}{2} \Rightarrow 42 = \frac{(6-1)\lambda}{2}$$

$\therefore \quad \lambda = 16.8\ \text{cm}$

**16.** (a) **17.** (b)

**18.** (b)

| Colour | | λ(nm) - range |
|---|---|---|
| Violet | – | 440-390 |
| Blue | – | 490-440 |
| Green | – | 565-490 |
| Yellow | – | 590-565 |
| Orange | – | 625-590 |
| Red | – | 740-625 |

Thus,

| | | |
|---|---|---|
| (i) (700 nm) | – | Red (u) |
| (ii) (600 nm) | – | Orange (t) |
| (iii) (580 nm) | – | Yellow (s) |
| (iv) (500nm) | – | Green (r) |
| (v) (460 nm) | – | Blue (q) |
| (vi) (400 nm) | – | Violet (p) |

**19.** (8) Charge on one electron $= -1.6022 \times 10^{-19}$ C

Let, number of electrons = *N*

Then total charge $= -N \times 1.6022 \times 10^{-19}$C

Given, total charge $= -1.282 \times 10^{-18}$ C

Thus, $-N \times 1.6022 \times 10^{-19}\ \text{C} = -1.282 \times 10^{-18}\ \text{C}$

$\therefore \quad N = 8$

**20.** (4) $= N_0 h\nu\ \text{mol}^{-1} = 6.02 \times 10^{23}\ \text{mol}^{-1} \times 6.62 \times 10^{-34}\ \text{Js} \times 1\ \text{s}^{-1}$

$= 4 \times 10^{-10}\ \text{J mol}^{-1}$

$x = 4$

## DPP-2 **Dual Nature of Electron** (de-Broglie's Equation), **Heisenberg's Uncertainty Principle and Photoelectric Effect**

**1. Plan** Photon moves with velocity of light

$$v = c = 3.0 \times 10^8\ \text{ms}^{-1}$$

$$\lambda = 3.6\ \text{Å} = 3.6 \times 10^{-10}\ \text{m}$$

By de-Broglie equation, $\lambda = \frac{h}{mv}$

$$m = \frac{h}{\lambda v} = \frac{6.62 \times 10^{-34}\ \text{Js}}{3.6 \times 10^{-10}\text{m} \times 3.0 \times 10^8\text{ms}^{-1}}$$

$$= 6.13 \times 10^{-33}\ \text{kg}$$

**Note** ($1\ \text{kg m}^2\text{s}^{-2} = 1\ \text{J}$)

**2. Plan**

(i) By de-Broglie equation

$$\lambda = \frac{h}{mv} = \frac{h}{\sqrt{2mKE}}$$

$m$ (electron) $= 9.1 \times 10^{-31}$ kg

$$\lambda = \frac{6.62 \times 10^{-34}\ \text{Js}}{\sqrt{2 \times 9.1 \times 10^{-31}\ \text{kg} \times 3.0 \times 10^{-25}\ \text{J}}}$$

$$= 8.96 \times 10^{-7}\ \text{m}$$

$$= 896 \times 10^{-9}\ \text{m} = 896\ \text{nm}$$

**Note** ($1\ \text{J} = \text{kg m}^2\text{s}^{-2}$)

(ii) Power = Energy time$^{-1}$

Power of 1 watt = 1 Js$^{-1}$

$$E = Nh\nu = \frac{Nhc}{\lambda}$$

$$= \frac{5.6\times 10^{24}\ s^{-1}\times 6.62\times 10^{-34}\ Js\times 3\times 10^{8}\ ms^{-1}}{337.1\times 10^{-9}\ m}$$

$$= 3.3\times 10^{6}\ Js^{-1} = 3.3\times 10^{6}\ watt$$

**3. Plan** By photoelectric effect

$$h\nu = h\nu_0 + \frac{1}{2}mv^2 = (W_0 + KE)$$

Since, $v$ (velocity) = zero

Hence, $KE = \frac{1}{2}mv^2 = 0$

$\therefore \quad h\nu = h\nu_0,\ \nu = \nu_0$

(i) $\nu_0 = \frac{c}{\lambda} = \frac{3\times 10^{8}\ ms^{-1}}{6800\times 10^{-10}\ m} = 4.4118\times 10^{14}\ s^{-1}$

(ii) Work function $W_0 = h\nu_0 = \phi$

$= 6.62\times 10^{-34}\ Js\times 4.4118\times 10^{14}\ s^{-1}$

$= 2.921\times 10^{-19}\ J$

**4. Plan** Voltage applied = eV = KE

Also $E = h\nu - \frac{hc}{\lambda}$ and $h\nu = W_0 + KE$

Then $W_0 = (h\nu - KE)$

$$h\nu = \frac{hc}{\lambda} = \frac{6.62\times 10^{-34}\ Js\times 3\times 10^{8}\ ms^{-1}}{256\times 10^{-9}\ m}$$

$= 7.76\times 10^{-19}\ J$

$KE = 0.35\ eV = 0.35\times 1.6\times 10^{-19}\ J = 0.56\times 10^{-19}\ J$

$\therefore W_0 = 7.76\times 10^{-19}\ J - 0.56\times 10^{-19}\ J = 7.20\times 10^{-19}\ J$

**5.** (d) $E = h\nu = \frac{hc}{\lambda}$

Thus, larger the value of wavelength smaller the energy.

Wavelength of

Cosmic rays < X-rays < traffic amber light < microwave < FM

IV V II I III

Thus, energy is in order

III < I < II < V < IV

**6.** (d) By photoelectric effect

$$h\nu = h\nu_0 + \frac{1}{2}mv^2$$

$\therefore \frac{1}{2}mv^2\ (KE) = h(\nu - \nu_0)$

$= 6.62\times 10^{-34}\ Js\ [1.0\times 10^{15} - 0.7\times 10^{15}]\ s^{-1}$

$= 6.62\times 10^{-34}\times 0.3\times 10^{15}\ J$

$= 1.986\times 10^{-19}\ J \approx 2.0\times 10^{-19}\ J$

**7.** (d) By Heisenberg's uncertainty principle, it is impossible to determine position and momentum simultaneously with same accuracy. Thus, existence of definite path is ruled out.

**8.** (c) By Heisenberg's uncertainty principle

$$\Delta x\cdot\Delta p \approx \frac{h}{4\pi}$$

Since, $\Delta p = m\Delta v$

$\therefore \quad \Delta x\cdot\Delta v\, m = \frac{h}{4\pi}$

$$\Delta x = \frac{h}{4\pi\ \Delta v\cdot m}$$

Given, $\Delta v = 600\ ms^{-1}\times\frac{0.005}{100} = 0.03\ ms^{-1}$

$\therefore \quad \Delta x = \frac{6.6\times 10^{-34}\ Js\ (kg\ m^2s^{-1})}{4\pi\times 0.03\ ms^{-1}\times 9.1\times 10^{-31}\ kg}$

$= 0.01923\ m = 1.92\times 10^{-3}\ m$

**9.** (b) By de-Broglie's equation

$$\lambda = \frac{h}{mv} = \frac{6.62\times 10^{-34}\ Js\ (kg\ m^2s^{-1})}{1.67\times 10^{-27}\ kg\times 1.0\times 10^{-3}\ ms^{-1}}$$

$= 4.0\times 10^{-10}\ m$

$= 0.4\times 10^{-9}\ m$

$= 0.4\ nm$

**Note** ($1\ J = 1\ kg\ m^2s^{-2}$)

**10.** (c) By Heisenberg's uncertainty principle, Satement I is correct. Since, path of electron is not clearly defined hence, Statement II is incorrect.

**11.** (b) Binding energy is the minimum energy to which electron is bound to nucleus or which is required to eject the electron.

Thus, binding energy = – threshold energy

$$E = h\nu = \frac{hc}{\lambda} = \frac{6.62\times 10^{-34}\ Js\times 3.0\times 10^{8}\ ms^{-1}}{150\times 10^{-12}\ m}$$

$= 1.324\times 10^{-15}\ J$

$KE = \frac{1}{2}mv^2 = \frac{1}{2}\times 9.11\times 10^{-31}\ kg\times(1.5\times 10^{7})^2\ m^2s^{-2}$

$= 1.025\times 10^{-16}\ J$ ($1\ kg\ m^2s^{-2} = 1\ J$)

Thus, $E = h\nu_0 + \frac{1}{2}mv^2 = W_0 + \frac{1}{2}mv^2 = W_0 + KE$

$\therefore \quad W_0 = (E - KE) = 1.324\times 10^{-15}\ J - 0.102\times 10^{-15}\ J$

$= 1.222\times 10^{-15}\ J$

$\therefore$ Binding energy = $-1.222\times 10^{-15}\ J$

**12.** (d) For He atom $(\Delta x\cdot\Delta p)_{He} \geq \frac{h}{4\pi}$

and for electron

$(\Delta x\cdot\Delta p)_{electron} \geq \frac{h}{4\pi}$

$(\Delta x)_{helium} = (\Delta x)_{electron} = \pm 1.00\ mm = \pm 1.00\times 10^{-3}\ m$

Hence, $(\Delta p)_{helium} = (\Delta p)_{electron} = 5.0\times 10^{-26}\ kg\ ms^{-1}$

**13.** (a) Kinetic energy = $kT$

where, $k$ = Boltzmann's constant = $\frac{R}{N_0}$

Kinetic energy at 300 K = $\frac{RT}{N_0} = \frac{8.314\ J\ mol^{-1}K^{-1}\times 300\ K}{6.02\times 10^{23}\ mol^{-1}}$

$= 4.143\times 10^{-21}\ J$

By de-Broglie equation, $\lambda = \frac{h}{\sqrt{2mKE}}$

$m$ (neutron) = $1.675\times 10^{-27}\ kg$

$$\lambda = \frac{6.62\times 10^{-34}\ Js}{\sqrt{2\times 1.675\times 10^{-27}\ kg\times 4.143\times 10^{-21}\ J}}$$

$= 1.78\times 10^{-10}\ m = 178\times 10^{-12}\ m$

$= 178\ pm$

**Note** ($1\ J = 1\ kg\ m^2s^{-2}$)

**14.** (b) By de-Broglie equation

$$\lambda = \frac{h}{mv}$$

For a matter wave

Frequency, $\nu = \frac{\text{velocity } (v)}{\lambda} = \frac{v}{h/mv}$

$$\nu = \frac{mv^2}{h}$$

Since, $KE = \frac{1}{2}mv^2$

$\therefore \quad mv^2 = \frac{2KE}{h}$

Frequency, $\nu = \left(\frac{2KE}{h}\right)$

**Explanations for Q. Nos. (15-16)**

Photoelectric effect is represented by

$$h\nu = h\nu_0 + \frac{1}{2}mv^2$$

$$h\nu = W_0 + \frac{1}{2}mv^2 = \text{(work function)} + \text{KE}$$

$$\therefore \quad \nu = \frac{\text{work function}}{h} + \frac{\text{KE}}{h}$$

Graph between $\nu$ (y-axis) and $\frac{\text{KE}}{h}$ (x-axis) is a straight line with

$$OA = 5 \times 10^{14}\ s^{-1}$$

Thus, $\frac{W_0}{h} = 5 \times 10^{14}\ s^{-1}$ atom$^{-1}$

$W_0 = h \times 5 \times 10^{14}\ s^{-1}$ atom$^{-1}$

$W_0 = 6.62 \times 10^{-34}$ Js $\times 5 \times 10^{14}\ s^{-1} \times 6.02 \times 10^{23}$ mol$^{-1}$

$= 199262$ J mol$^{-1}$ $= 199.262$ kJ mol$^{-1}$

$$h\nu = W_0 + \frac{1}{2}mv^2$$

$\therefore \quad \frac{1}{2}mv^2 = h\nu - W_0 = (h\nu - h\nu_0)$ for one electron

$= N_0(h\nu - h\nu_0)$ for one mole

$$= \frac{N_0 hc}{\lambda} - N_0 h\nu_0$$

$$= \frac{6.02 \times 10^{23} \times 6.62 \times 10^{-34} \times 3 \times 10^8 \text{kJ}}{2.5 \times 10^{-7} \times 1000} - 199.262 \text{ kJ}$$

$= 278.97$ kJ mol$^{-1}$

**15.** (a) **16.** (b)

**17.** (b)

(i) $\nu = \frac{c}{\lambda} = \frac{3 \times 10^8\ \text{ms}^{-1}}{616 \times 10^{-9}\ \text{m}} = 4.87 \times 10^{14}\ s^{-1}$

**(i) → (s)**

(ii) Time = 30 s

$v = 3.0 \times 10^8$ ms$^{-1}$

Distance travelled $= v \times t = 3.0 \times 10^8$ ms$^{-1}$ $\times 30$ s

$= 9.0 \times 10^9$ m

$= 9.0 \times 10^6$ km

**(ii) → (r)**

(iii) $E = h\nu = 6.62 \times 10^{-34}$ Js $\times 4.87 \times 10^{14}\ s^{-1}$

$= 3.22 \times 10^{-19}$ J $= 3.22 \times 10^{-22}$ kJ

**(iii) → (q)**

(iv) $E = Nh\nu$

$$N = \frac{E}{h\nu} = \frac{2.0\ \text{J}}{3.22 \times 10^{-19}\ \text{J}} = 6.20 \times 10^{18} \text{ quanta}$$

**(iv) → (p)**

**18.** (4)

Energy of photon $= h\nu$

$$= \frac{hc}{\lambda} = \frac{6.62 \times 10^{-34}\ \text{Js} \times 3 \times 10^8\ \text{ms}^{-1}}{300 \times 10^{-9}\ \text{m}} = 6.62 \times 10^{-19}\ \text{J}$$

$$= \frac{6.62 \times 10^{-19}\ \text{J}}{1.60 \times 10^{-19}\ \text{J/eV}} = 4.14 \text{ eV}$$

Thus, any metal with energy less than 4.14 eV will emit photons thus, showing photoelectric effect.

(Li, Na, K, Mg)

**19.** (6) By Heisenberg's uncertainty principle

$$\Delta x \cdot \Delta p \approx \frac{h}{4\pi}$$

Given $\Delta x$ (in m) $= \Delta p$ (in kg ms$^{-1}$)

Thus, $\Delta p = \Delta x = \sqrt{\frac{h}{4\pi}}$

$$m\,\Delta v = \frac{1}{2}\sqrt{\frac{h}{\pi}}$$

$$\therefore \quad \Delta v = \frac{1}{2m}\sqrt{\frac{h}{\pi}}$$

$$\Delta v = \frac{1}{2 \times 1.21 \times 10^{-18}\ \text{kg}}\sqrt{\frac{6.62 \times 10^{-34}\ \text{Js}}{\pi}} = 6.00\ \text{ms}^{-1}$$

**20.** (9) By de-Broglie equation

$$\lambda = \frac{h}{mv}$$

$m = 9.1 \times 10^{-31}$ kg (electron)

$$= \frac{h}{\sqrt{2m(\text{KE})}} = \frac{6.62 \times 10^{-34}}{\sqrt{2 \times 9.1 \times 10^{-31} \times 3.0 \times 10^{-25}}}$$

$= 8.96 \times 10^{-7}$m $\approx 9.0 \times 10^{-7}$m

Thus, $x = 9$

## DPP-3 Bohr's Theory of H-Atom

**1. Plan** $\Delta E = E_{n_2} - E_{n_1}$

Also $\Delta E = h\nu = \frac{hc}{\lambda}$

Thus, $\lambda = \frac{hc}{\Delta E}$

$\Delta E = (E_3 - E_2) = -2.42 \times 10^{-12} - (-5.42 \times 10^{-12})$

$= 3.0 \times 10^{-12}$ ergs

$$\lambda = \frac{6.62 \times 10^{-27}\ \text{ergs} \times 3 \times 10^{10}\ \text{cm s}^{-1}}{3.0 \times 10^{-12}\ \text{ergs}} = 6.62 \times 10^{-5}\ \text{cm}$$

**2. Plan** By Bohr's theory

$$\bar{\nu} = \frac{1}{\lambda} = \frac{(\text{IE})_H Z^2}{ch}\left[\frac{1}{n_1^2} - \frac{1}{n_2^2}\right]$$

Given, $(\text{IE})_H = 2.17 \times 10^{-11}$ erg atom$^{-1}$ $= 2.17 \times 10^{-18}$ J atom$^{-1}$

$Z = 1$ (H-atom)

$$\therefore \quad \frac{1}{\lambda} = \frac{2.17 \times 10^{-18}\ \text{J} \times 1^2}{3 \times 10^8\ \text{ms}^{-1} \times 6.62 \times 10^{-34}\ \text{Js}}\left[\frac{1}{1^2} - \frac{1}{2^2}\right]$$

$= 8.195 \times 10^6$ m$^{-1}$

$\therefore \quad \lambda = 1.22 \times 10^{-7}$m $= 122 \times 10^{-9}$ m $= 122$ nm

**3. Plan** (i) $(\text{IE})_Z$ of atomic number ($Z$) is related to $(\text{IE})_H$ by equation

$$(\text{IE})_Z = (\text{IE})_H \frac{Z^2}{n^2}$$

Electron in $He^+$ is in first orbit ($n = 1$)

$$(\text{IE})_{He^+} = 2.18 \times 10^{-18}\ \text{J} \times \frac{(2)^2}{(1)^2} = 8.72 \times 10^{-18}\ \text{J atom}^{-1}$$

$= 8.72 \times 10^{-18} \times 6.02 \times 10^{23}$ J mol$^{-1}$

$= 5.25 \times 10^6$ J mol$^{-1}$

(ii) For Balmer series, $n_1 = 2$ and $n_2 = 3, 4, \ldots, \infty$

$$\bar{\nu} = \frac{1}{\lambda} = \bar{R}_H\left[\frac{1}{n_1^2} - \frac{1}{n_2^2}\right]$$

$$\bar{R}_H = 1.09677 \times 10^7\ \text{m}^{-1}$$

If $\bar{\nu}$ is minimum, then $\lambda$ is maximum.

It is when $n_2 = (n_1 + 1) = 3$

$$\therefore \quad \bar{\nu}_{min} = \frac{1}{\lambda_{max}} = 1.09677 \times 10^7\left[\frac{1}{4} - \frac{1}{9}\right]\text{m}^{-1}$$

$$= 1.09677 \times 10^7 \times \frac{5}{36}\ \text{m}^{-1}$$

$\therefore \quad \lambda_{max} = 6.565 \times 10^{-7}$ m $= 656.5$ nm

**4.** (i) **Plan** $\Delta E = (\text{IE})_H Z^2 \left[\frac{1}{n_1^2} - \frac{1}{n_2^2}\right] = \frac{hc}{\lambda}$

For Balmer series (given),

$n_1 = 2, n_2 = 4, Z = 2$ for $He^+$

$$\therefore \Delta E (He^+) = (\text{IE})_H (2)^2 \left[\frac{1}{2^2} - \frac{1}{4^2}\right]$$

$$\Delta E (H) = (\text{IE})_H (1)^2 \left[\frac{1}{n_1^2} - \frac{1}{n_2^2}\right]$$

Wavelengths are equal, hence

$$\Delta E (He^+) = \Delta E (H)$$

$$\therefore \quad 4\left[\frac{1}{4} - \frac{1}{16}\right] = \left[\frac{1}{n_1^2} - \frac{1}{n_2^2}\right]$$

$$\frac{3}{4} = \left[\frac{1}{n_1^2} - \frac{1}{n_2^2}\right]$$

This gives $n_1 = 1, n_2 = 2$

(ii) **Plan** $\nu$ (frequency) is maximum when $n_2 \to \infty$

and is minimum when $n_2 = (n_1 + 1) = 2$, (i.e. $n = 2$)

$$\therefore \quad \nu_{max} = 3.2881 \times 10^{15} \text{ s}^{-1}$$

$$\nu_{min} = 3.2881 \times 10^{15} \times \frac{3}{4}$$

$$= 2.4661 \times 10^{15} \text{ s}^{-1}$$

**5. Plan** Energy in the $n$th state is given by

$$E_n = -\frac{13.6 Z^2}{n^2} \text{ eV}$$

———— $n=3$

———— $n=2$ (first excited state)

———— $n=1$ (ground)

For H-atom, $Z = 1$, first excited state is $n = 2$

Then, $E_2 = -\frac{13.6}{4} = -3.4$ eV

**6.** (d) $\bar{\nu} = \bar{R}_H Z^2 \left[\frac{1}{n_1^2} - \frac{1}{n_2^2}\right]$

$$\bar{R}_H = \frac{2\pi^2 me^4 k^2}{ch^3}$$

$$= x \text{ m}^{-1} \quad \text{(For H-atom)}$$

$$\bar{R}_Z = \frac{2\pi^2 me^4 k^2}{ch^3} (Z^2)$$

For a species of atomic number $Z$ (with one electron only)

$$\frac{\bar{R}_Z}{\bar{R}_H} = Z^2$$

$$\therefore \quad \bar{R}_Z = (2)^2 \bar{R}_H = 4x \text{ m}^{-1} = \frac{4x}{100} (\text{cm}^{-1}) = 0.04x \text{ cm}^{-1}$$

**7.** (b) The potential energy of an electron in the first Bohr's orbit in the H-atom is

$$= 2 \times (-E_1) \text{ (H-atom)}$$

$$= 2 \times (-13.6 \text{ eV}) \text{ atom}^{-1}$$

$$= -27.2 \text{ eV atom}^{-1}$$

This energy is defined as an atomic unit of energy called **HARTREE.**

**8.** (b) $\bar{\nu} = \bar{R}_H Z^2 \left[\frac{1}{n_1^2} - \frac{1}{n_2^2}\right]$

$$x = \bar{\nu} (He^+) = \bar{R}_H (2)^2 \left[\frac{1}{n_1^2} - \frac{1}{n_2^2}\right]$$

$$\bar{\nu} (Be^{3+}) = \bar{R}_H (4)^2 \left[\frac{1}{n_1^2} - \frac{1}{n_2^2}\right]$$

$$\frac{\bar{\nu} (Be^{3+})}{\bar{\nu} (He^+)} = \frac{16}{4} = 4$$

$$\therefore \bar{\nu} (Be^{3+}) = 4\bar{\nu} (He^+) = 4x \text{cm}^{-1}$$

**9.** (a) $\Delta E = +2.18 \times 10^{-18} \left[\frac{1}{n_1^2} - \frac{1}{n_2^2}\right] = k\left[\frac{1}{n_1^2} - \frac{1}{n_2^2}\right] = kx$

where, $x = \left[\frac{1}{n_1^2} - \frac{1}{n_2^2}\right]$

Greater the value of $x$, greater the energy absorbed

(a) $x = \left[\frac{1}{1^2} - \frac{1}{2^2}\right] = \frac{3}{4} = 0.75$

(b) $x = \left[\frac{1}{2^2} - \frac{1}{4^2}\right] = \frac{12}{64} = 0.1875$

(c) $x = \left[\frac{1}{3^2} - \frac{1}{6^2}\right] = \frac{27}{324} = 0.0833$

(d) $x = \left[\frac{1}{9} - \frac{1}{0}\right] = 0.111$

Therefore, electronic transition (a) requires greatest quantity of energy.

**10.** (a) $E_1$ = Energy of H-atom in the ground state = 13.6 eV

$E_2$ ——— 13.6 + (13.6 × 1.5) eV

$E_1$ ——— 13.6 eV

Energy absorbed = (13.6 × 1.5) = 20.4 eV

$E_2$ = Energy of the excited state

= 13.6 + 20.4 = 34.0 eV

$\Delta E = \text{KE} = (E_2 - E_1)$

= 34.0 − 13.6 = 20.4 eV

**11.** (b) Ionisation energy = − Energy of the electron

$$(\text{IE})_Z = (\text{IE})_H \left(\frac{Z^2}{n^2}\right)$$

$He^+$ $(1s^1)$, $Z = 2, n = 1$

$$(\text{IE})_{He^+} = (\text{IE})_H \left(\frac{2^2}{1^2}\right) = 4(\text{IE})_H$$

$Li^{2+}$ $(1s^1)$, $Z = 3, n = 1$

$$(\text{IE})_{Li^{2+}} = (\text{IE})_H \left(\frac{3^2}{1^2}\right) = 9(\text{IE})_H$$

$$(\text{IE})_{Li^{2+}} = \frac{9}{4} (\text{IE})_{He^+}$$

$$= \frac{9}{4} \times 19.6 \times 10^{-18} \text{ J atom}^{-1}$$

$$= 4.41 \times 10^{-17} \text{ J atom}^{-1}$$

$\therefore$ Energy of electron = $-4.41 \times 10^{-17}$ J

**12.** (c) $\text{KE} = \frac{1}{2} mv^2$

Also, angular momentum is quantised

$$mvr_n = \frac{nh}{2\pi}$$

$$\therefore \quad mv = \frac{nh}{2\pi r_n}$$

$$m^2v^2 = \frac{n^2h^2}{4\pi^2 r_n^2}$$

$$\therefore \quad mv^2 = \frac{n^2h^2}{4\pi^2 r_n^2 m}$$

For $(n = 2)$

$$mv^2 = \frac{4h^2}{4\pi^2 r_n^2 m}$$

$$\therefore \quad mv^2 = \frac{h^2}{\pi^2 r_n^2 m}$$

$$\therefore \quad \text{KE} = \frac{1}{2} mv^2 = \frac{h^2}{2\pi^2 r_n^2 m}$$

Also $r_n = \frac{n^2a_0}{Z} = \frac{n^2a_0}{1}$ for H-atom $= 4a_0$

In second orbit

$\therefore \quad r_n^2 = 16\, a_0^2$

$$KE = \frac{h^2}{32\pi^2 m a_0^2}$$

**13.** (a) For H-atom, $Z = 1$

$$E_2 = -2.178 \times 10^{-18}\ J\left(\frac{1}{2^2}\right)$$

$$E_1 = -2.178 \times 10^{-18}\ J\left(\frac{1}{1^2}\right)$$

$$\Delta E = E_2 - E_1 = 2.178 \times 10^{-18}\left[\frac{1}{1^2} - \frac{1}{2^2}\right]$$

$$= 1.6335 \times 10^{-18}\ J$$

Also, $\Delta E = h\nu = \frac{hc}{\lambda}$

$$\therefore \lambda = \frac{hc}{\Delta E} = \frac{6.62 \times 10^{-34}\ Js \times 3 \times 10^8\ ms^{-1}}{1.6335 \times 10^{-18}\ J} = 1.216 \times 10^{-7}\ m$$

**14.** (b) Potential energy of the electron $= -\frac{4\pi^2 mZ^2e^4k^2}{n^2h^2}$

(PE) of H-atom in ground state ($n = 1$)

$$= -\frac{4\pi^2 me^4k^2}{h^2} = 2\left(-\frac{2\pi^2 me^4k^2}{h^2}\right)$$

$$= 2(-13.6\ eV) = -27.2\ eV$$

**Note** $E_1$(H) *in ground state* $= -\frac{2\pi^2 me^4k^2}{h^2} = -13.6$ eV

For $He^+$ in second Bohr's orbit ($n = 2$, $Z = 2$)

$$PE = -\frac{4\pi^2 me^4k^2}{h^2}\left(\frac{Z^2}{n^2}\right) = -27.2\ eV$$

**15.** (a) For the spectral line in H-atom and H-like species (one electron)

$$\frac{1}{\lambda} = \bar{\nu} = \bar{R}_H Z^2\left[\frac{1}{n_1^2} - \frac{1}{n_2^2}\right]$$

For Lyman series

For shortest wavelength (maximum wave number) $n_2 \to \infty$

$$\therefore \quad \frac{1}{\lambda_{min}} = \bar{\nu}_{max} = \bar{R}_H(1)^2\left[\frac{1}{1^2}\right] = \bar{R}_H$$

$$\therefore \quad \frac{1}{x} = \bar{R}_H$$

For longest wavelength (minimum wave number), $n_2 = (n_1 + 1)$.
For Balmer series, $n_1 = 2$

$$\therefore \quad \frac{1}{\lambda_{max}} = \bar{\nu}_{min} = \bar{R}_H(Z)^2\left[\frac{1}{2^2} - \frac{1}{3^2}\right]$$

For $He^+$, $Z = 2$

$$\frac{1}{\lambda_{max}} = \frac{1}{x}(2)^2\left[\frac{5}{36}\right] \Rightarrow \frac{1}{\lambda_{max}} = \frac{5}{9x}$$

$$\therefore \quad \lambda_{max} = \frac{9x}{5}$$

**16.** (b) Angular momentum is quantised, hence

$$mvr_n = \frac{nh}{2\pi}$$

By Bohr's theory ($a_0$ = Bohr's radius = $x$)

$$r_n = \frac{n^2a_0}{Z} = \frac{9a_0}{1} = 9x$$

$$\therefore \quad mv(9x) = \frac{3h}{2\pi}$$

$$\therefore \quad mv = \frac{h}{6\pi x}$$

By de-Broglie's equation

$$\lambda = \frac{h}{mv} = \frac{h}{h/6\pi x} = 6\pi x$$

**17.** (a) $r_n = \frac{n^2a_0}{Z} = n^2a_0$ for H-atom

$1.3225\ nm = 1.3225 \times 10^{-9}\ m$

$= 1322.5 \times 10^{-12}\ m$

$= 1322.5\ pm$

For $r_n = 1.3225$ nm(starting of emission)

$1322.5\ pm = n^2 \times 52.9\ pm$

$n^2 = 25$

$\Rightarrow \quad n = 5$

For $r_n = 211.6$ pm (where electron falls)

$211.6\ pm = n^2 \times 52.9\ pm$

$n^2 = 4$

$n = 2$

Thus, $n_2 = 5$, $n_1 = 2$

$$\bar{\nu} = \frac{1}{\lambda} = \bar{R}_H Z^2\left[\frac{1}{n_1^2} - \frac{1}{n_2^2}\right]$$

$$= 1.09677 \times 10^7\ m^{-1}\ (1)^2\left[\frac{1}{2^2} - \frac{1}{5^2}\right]$$

$$\bar{\nu} = 2.3032 \times 10^6\ m^{-1}$$

$\therefore \quad = 4.3417 \times 10^{-7}\ m = 434.17 \times 10^{-9}\ m$

$= 434.17$ nm

**18.** (b) Emission ($n_2 = 5$ to $n_1 = 2$) is called Balmer series and appears in **visible region.**

**19.** (a) (i) $\to$ s, (ii) $\to$ u, (iii) $\to$ t, (iv) $\to$ r, (v) $\to$ p, (vi) $\to$ q

**20.** (b) $E_n = \frac{-2\pi^2 mZ^2e^4}{n^2h^2} = \frac{-Ze^2}{8\pi\varepsilon_0 r_n}$

$$K_n = \frac{Zke^2}{2r_n}, \quad V_n = \frac{-Zke^2}{r_n},$$

$$r_n = \frac{n^2a_0}{Z}, \quad \frac{V_n}{K_n} = -2$$

Thus, **(i)→(r)** $\quad r_n \propto E_n^{-1}$

**(ii)→(q)**

$$mvr \text{ (angular momentum)} = \sqrt{l(l+1)}\frac{h}{2\pi} = 0$$

**(iii)→(p)** $\quad r_n \propto \frac{1}{Z}$ ($n = 1$)

**(iv)→(s)**

**21.** (2) For atomic number $Z$, energy of the $n$th orbit is

$$E_n = -\frac{13.6Z^2}{n^2}\ eV$$

$E_1(H) = -13.6$ eV

$E_n(H) = -3.4\ eV = -\frac{13.6}{n^2}$

$$n^2 = \frac{13.6}{3.4} = 4$$

$\therefore \quad n = 2$

**22.** (4)

$r_n = n^2a_0/Z$

$r_n = 2.116$ A°

$a_0 = 0.529$ A°

$Z = 4(Be^{3+})$

$$\therefore \quad n^2 = \frac{r_n Z}{a_0}$$

$$n^2 = \frac{2.116 \times 4}{0.529} = 16$$

$\therefore \quad n = 4$

**23.** (2)

By Bohr's theory, $\frac{1}{\lambda} = \bar{R}_H Z^2 \left[\frac{1}{n_1^2} - \frac{1}{n_2^2}\right]$

$$\frac{1}{\lambda} = 1.09679 \times 10^7 \text{m}^{-1}\ Z^2 \left[\frac{1}{n_1^2} - \frac{1}{n_2^2}\right]$$

$$\frac{1}{3.0 \times 10^{-8}\ \text{m}} = 1.09679 \times 10^7\ \text{m}^{-1}\ (Z^2)\left[\frac{1}{1^2} - \frac{1}{2^2}\right]$$

$$= 1.09679 \times 10^7\ \text{m}^{-2}\ (Z^2)\left(\frac{3}{4}\right)$$

$$Z^2 = \frac{4}{3 \times 3.0 \times 10^{-8} \times 1.09679 \times 10^7}$$

$$= 4.05$$

$Z = 2.01 = 2$ ($Z$, being whole number)

**24.** (3)

For radius $r$,

circumference of the orbit $= 2\pi r_n$

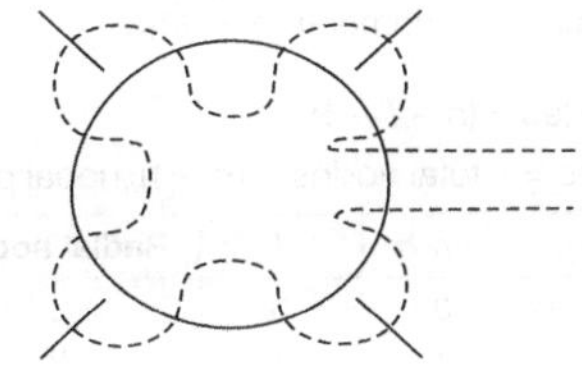

Number of waves in one complete revolution $= n\lambda$

For third orbit $= n\lambda = 3\lambda$

Thus, three waves are formed in one revolution.

## DPP-4 Quantum Numbers and Electronic Configuration

**1. Plan** V ($Z = 23$)

(i) Magnetic moment of $N$-unpaired electrons

$$= \sqrt{N(N+2)}\ \text{BM}$$

$$N(N+2) = (1.73)^2$$

$$N(N+2) = 3$$

$\therefore$ $N^2 + 2N - 3 = 0$

$(N-3)(N+1) = 0$

Thus, $N = 3$, ($N = -1$ is not possible)

(ii) **Plan** V: [Ar] $4s^2 3d^3$

$V^{2+}$ = [Ar] $3d^3$ [↿][↿][↿][ ][ ] ($3d^3$)

$V^{2+}$ = [Ar] $3d^3$

**2. Plan** Suborbits are based on numbers ($l$)

$l \rightarrow 0\ 1\ 2\ 3\ 4$

Suborbit $s\ p\ d\ f\ g$

Thus, $g$-suborbit will correspond to $l = 4$

$l = 4$ will be first in $n = 5$

Thus, 5 $g$ suborbit will first appear

Suborbits = one

Orbitals = $(2l + 1) = 9$

Electrons = $(2l + 1) \times 2 = 18$

**3. Plan**

Before you decide, check whether Aufbau rule is followed

$1s < 2s < 2p < 3s$

Hund's rule is followed (pairing of electrons takes place only after each orbital is half-filled)

Pauli's exclusion principle

(i) forbidden : Hund's rule and Aufbau rule are violated

(ii) ground state

(iii) first excited state

(iv) forbidden : Hund's rule is violated

(v) forbidden : Pauli's exclusion principle is violated

(vi) ground state

**4. Plan** For a given value of $n$,

$$l = 0, 1, 2, \ldots, n-1$$

$$m_l = -l, 0, +l$$

$$m_s = +\frac{1}{2}, 0, -\frac{1}{2}$$

Thus, each orbital can take three electrons instead of two.

For second period $n = 2$

| $l$ | $m_l$ | $m_s$ | Elements |
|---|---|---|---|
| 0 | 0 | $+\frac{1}{2}, 0, -\frac{1}{2}$ | Three |
| 1 | −1 | $+\frac{1}{2}, 0, -\frac{1}{2}$ | Three |
| | 0 | $+\frac{1}{2}, 0, -\frac{1}{2}$ | Three |
| | +1 | $+\frac{1}{2}, 0, -\frac{1}{2}$ | Three |

Thus, total = 12 elements

**5.** (d) $(P)_{15}$ : $1s^2 2s^2 2p^6 3s^2 3p^3$

Valence unpaired electrons are in $3p$

[⥮] [↿][↿][↿]

Paired 3s valence electrons — Unpaired 3p valence electrons

Valence unpaired electrons are in $3p$.

Thus, $n = 3$

$l = 1$

$m = -1, 0, +1$

$s = +\frac{1}{2}$ or $-\frac{1}{2}$

Thus, (d).

**6.** (c) For an atom, orbitals are $1s, 2s, 2p, 3s, \ldots$

$2s$ is completely occupied hence next electron is in $2p$

Thus, $n = 2$

$l = 1$

**7.** (c)

Fe (26) = [Ar] $4s^2 3d^6$ — $4s^2$ [⥮] $3d^6$ [⥮][↿][↿][↿][↿]

$Fe^{2+}$ (24) = [Ar] $4s^0 3d^6$ [⥮][↿][↿][↿][↿]

unpaired electrons, $N$=4

Cr (24) = [Ar] $4s^1 3d^5$ — $4s^1$ [↿] $3d^5$ [↿][↿][↿][↿][↿]

unpaired electrons, $N$=6

Mn (25) [Ar] $4s^2\ 3d^5$

$Mn^+$ ($24e^-$) as in Cr $N$=6

Co (27) [Ar] $4s^2 3d^7$

$Co^{3+}$ ($24e^-$) [as] in $Fe^{2+}$ $N$=4

Magnetic moment $= \sqrt{N(N+2)}$ BM

Thus, $Fe^{2+} = Co^{3+} < Mn^+ = Cr$

**8.** (a) Orbital angular momentum is $= \sqrt{l(l+1)}\,\frac{h}{2\pi}$

For a $d$-orbital electron, $l = 2$

Thus, orbital angular momentum $= \sqrt{2(3)}\,\frac{h}{2\pi} = \sqrt{6}\,\frac{h}{2\pi}$

**9.** (a) Radial nodes $= (n - l - 1)$

Angular nodes $= l$, total nodes $= (n-1)$, nodal plane $= l$

| Suborbit | $n$ | $l$ | Radial nodes |
|---|---|---|---|
| 3s | 3 | 0 | 2 |
| 2p | 2 | 1 | 0 |

**10.** (c) $r_n = \frac{n^2 a_0}{Z}$

$$Z = \frac{n^2 a_0}{r_n} = \frac{16 a_0}{1.5 a_0} = 10.67 \approx 11$$

Thus, $Z = 11$

$1s^2 2s^2 2p^6 3s^1$ valence electron in 3s.

**11.** (c)

| $n$ | $l$ | Suborbit | $(n+l)$ |
|---|---|---|---|
| I. | 4 | $4p$ | 5 |
| II. | 4 | $4s$ | 4 |
| III. | 3 | $3d$ | 5 |
| IV. | 3 | $3p$ | 4 |

By Aufbau rule, smaller the value of $(n + l)$, smaller the energy. If $(n + l)$ is same for two or more orbits, suborbit with lower value of $n$, has smaller energy.

$$4s,\ 3p < 4p,\ 3d$$
$$3p < 4s < 3d < 4p$$

Thus, IV < II < III < I.

**12.** (a) Electronic configuration of the element with ($Z = 37$) is

$[Ar]\ 4s^2\ 3d^{10}\ 4p^6\ 5s^1$ [↑] ($5s^1$)

Last filling (valence electron) goes into 5s

Thus, $n = 5, l = 0, m = 0$

$s = +\frac{1}{2}$

**13.** (b) EC of Cr (24) is

| | $1s^2$ | $2s^2$ | $2p^6$ | $3s^2$ | $3p^6$ | $4s^1$ | $3d^5$ |
|---|---|---|---|---|---|---|---|
| $l$ | 0 | $\phi$ | 1 | 0 | 1 | 0 | 2 |

Thus, electrons with ($l = 1$) = six in $2p$ + six in $3p$ = 12

Electrons with ($l = 2$) = Five in $3d$ = 5

**14.** (c) 4$f$-orbital

Thus, $n = 4$, $l = 3$

$m_l$ (anyone) $= -3\ -2\ -1\ 0\ +1\ +2\ +3$

$m_s = +\frac{1}{2} = -\frac{1}{2}$

(a) $m_l$-incorrect (b) $l$-incorrect

(c) Correct (d) $l$-incorrect

**15.** (a, d)

(a) Three-dimensional shape of orbital is derived from angular quantum number : True

(d) Spin of the electron is derived from spin quantum number : True

**16.** (a, b) Greater the value of $n$, smaller the effective nuclear charge For different subshells ($l$) of same shell ($n$), smaller the energy, greater the effective nuclear charge.

| | Energy | Result |
|---|---|---|
| (a) 2s and 3s | 2s < 3s | II |
| (b) 4d and 4f | 4d < 4f | II |
| (c) 3d and 3p | 3p < 3d | I |
| (d) 3d and 4s | 4s < 3d | I |

**17.** (c, d) (a) 2s (b) 2p (c) 3d (d) 3d

Thus, (c) and (d), both 3d, have equal energy.

**18.** (a, b, c, d)

(a) Cr (24) $= [Ar]\ 4s^1 3d^5$ Correct

(b) $K = 1s^2 2s^2 2p^6 3s^2 3p^6 4s^1$

[↑] or [↓]

Spin can be $+\frac{1}{2}$ or $-\frac{1}{2}$ thus, correct.

(c) $p$-suborbit has three orbitals, each orbital can have maximum of two electrons thus, correct.

(d) $s$-suborbit has one orbital and this orbital can have maximum of two electrons thus, correct.

**19.** (b, c, d)

(a) $M^{2+}$ (electrons = 24, $Z$ = 26)

Thus, EC of $M = [Ar]\ 4s^2 3d^6$

In the formation of $M^{2+}$, electrons are taken out from valence shell (4s). Thus, $M^{2+}$ has EC is $[Ar]\ 3d^6$ : Incorrect

(b) M (electrons = 24, $Z$ = 24)

This represents Cr.

Thus $[Ar]\ 3d^5 4s^1$ : Correct

(c) $M^+$ (electrons = 24, $Z$ = 25)

$M : [Ar]\ 4s^2 3d^5$

$M^+ = [Ar]\ 4s^1 3d^5$ : Correct

(d) If Aufbau rule is not followed, electrons are filled in available orbitals.

Thus, $1s^2 2s^2 2p^6 3s^2 3p^6 3d^1$

Thus, correct.

**20.** (c) If $l$ can have value $n$ also, then $l = 0, 1, 2, 3, \ldots, n$

| $n$ | $l$ | Notation | Orbital $= (2l+1)$ | Electrons $= 2(2l+1)$ | |
|---|---|---|---|---|---|
| 1 | 0 | 1s | one | 2 | $1s^2$ |
| | 1 | 1p | three | 6 | $1p^6$ |
| 2 | 0 | 2s | one | 2 | $2s^2$ |
| | 1 | 2p | three | 6 | $2p^6$ |
| | 2 | 2s | five | 10 | $2d^{10}$ |
| 3 | 0 | 3s | one | 2 | $3s^2$ |
| | 1 | 3p | three | 6 | $3p^6$ |
| | 2 | 3d | five | 10 | $3d^{10}$ |
| | 3 | 3f | seven | 14 | $3f^{14}$ |

By Aufbau rule, energy order is

$$1s < 1p < 2s < 2p < 3s < 2d \ldots$$

Thus, vanadium ($Z$ = 23) will have EC

$$1s^2\ 1p^6\ 2s^2\ 2p^6\ 3s^2\ 2d^5$$

[⥮] [⥮|⥮|⥮] [⥮] [⥮|⥮|⥮] [⥮] [↿|↿|↿|↿|↿]

$1s^2$ $1p^6$ $2s^2$ $2p^6$ $3s^2$ $3d^5$

**21.** (b) Ti ($Z$ = 22) based on the above assumption would have EC

$$1s^2 1p^6 2s^2 2p^6 3s^2 2d^4$$

[⥮] [↿|↿|↿|↿| ]

$3s^2$ $2d^4$

which changes to [↿] [↿|↿|↿|↿|↿]

$3s^1$

to stabilise it. (Hund's rule)

Thus, unpaired electrons $N = 6$

Magnetic moment $= \sqrt{N(N+2)}$ BM

$= \sqrt{48}$ BM

**22.** (b)

(i) Orbital angular momentum $L = \sqrt{l(l+1)}\,\frac{h}{2\pi}$

$L$ depends on the value of $l$ (azimuthal quantum number)

(ii) To describe wave function ($\psi$), $n$, $l$ and $m$ are needed. If, it obeys Pauli's exclusion principle, then $s$ is also needed.

(iii) Value of $n$, $l$ and $m$ are needed to determine size, shape and orientation.

(iv) Probability density ($\psi^2$) is based on $n$, $l$ and $m$

Thus, (i)→ q; (ii)→ p, q, r, s; (iii)→ p, q, r; (iv)→ (p, q, r)

**23.** (6)

$|m_l| = 1$ thus, magnetic quantum number = + 1 and −1

$$(m_s) = -\frac{1}{2}$$

$n = 4$ can have suborbits

| | $l$ | $m_l$ | $m_s$ |
|---|---|---|---|
| 4s | 0 | 0 | $+\frac{1}{2}\,-\frac{1}{2}$ |
| 4p | 1 | $\underline{-1}, 0, \underline{+1}$ | $+\frac{1}{2}\,-\frac{1}{2}$ |
| 4d | 2 | $-2, \underline{-1}, 0, \underline{+1}, +2$ | |
| 4f | 3 | $-3, -2, -1, 0, +1 +2, +3$ | |

Thus underlined electrons with ($m_l$ = − 1, + 1) and spin $-\left(\frac{1}{2}\right)$ are six.

**24.** (9) $n = 3$ can have suborbits

| | $l$ | $m_l$ | Spin ($m_s$) | Electron with $-\frac{1}{2}$ spin |
|---|---|---|---|---|
| 3s | 0 | 0 | $+\frac{1}{2}$, $-\frac{1}{2}$ | 1 |
| 3p | 1 | −1 | $+\frac{1}{2}$, $-\frac{1}{2}$ | 3 |
| | | 0 | $\pm\frac{1}{2}$ | |
| | | +1 | $\pm\frac{1}{2}$ | |
| 3d | 2 | −2 | $\pm\frac{1}{2}$ | 5 |
| | | −1 | $\pm\frac{1}{2}$ | |
| | | 0 | $\pm\frac{1}{2}$ | |
| | | +1 | $\pm\frac{1}{2}$ | |
| | | +2 | $\pm\frac{1}{2}$ | |

Total nine electrons.

## DPP-5 Quantum Mechanical Model of an Atom

**1.** (d) For 1s-orbital radial wave function ($R$) is maximum at $r = 0$, and falls rapidly as $r$ increases thus, (a) correct.

For 2s-orbital, radial wave function ($R$) is maximum at ($r = 0$), falls to zero and further decreases with $r$. There appears radial nodes. Thus (b) correct.

For 2p-orbital, radial wave function is zero at $r = 0$, reaches maximum value

(at $r = a_0$) and then falls thus (c) is correct.

**2.** (b)

(a) It represents $R^2$ vs $r$ for 1s

(b) It represents $R^2$ vs $r$ for 2s

(c) It represents $R^2$ vs $r$ for 2p

(d) It represents $R$ vs $r$ for 2s

**3.** (d)

(a) Describes radial wave function as a function of $r$ for 1s

(b) Describes radial probability function as a function of $r$ for 2s

(c) Describes radial wave function as a function $r$ for 2s

(d) Describes radial probability function as a function of $r$ for 2p

**4.** (d) In case of 1s, $\psi^2$ falls as r increases thus, (A) is for 1s.

In case of 2s, $\psi^2$ is maximum in the vicinity of the nucleus, falls to zero giving spherical nodes and then rises to second highest value. Thus, (B) is for 2s.

**5.** (d) (a) Thus, $\psi^2$ is dependent on $r$ true

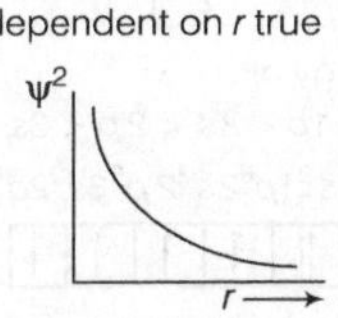


(b) $\psi$ and $|\psi^2|$ are dependent on $r$, $\theta$ and $\phi$ : true

(c) Angular wave function is determined by $l$ and $m_l$ and not by $n$: true

**6.** (c) 3s is spherically symmetrical and its electron density is maximum at the nucleus. It decreases with $r$.

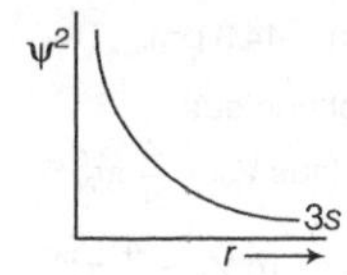


**7.** (c) $H = T + V$

where, $T$ is called kinetic energy $= \frac{1}{2}mv^2$

$$= \frac{p^2}{2m}$$

$V$ is called potential energy $= -\frac{Ze^2}{r}$

**8.** (c) Angular nodes in 4s = 0

| Sub orbit | $n$ | $l$ | Angular nodes = ($l$) | Radial nodes = ($n - l - 1$) |
|---|---|---|---|---|
| 3d | 3 | 2 | 2 | 0 |
| 2p | 2 | 1 | 1 | 0 |

**9.** (a) Radial probability density $= 4\pi r^2\psi^2$ of finding the electron within the volume of a very thin spherical shell at a distance $r$ from nucleus.

**10.** (a) For $n$ nodes in a string

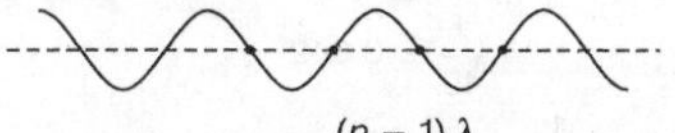

$$l = \frac{(n-1)\lambda}{2}$$

$$12 = \frac{(5-1)\lambda}{2}$$

$$\lambda = 6.0 \text{ cm}$$

**11.** (c)

$$\psi = A\sin\left(\frac{2\pi x}{\lambda}\right)$$

$$\frac{d\psi}{dx} = \left(\frac{2\pi}{\lambda}\right)A\cos\left(\frac{2\pi x}{\lambda}\right)$$

$$\frac{d^2\psi}{dx^2} = -\left(\frac{2\pi}{\lambda}\right)^2 A\sin\left(\frac{2\pi x}{\lambda}\right) = -\frac{4\pi^2}{\lambda^2}\psi$$

$$\therefore \quad \frac{d^2\psi}{dx^2} + \frac{4\pi^2}{\lambda^2}\psi = 0$$

**12.** (b) At radial node, wave function $\psi_{2s} = 0$

$$\therefore \quad \left(2 - \frac{r_0}{a_0}\right) = 0$$

$$r_0 = 2a_0$$

$$a_0 = \frac{r_0}{2}$$

**13.** (a, b, c, d) The wave function describes properties of the orbital and the electron that occupies the orbital

(a) Type of orbital is described : true

(b) Energy of electron : true

(c) Shape of the orbital : true

(d) Probability : true

**14.** (a, b, c, d) For $2s$-electron of H-atom (first excited state)

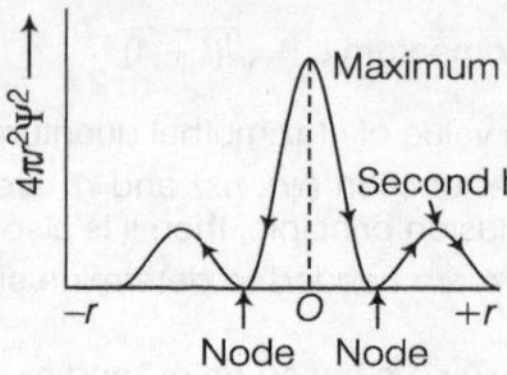


For $1s$-electron of H-atom in ground state

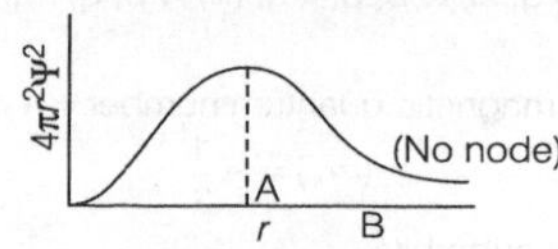


At point $A$ $r = a_0$ $r$ at point $B$ $r > a_0$

(a) True (b) True (c) True (d) True

**15.** (a, c) Radial probability increases as $r$ increase reaches a maximum value when $r = a_0$ (Bohr's radius) and then falls. When $r > 1$ Å, radial probability is very small.

Thus, (a) and (c) are true.

## Revisal Problems (JEE Main)

**1.** (a) $\lambda = \dfrac{h}{mv}$

Mass of one molecule of $H_2O$

$$= \frac{18 \times 10^{-3}\ \text{kg}}{6.02 \times 10^{23}}$$

$$\lambda = \frac{6.62 \times 10^{-34} \times 6.02 \times 10^{23}}{18 \times 10^{-13} \times 500}$$

$$= 4.43 \times 10^{-11}\ \text{m}$$

$$= 44.3 \times 10^{-12}\ \text{m} = 44.3\ \text{pm}$$

**2.** (a) Proton is a part of nucleus

Thus, volume of nucleus $V_N = \dfrac{4}{3}\pi r_N^3$

Volume of hydrogen atom $V_H = \dfrac{4}{3}\pi r_H^3$

Fraction of the volume of hydrogen occupied by nucleus

$$= \frac{V_N}{V_H}$$

$$\frac{V_N}{V_H} = \left(\frac{r_N}{r_H}\right)^3 = \frac{(1.5 \times 10^{-15}\ \text{m})^3}{(0.0529 \times 10^{-9}\ \text{m})^3} = 2.8355 \times 10^{-5})^3$$

$$= 2.28 \times 10^{-14}$$

**3.** (b) $\bar{\nu} = \dfrac{1}{\lambda} = \bar{R}_H Z^2\left[\dfrac{1}{n_1^2} - \dfrac{1}{n_2^2}\right]$

$\lambda = 2170$ nm

For H-atom, $Z = 1$ and $n_2 = 7$ (given)

$$\therefore \quad \frac{1}{2170 \times 10^{-9}\ \text{m}} = 1.09677 \times 10^7\ \text{m}^{-1}\left[\frac{1}{n_1^2} - \frac{1}{7^2}\right]$$

$$\therefore \quad \frac{1}{n_1^2} = \frac{1}{49} + \frac{1}{\left[\begin{array}{l}2170 \times 10^{-9}\text{m} \\ \times 1.09677 \times 10^7\text{m}^{-1}\end{array}\right]}$$

$$= \frac{1}{49} + 0.042$$

$$\frac{1}{n_1^2} = 0.0624$$

$$n_1^2 = 16.02$$

$$n_1 = 4$$

**4.** (d)

| $n$ | $l$ | $m_l$ | $m_s$ | | Electron |
|---|---|---|---|---|---|
| 1 | 0 | 0 | $\pm\frac{1}{2}$ | 2 | $1s^2$ |
| | 1 | −1 | $\pm\frac{1}{2}$ | | |
| | | 0 | $\pm\frac{1}{2}$ | 6 | $1p^6$ |
| | | +1 | $\pm\frac{1}{2}$ | | |
| 2 | 0 | 0 | $\pm\frac{1}{2}$ | | $2s^2$ |
| | 1 | −1 | $\pm\frac{1}{2}$ | | |
| | | 0 | $\pm\frac{1}{2}$ | 6 | $2p^6$ |
| | | +1 | $\pm\frac{1}{2}$ | | |
| | 2 | −2 | $\pm\frac{1}{2}$ | | |
| | | −1 | $\pm\frac{1}{2}$ | | |
| | | 0 | $\pm\frac{1}{2}$ | 10 | $2d^{10}$ |
| | | +1 | $\pm\frac{1}{2}$ | | |
| | | +2 | $\pm\frac{1}{2}$ | | |

By Aufbau rule, energy of

$$1s < 1p < 2s < 2p < 3s < 2d$$

Cr (24) has EC $= 1s^2 1p^6 2s^2 2p^6 3s^2 2d^6$

| ⇅ | ↑ | ↑ | ↑ | ↑ |
|---|---|---|---|---|

Unpaired electron $(N) = 4$

Magnetic moment $= \sqrt{N(N+2)}$ BM $= \sqrt{24}$ BM

**5.** (c) Angular momentum is quantised, thus

$$mvr_n = \frac{nh}{2\pi}$$

$\therefore$ elocity, $v = \dfrac{nh}{2\pi m r_n}$

But $r_n = \dfrac{n^2 a_0}{Z} = \dfrac{n^2 a_0}{1}$ (for H-atom)

$\therefore$ $v = \dfrac{nh}{2\pi m\,(n^2 a_0)} = \dfrac{h}{2\pi m n a_0}$

In $n = 2$, $v = \dfrac{h}{4\pi m a_0}$

**6.** (a) By Heisenberg's uncertainty principle

$$\Delta x \cdot \Delta p \approx \frac{h}{4\pi}$$

Given, $\Delta x = \Delta p$

Thus, $(\Delta x)^2 = (\Delta p)^2 = \dfrac{h}{4\pi}$

$\therefore$ $\Delta p = \sqrt{\dfrac{h}{4\pi}} = \dfrac{1}{2}\sqrt{\dfrac{h}{\pi}}$

But $\Delta p = m\,\Delta v$

$\therefore$ $m\,\Delta v = \dfrac{1}{2}\sqrt{\dfrac{h}{\pi}}$

$\therefore$ $\Delta v = \dfrac{1}{2m}\sqrt{\dfrac{h}{\pi}}$

**7.** (d) de-Broglie wavelength is given by $\lambda = \dfrac{h}{mv}$

$$\lambda_A = \frac{h}{m \times 0.1} \quad \text{and} \quad \lambda_B = \frac{h}{5m \times 0.05}$$

$\therefore$ $\dfrac{\lambda_A}{\lambda_B} = \dfrac{5 \times 0.05}{0.10} = \dfrac{5}{2}$

**8.** (a) Wave number of the spectral line in H-atom is given by

$$\bar{\nu} = \bar{R}_H Z^2 \left[\frac{1}{n_1^2} - \frac{1}{n_2^2}\right] = \bar{R}_H \left[\frac{1}{n_1^2} - \frac{1}{n_2^2}\right]$$

For H-atom, $Z = 1$

(a) $\bar{\nu} = \bar{R}_H \left[\dfrac{1}{1^2} - \dfrac{1}{3^2}\right] = \bar{R}_H \left[\dfrac{9-1}{9}\right] = \dfrac{8}{9}\bar{R}_H$

**9.** (b) $C_2^{2-}$ $6 \times 2 + 2 = 14$; $O_2^-$ $16 + 1 = 17$

CO $6 + 8 = 14$; NO $7 + 8 = 15$

$NO^+$ $7 + 8 - 1 = 14$; $CN^-$ $7 + 6 + 1 = 14$

$N_2$ $7 \times 2 = 14$; $O_2^{2-}$ $8 \times 2 + 2 = 18$

$CO_3^{2-}$ $6 + 24 + 2 = 32$

Then $C_2^{2-}$, CO, $NO^+$, $N_2$ are *iso*-electronic.

**10.** (c) $22 \times 10^{-3}$ mg per year $= 22 \times 10^{-6}$ g per year

$$= \frac{22 \times 10^{-6}\ \text{g}}{365 \times 24 \times 3600}\ \text{g per second}$$

$$= 6.976 \times 10^{-13}\ \text{g per second}$$

Thus, mass of $1.06 \times 10^{11}$ $\alpha$-particles $= 6.976 \times 10^{-13}$ g

Thus, 4 g mol$^{-1}$ ${}_2^4He^{2+}$ ($\alpha$-particles) $= \dfrac{1.06 \times 10^{11} \times 4}{6.976 \times 10^{-13}}$

$$= 6.08 \times 10^{23} = N_0$$

**11.** (b) 1 dL $= \dfrac{1}{10}$ L $= 0.1$ L $= 100\ \text{cm}^3$

100 cm$^3$ of blood has $= 30$ mg Pb $= 30 \times 10^{-6}$ g Pb

$$= \frac{30 \times 10^{-6}}{208}\ \text{mol Pb} = \frac{30 \times 10^{-6} \times N_0}{208}\ \text{atoms Pb}$$

$= 8.6827 \times 10^{16}$ atoms Pb

1 cm$^3$ of blood has $= 8.6827 \times 10^{14}$ atoms Pb

**12.** (b) $\nabla^2$ is called Laplacian operator

$$\nabla^2\psi = \frac{d^2\psi}{dx^2} + \frac{d^2\psi}{dy^2} + \frac{d^2\psi}{dz^2}$$

and $U(\text{PE}) = -\dfrac{Ze^2}{r}$

**13.** (b) Radius $(r_n) = \dfrac{n^2 a_0}{Z}$

For Bohr's orbit of H-atom, $Z = 1, n = 1$

$\therefore$ $r_1 = a_0$

(a) $n = 2$, $Li^{2+}$ $(Z = 3)$

$$r_2 = \frac{4a_0}{3}$$

(b) $n = 2$, $Be^{3+}$ $(Z = 4)$

$$r_2 = \frac{4a_0}{4} = a_0$$

(c) $n = 2$, $He^+$ $(Z = 2)$

$$r_2 = \frac{4a_0}{2} = 2a_0$$

(d) $n = 3$, $Li^{2+}$ $(Z = 3)$

$$r_3 = \frac{9a_0}{3} = 3a_0$$

**14.** (b)

(a) 2*p* 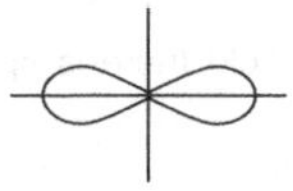

Symmetric about *x*, *y* or *z*-axis

(b) 3*s* 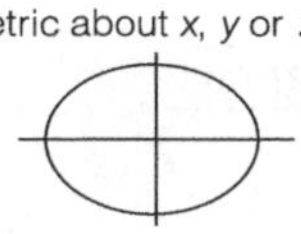

Spherically symmetric about the nucleus

(c) 3*p* as (a)

(d) 3*d* 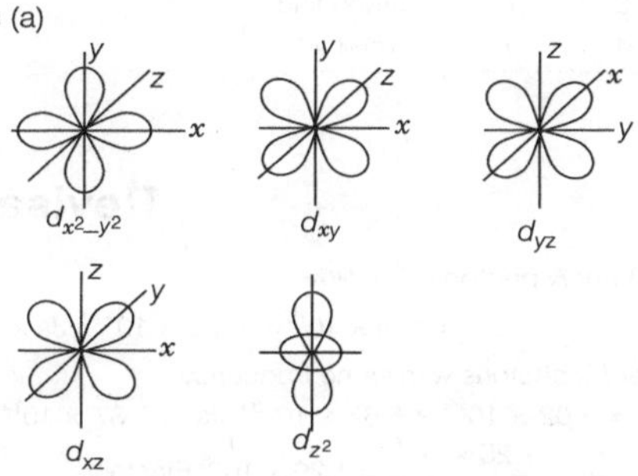


**15.** (a) H (1) has EC $= 1s^1$

Thus, w.r.t. (1*s*) orbital, 3*s*, 3*p* and 3*d* orbitals will have same energy.

**16.** (b) Due to extra stability of completely filled suborbit or half-filled suborbit

$4s^1 3d^{10}$ is more stable than $4s^2 3d^9$. Thus, [Ar] $4s^2 3d^9$ is not the ground state electronic configuration.

**17.** (a) N (7) has electronic configuration as

$$1s^2\, 2s^2\, 2p^3 \quad [\uparrow\downarrow] \quad [\uparrow\downarrow] \quad [\uparrow|\uparrow|\uparrow]$$

Thus, Statement I is correct.

By Hund's rule of maximum multiplicity, pairing of electrons takes place only after each orbital is half-filled, i.e. maximum multiplicity, and thus, has stable configuration.

Thus, Statement II is also correct and is the correct explanation of Statement I.

**18.** (d) For a given value of $n$

$$l = 0, 1, 2, \ldots, (n-1)$$

Total orbitals = $(2l + 1)$

Total electrons = $2(2l + 1)$

Thus, total electrons are obtained by taking sum of all values of $l$ from 0 to $(n - 1)$

$$\sum_{l=0}^{l=(n-1)} 2(2l+1)$$

**19.** (d) For a given orbit, ($3d$, $n = 3$, $l = 2$)

Number of angular nodes = $l = 2$ for $3d$ number of radial nodes

$$= (n - l - 1) = (3 - 2 - 1) = 0$$

Total nodes = $(n - l - 1 + l) = (n - 1) = 2$

**20.** (a) Velocity = $v$ cm $s^{-1}$ (say)

Distance travelled in 1s = $v$ cm $s^{-1} \times 1\, s$ = $v$ cm

$\therefore$ $\lambda$ (wavelength) = $v$ cm

$$\lambda = \frac{h}{mv} \text{ (de-Broglie equation)}$$

$$\therefore \quad v = \frac{h}{mv}$$

$$\therefore \quad mv^2 = h$$

$$v^2 = \frac{h}{m} \Rightarrow v = \sqrt{\frac{h}{m}}$$

**21.** (a) In the presence of magnetic field there is splitting of lines (Zeeman's effect)

Total lines due to splitting = $(2l + 1)$

Total orbitals = $(2l + 1)$

| $l$ | Suborbit | Degenerate = $(2l + 1)$ |
|---|---|---|
| 0 | $s$ | one |
| 1 | $p$ | three-fold |
| 2 | $d$ | five-fold |
| 3 | $f$ | seven-fold |
| 4 | $g$ | nine-fold |

**22.** (d) $\lambda = \dfrac{h}{mv} = \dfrac{h}{\sqrt{2m\,(\text{KE})}}$

$$\therefore \quad \lambda^2 = \frac{h^2}{2m\,(\text{KE})}$$

$$\therefore \quad \text{KE} = \frac{h^2}{2m\lambda^2} = \frac{(6.62 \times 10^{-34}\text{ J s})^2}{(2 \times 9.1 \times 10^{-31}\text{ kg})\,(0.3 \times 10^{-9}\text{ m})^2}$$

$$= 2.67 \times 10^{-18}\text{ J}^2\text{s}^2\text{kg}^{-1}\text{m}^{-2} = 2.67 \times 10^{-18}\text{ J}$$

$$= \frac{2.67 \times 10^{-18}\text{ J}}{1.6 \times 10^{-19}\text{ J/eV}} = 16.72\text{ eV} \qquad (1\text{J} = 1\text{ kg m}^2\text{ s}^{-2})$$

**23.** (d) Fe = [Ar] $4s^2 3d^6$

$Fe^{2+}$ = [Ar] $4s^0 3d^6$

[Electrons of the outermost shell are taken out in the formation of $Fe^{2+}$ (i.e. $4s$)]

Electrons in $3d$ in $Fe^{2+}$ = 6

Ne (10) = $1s^2, 2s^2, 2p^6$

$p$-electrons = 6 : Correct

Mg (12) = $1s^2 2s^2 2p^6 3s^2$

$s$-electrons = 6 : Correct

Fe (26) = $1s^2 2s^2 2p^6 3s^2 3p^6 4s^2 3d^6$

$d$-electrons = 6 : Correct

Cl (17) = $1s^2 2s^2 2p^6 3s^2 3p^5$

$p$-electrons 11 : Incorrect

**24.** (c) Bohr's theory is applicable to H-atom and H-like species with one electron.

H ($Z$ = 1) 1 electron

$He^+$ ($Z$ = 2) 1 electron

$Li^{2+}$ ($Z$ = 3) 1 electron

$Be^{3+}$ ($Z$ = 4) 1 electron

## Revisal Problems (JEE Advanced)

**1.** (a) For $N$ photons $E = Nh\nu$

$$= 7.25 \times 10^{15} \times 6.62 \times 10^{-34}\text{ Js} \times 5.37 \times 10^{14}\text{ s}^{-1}$$

For $N_0$ photons with same frequency

$E' = 6.02 \times 10^{23} \times 6.62 \times 10^{-34}$ Js $\times 5.37 \times 10^{14}$ $s^{-1}$

$$\therefore \quad \frac{E}{E'} = \frac{7.25 \times 10^{15}}{6.02 \times 10^{23}} = 1.20 \times 10^{-8}\text{ Einstein}$$

**2.** (d) Velocity varies within $\pm 1$ $ms^{-1}$

Thus, it can be 99 $ms^{-1}$ to 101 $ms^{-1}$

$$(\text{KE})_{99} = \frac{1}{2} m\,(99)^2$$

$$(\text{KE})_{101} = \frac{1}{2} m\,(101)^2$$

$$\therefore \quad \Delta E = \frac{1}{2} m\,[(101)^2 - (99)^2]$$

$$= \frac{1}{2} \times 9.1 \times 10^{-31}\text{ kg}\,[(101)^2 - (99)^2]$$

$$= 1.82 \times 10^{-28}\text{ kg m}^2\text{s}^{-2} = 1.82 \times 10^{-28}\text{ J}$$

By Heisenberg's uncertainty principle

$$\Delta E \times \Delta t \approx \frac{h}{4\pi}$$

$$\therefore \quad \Delta t = \frac{h}{4\pi\,\Delta E} = \frac{6.62 \times 10^{-34}\text{ Js}}{4\pi \times 1.82 \times 10^{-28}\text{ J}}$$

$$= 2.89 \times 10^{-7}\text{ s} = 0.289 \times 10^{-6}\text{ s}$$

$$= 0.289\,\mu\text{s} \approx 0.3\,\mu\text{s}$$

**3.** (c) For photoelectric effect

$$h\nu = h\nu_0 + \text{KE}$$

$$\text{KE} = h(\nu - \nu_0)$$

$$(\text{KE})_1 = h\,[1.6 \times 10^{16} - \nu_0]$$

$$(\text{KE})_2 = h\,[1.0 \times 10^{16} - \nu_0]$$

Given, $(\text{KE})_1 = 2(\text{KE})_2$

$$h(1.6 \times 10^{16} - \nu_0) = 2h\,[1.0 \times 10^{16} - \nu_0]$$

$$\nu_0 = 4 \times 10^{15}\text{ Hz}$$

**4.** (a) $O_2 \longrightarrow O + O^*$

$$\Delta H^\circ = 498\text{ kJ mol}^{-1}$$

$$= \frac{498 \times 10^3\text{ J}}{6.02 \times 10^{23}}\text{ J molecule}^{-1}$$

$$= 8.27 \times 10^{-19}\text{ J molecule}^{-1}$$

Energy required to convert normal (O) into (O*) = 1.967 eV

$$= 1.967 \times 1.6 \times 10^{-19}\text{ J} = 3.147 \times 10^{-19}\text{ J}$$

Total energy = $(8.27 \times 10^{-19} + 3.15 \times 10^{-19})$

$$= 1.142 \times 10^{-18}\text{ J}$$

Thus, $\lambda = \dfrac{hc}{E} = \dfrac{6.62 \times 10^{-34}\text{ Js} \times 3 \times 10^8\text{ ms}^{-1}}{1.142 \times 10^{-18}\text{ J}}$

$$= 1.739 \times 10^{-7} = 173.9 \times 10^{-9}$$

$$= 174\text{ nm}$$

**5.** (b) Electron-density along axis.

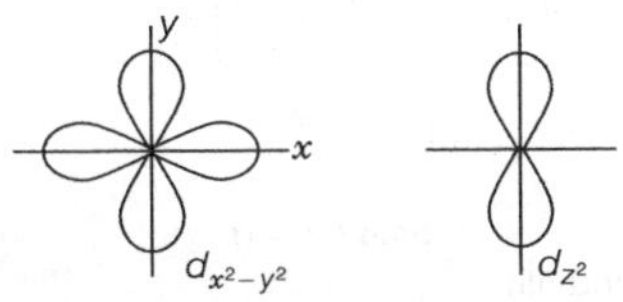


**6.** (b) Most probable distance is when

$$\frac{d}{dr}[4\pi r^2\psi^2] = 0$$

$$\frac{d}{dr}[4\pi r^2(\sqrt{N}\, e^{-Zr/a_0})^2] = 0$$

$$\frac{d}{dr}[4\pi r^2 N e^{-2Zr/a_0}] = 0$$

$$\left[4\pi(2r)Ne^{-2Zr/a_0} + 4\pi r^2 N\left(\frac{-2Z}{a_0}\right)e^{-2Zr/a_0}\right] = 0$$

$$8\pi r N e^{-2Zr/a_0}\left[1 - r\left(\frac{Z}{a_0}\right)\right] = 0$$

Thus, $$r\left(\frac{Z}{a_0}\right) = 1$$

$\therefore$ $$r = \left(\frac{a_0}{Z}\right)$$

**7.** (a) If $\psi = A\sin\left(\frac{2\pi x}{\lambda}\right)$

$$\frac{d\psi}{dx} = \left(\frac{2\pi}{\lambda}\right)A\cos\left(\frac{2\pi x}{\lambda}\right)$$

$$\frac{d^2\psi}{dx^2} = -\left(\frac{2\pi}{\lambda}\right)^2 A\sin\left(\frac{2\pi x}{\lambda}\right) = -\frac{4\pi^2}{\lambda^2}\psi$$

$$\therefore \quad \frac{d^2\psi}{dx^2} + \frac{4\pi^2}{\lambda^2}\psi = 0$$

Thus, true.

(b) If $\psi = Ae^{-2\pi x/\lambda}$

$$\frac{d\psi}{dx} = \left(-\frac{2\pi}{\lambda}\right)Ae^{-2\pi x/\lambda}$$

$$\frac{d^2\psi}{dx^2} = \left(-\frac{2\pi}{\lambda}\right)^2 Ae^{-2\pi x/\lambda} = \left(\frac{4\pi^2}{\lambda^2}\right)\psi$$

$$\frac{d^2\psi}{dx} - \frac{4\pi^2}{\lambda^2}\psi = 0$$

Thus, incorrect.

**8.** (a) There is uncertainty with 1 cm $s^{-1}$, hence, velocity varies between 99 cm $s^{-1}$ to 101 cm$s^{-1}$.

Thus, $\Delta v$ (uncertainty in velocity) = 101 − 99 = 2 cm $s^{-1}$

= 0.02 m $s^{-1}$

By Heisenberg's uncertainty principle

$$\Delta x \cdot \Delta p \approx \frac{h}{4\pi}$$

$$\Delta x \cdot m\,\Delta v \approx \frac{h}{4\pi}$$

$\therefore$ Uncertainty in position, $\Delta x = \dfrac{h}{4\pi m\,\Delta v}$

$$\Delta x = \frac{6.62 \times 10^{-34}\ \text{Js (kg m}^2\text{s}^{-2})}{4\pi \times 1.00 \times 10^{-3}\ \text{kg} \times 0.02\ \text{ms}^{-1}} = 2.63 \times 10^{-30}\ \text{m}$$

**9.** (b) KE given to iodine molecule = [Energy absorbed − energy required to dissociate $I_2$]

Bond energy = 240 kJ $mol^{-1}$ = 240 × $10^3$ J $mol^{-1}$

$$= \frac{240 \times 10^3}{6.02 \times 10^{23}}\ \text{J molecule}^{-1} = 3.987 \times 10^{-19}\ \text{J}$$

Energy absorbed (one photon) = $h\nu = \dfrac{hc}{\lambda}$

$$= \frac{6.62 \times 10^{-34}\ \text{Js} \times 3 \times 10^8\ \text{ms}^{-1}}{4500 \times 10^{-10}\ \text{m}}$$

$= 4.413 \times 10^{-19}$ J

KE of $I_2$ molecule = $4.413 \times 10^{-19} - 3.987 \times 10^{-19}$

$= (4.413 - 3.987) \times 10^{-19}$ J

$= 4.26 \times 10^{-20}$ J

KE of each I atom = $\dfrac{4.26 \times 10^{-20}}{2} = 2.13 \times 10^{-20}$ J

**10.** (d) I : $n = 4$, $m_s = -\frac{1}{2}$

Maximum electrons = $2n^2 = 32$

$m_s = +\frac{1}{2}$ for 16 electrons

$= -\frac{1}{2}$ for 16 electrons

Thus, 16 electrons will have $m_s = -\frac{1}{2}$

II : $n = 3$, $l = 0$

3s-suborbit can have $2(2l + 1) = 2$ electrons

Only two electrons

III : $n = 2$, $l = 1$

2p-suborbit can have $2(2l + 1) = 6$ electrons

Thus, six electrons.

IV : $n = 2$, $m_l = 0$

$n = 2, l = 0, m_l = 0$

$n = 2, l = 1, m_l = 1 - 1, 0, + 1$

Thus, four electrons.

**11.** (a) Colour of the emitted radiations change as temperature increases

$E = h\nu$

Also, $E \propto$ temperature

Thus, as temperature changes frequency also changes and colour changes.

Thus,both Statements I and II are correct and Statement II is the correct explanation of Statement I.

**12.** (b) Both Statements I and II are correct but Statement II is not the correct explanation of Statement I.

**13.** (a) Electrons are negatively charged thus, Statement I is correct. In cathode ray tube, these rays are deflected towards positive plate thus, Statement II is also correct and also the correct explanation of Statement I.

**14.** (a,b,c,d)

$$\nu = \bar{R}_H\left[\frac{1}{n_1^2} - \frac{1}{n_2^2}\right]$$

$$\bar{\nu}_{max} = \frac{\bar{R}_H}{n_1^2} \quad \text{(when } n_2 \to \infty\text{)}$$

$= \bar{R}_H$ for $n_1 = 1$ (Lyman) $= \dfrac{\bar{R}_H}{4}$ for $n_1 = 2$ (Balmer)

$= \dfrac{\bar{R}_H}{9}$ for $n_1 = 3$ (Paschen) $= \dfrac{\bar{R}_H}{10}$ for $n_1 = 4$ (Brackett)

$\bar{\nu}_{min}$ when $n_2 = (n_1 + 1)$

Thus, $\bar{\nu}_{min} = \dfrac{3}{4}\bar{R}_H$ for Lyman

$= \dfrac{5}{36}\bar{R}_H$ for Balmer

$= \dfrac{7}{144}\bar{R}_H$ for Paschen

$= \dfrac{9}{400}\bar{R}_H$ for Brackett

**15.** (a, b) Velocity of the electron of the species with atomic number $Z$ in $n$th orbit is

$$(2.182 \times 10^6) \frac{Z}{n} \text{ ms}^{-1}$$

| | $n$ | Species | $Z$ | $v$ |
|---|---|---|---|---|
| | 3 | $Li^{2+}$ | 3 | $2.182 \times 10^6$ ms$^{-1}$ |
| (a) | 2 | $He^+$ | 2 | $2.182 \times 10^6$ ms$^{-1}$ |
| (b) | 1 | H | 1 | $2.182 \times 10^6$ ms$^{-1}$ |
| (c) | 3 | $Be^{3+}$ | 4 | $\frac{4}{3}$ $v$ of $Li^{2+}$ |
| (d) | 1 | $He^+$ | 2 | $2v$ of $Li^{2+}$ |

**16.** (a, b, c) If $n = n$ then $l = 0, 1, 2, 3, \ldots, n$

(a, b, c)

If $n = n$ then $l = 0, 1, 2, 3, \ldots n$

| $n$ | $l$ | $m$ | Spin | Electron | Notation |
|---|---|---|---|---|---|
| 1 | 0 | 0 | $\pm\frac{1}{2}$ | 2 | $1s^2$ |
| | 1 | $-1$ | $\pm\frac{1}{2}$ | | |
| | | 0 | $\pm\frac{1}{2}$ | 6 | $1p^6$ |
| | | +1 | $\pm\frac{1}{2}$ | | |
| 2 | 0 | 0 | $\pm\frac{1}{2}$ | 2 | $1s^2$ |
| | 1 | $-1$ | $\pm\frac{1}{2}$ | | |
| | | 0 | $\pm\frac{1}{2}$ | 6 | $2p^6$ |
| | | +1 | $\pm\frac{1}{2}$ | | |
| | 2 | $-2$ | $\pm\frac{1}{2}$ | | |
| | | $-1$ | $\pm\frac{1}{2}$ | | |
| | | 0 | $\pm\frac{1}{2}$ | 10 | $2d^{10}$ |
| | | +1 | $\pm\frac{1}{2}$ | | |
| | | +2 | $\pm\frac{1}{2}$ | | |

Energy order, $1s < 1p < 2s < 2p < 3s < 2d$

(a) ($Z = 21$), $1s^2 1p^6 2s^2 2p^6 3s^2 2d^3$ : Correct

(b) (Fe = 26), $1s^2 1p^6 2s^2 2p^6 3s^2 2d^8$

| ⥮ | ⥮ | ⥮ | ↿ | ↿ |
|---|---|---|---|---|

Two unpaired electrons : correct

(c) $Fe^{2+} = 1s^2 1p^6 2s^2 2p^6 2d^8$

(Valence shell 3s electrons are taken out)

$Fe^{3+} = 1s^2 1p^6 2s^2 2p^6 2d^7$

Both have unpaired electrons in $d$-orbital thus, coloured : Correct

(d) $Cu = 1s^2 1p^6 2s^2 2p^6 3s^2 2d^{10} 3p^1$

$Cu^+ = 1s^2 1p^6 2s^2 2p^6 3s^2 2d^{10}$

Colourless : Incorrect

### Explanations for Q. Nos (17-18)

(I) $H(g) \longrightarrow H^+(g) + e^-$ $\quad$ $(IE)_H = 13.6$ eV

(II) $He(g) \longrightarrow He^+(g) + e^-$ $\quad$ $(IE)_{He}$ (first) = 24.6 eV

(III) $He^+(g) \longrightarrow He^{2+}(g) + e^-$ $\quad$ $(IE)_{He}$ (second) = ? = $q$

By Bohr's theory of H-atom

$$(IE)_Z = (IE)_H = \left(\frac{Z^2}{n^2}\right)$$

$$(IE)_{He^+} = (13.6)\left(\frac{2^2}{1^2}\right)$$

$$= 54.4 \text{ eV} = q$$

Adding (II) and (III)

$$He(g) \longrightarrow He^{2+}(g) + 2e^-$$

$$\Delta H = (24.6 + q) = 79.0 \text{ eV}$$

Hence for $\quad He^{2+}(g) + 2e^- \longrightarrow He(g)$

$$\Delta H = -79.0 \text{ eV}$$

**Note** *Second ionisation energy of He(g) is the first ionisation of $He^+(g)$.*

$He(g) \longrightarrow He^+(g) + e^-$ $(IE)_1$ of $He(g)$

$He^+(g) \longrightarrow He^{2+}(g) + e^-$ $(IE)_2$ of $He(g)$

$(IE)_1$ of $He^+(g)$

**17.** (d) $\quad$ **18.** (c)

### Explanations for Q. Nos. (19-20)

Energy of electron = 0.5 keV

$= 0.5 \times 10^3$ eV

$= 0.5 \times 10^3 \times 1.6 \times 10^{-19}$ J

$= 8.0 \times 10^{-17}$ J

Momentum $p = \sqrt{2m\,KE}$

$= \sqrt{2 \times 9.1 \times 10^{-31} \text{ kg} \times 8.0 \times 10^{-17} \text{ J}}$

$= 1.20 \times 10^{-23}$ kg ms$^{-1}$

**Note** $1\,J = 1\,kg\,m^2s^{-1}$

$$\Delta x = 0.2 \text{ nm} = 0.2 \times 10^{-9} \text{ m}$$

$$\Delta x \cdot \Delta p \approx \frac{h}{4\pi}$$

$$\therefore \quad \Delta p = \frac{h}{4\pi\, \Delta x}$$

$$= \frac{6.62 \times 10^{-34} \text{ Js}}{4\pi \times 0.2 \times 10^{-9}}$$

$$= 2.6 \times 10^{-26} \text{ kg ms}^{-1}$$

$$\text{Percentage uncertainty} = \frac{\Delta p}{p} \times 100$$

$$= \frac{2.6 \times 10^{-25}}{1.2 \times 10^{-23}} \times 100 = 2.2\%$$

**19.** (c) $\quad$ **20.** (a)

### Explanations for Q. Nos. (21-22)

$100 \text{ W} = 100 \text{ Js}^{-1}$

$E = 1000$ J in 10 s

$E = Nh\nu$

$$E = \frac{Nhc}{\lambda}$$

$$1000 = \frac{N \times 6.62 \times 10^{-34} \text{ J} \times 3 \times 10^8 \text{ ms}^{-1}}{560 \times 10^{-9} \text{ m}}$$

$\therefore N = 2.82 \times 10^{21}$ photons in 10s

Time to produce 1 mole ($6.02 \times 10^{23}$ photons)

$$= \frac{10 \times 6.02 \times 10^{23}}{2.82 \times 10^{21}}$$

$= 2135$ s

$= 36$ min

**21.** (a) $\quad$ **22.** (b)

**23.** (6) Total lines $= \frac{n(n-1)}{2} = \frac{4(3)}{2} = 6$

| Level | Falls to | Lines |
|---|---|---|
| $n = 4$ to | | |
| | $n = 3$ | |
| | $n = 2$ | Three |
| | $n = 1$ | |
| $n = 3$ to | $n = 2$ | |
| | $n = 1$ | Two |
| $n = 2$ to | $n = 1$ | One |

**24.** (5)
$$(\text{IE})_Z = (\text{IE})_H \left(\frac{Z^2}{n^2}\right)$$
$$(\text{IE})_H\ (n = 1) = (\text{IE})$$
$$(\text{IE})_H\ (n = n) = \frac{(\text{IE})_H}{n^2}$$
$$\frac{(\text{IE})_H\ (n = 1)}{(\text{IE})_H\ (n = n)} = n^2 = 25$$

Thus, $n = 5$

**25.** (5)
$$\nu = 3.29 \times 10^{15}\ \text{s}^{-1}\left[\frac{1}{9} - \frac{1}{n^2}\right]$$
$$\nu = \frac{c}{\lambda}$$
$$\therefore \quad \frac{3 \times 10^8 \text{ms}^{-1}}{1285 \times 10^{-9} \text{m}} = 3.29 \times 10^{15} \text{s}^{-1}\left[\frac{1}{9} - \frac{1}{n^2}\right]$$
$$\frac{3 \times 10^8 \text{ms}^{-1}}{1285 \times 10^{-9}\ \text{m} \times 3.29 \times 10^{15}\ \text{s}_{-1}} = \left[\frac{1}{9} - \frac{1}{n^2}\right]$$
$$0.071 = \frac{1}{9} - \frac{1}{n^2}$$
$$\therefore \quad \frac{1}{n^2} = \left[\frac{1}{9} - 0.071\right] = 0.04$$
$$n^2 = 25$$
$$n = 5$$

**26.** (i) (5)
$$\lambda = \frac{h}{mv} = \frac{h}{\sqrt{2m\,\text{KE}}}$$
$$\lambda = \frac{h}{\sqrt{2m\left(\frac{3}{2}RT\right)}}$$
$$\lambda\ (\text{He})\ \text{at 200 K} = \frac{h}{\sqrt{2 \times 4\left(\frac{3}{2}R \times 200\right)}}$$
$$\lambda\ (\text{Ne})\ \text{at 1000 K} = \frac{h}{\sqrt{2 \times 20\left(\frac{3}{2}R \times 1000\right)}}$$
$$\frac{\lambda_{(\text{He})}}{\lambda_{(\text{Ne})}} = \frac{\sqrt{20 \times 1000}}{\sqrt{4 \times 200}} = \sqrt{25} = 5$$

(ii) (8)
$$\lambda = \frac{h}{mv} = \frac{h}{\sqrt{2m\,(\text{KE})}}$$
$$= \frac{h}{\sqrt{2m\left(\frac{3}{2}RT\right)}}$$
$$\lambda_{(\text{He})} = \frac{h}{\sqrt{2 \times 4\left(\frac{3}{2}R \times 100\right)}}$$
$$\lambda_{(\text{X})} = \frac{h}{\sqrt{2 \times m\left(\frac{3}{2}R \times 1250\right)}}$$
$$\frac{\lambda_{(\text{He})}}{\lambda_{(\text{X})}} = \sqrt{\frac{m \times 1250}{4 \times 100}} = 5$$
$$\therefore \quad \frac{m \times 1250}{4 \times 100} = 25$$
$$m = 8$$

**27.** (a) (i)→ t; (ii)→ s; (iii)→ r ; (iv)→ q; (v)→ p

**28.** (c) N (7) $= 1s^2 2s^2 2p^3$

[↿ | ↿ | ↿]

Paramagnetic due to unpaired electron and sum of spin
$$= 3 \times \frac{1}{2} = 1.5$$

Thus, **(i)→ (p,r)**

Co(27) = [Ar] [⇅] [⇅ | ⇅ | ↿ | ↿ | ↿]
$4s^2$ $3d^7$

$Co^{2+}$ = [Ar] [⇅ | ⇅ | ↿ | ↿ | ↿]

Paramagnetic and coloured due to unpaired electrons in $3d$

Sum of spin = 1.5

Thus, **(ii) → (p,q,r)**

Mn (25) = [Ar] $4s^2 3d^5$ [⇅] [↿ | ↿ | ↿ | ↿ | ↿]

$Mn^{2+}$(25) = [Ar] $s^2 3d^5$ [↿ | ↿ | ↿ | ↿ | ↿]

Coloured and paramagnetic

Sum of spin = 2.5

Thus, **(iii)→ (p,q)**

Fe (26) = [Ar] $4s^2 3d^6$ [⇅] [⇅ | ↿ | ↿ | ↿ | ↿]

$Fe^{2+}$(26) = [Ar] $3d^6$ [⇅ | ↿ | ↿ | ↿ | ↿]

Coloured and paramagnetic

Sum of spin = 2.0

$$\text{Magnetic moment} = \sqrt{N(N+2)}$$
$$= \sqrt{4(6)}$$
$$= \sqrt{24}\ \text{BM}$$

Thus, **(iv)→ (p,q,s)**

## JEE Main & AIEEE Archive

**1.** (a) The electronic configuration of rubidium atom ($Z = 37$) is given by
$$\text{Rb} = [\text{Kr}]\ 5\,s^1$$
Hence, the quantum numbers for $5\,s^1$ electron is given by

$n = 5,\ l = 0,\ m = 0$
$$s = +\frac{1}{2} \text{ or } -\frac{1}{2}$$

**2.** (a) Energy change $\Delta E = 2.178 \times 10^{-18}\left(\frac{1}{1^2} - \frac{1}{2^2}\right) = \frac{hc}{\lambda}$
$$= 2.178 \times 10^{-18}\left(\frac{1}{1^2} - \frac{1}{2^2}\right)$$
$$= \frac{6.62 \times 10^{-34} \times 3.0 \times 10^8}{\lambda}$$
$$\therefore \lambda \approx 1.214 \times 10^{-7}\ \text{m}$$

**3.** (b)

| | $n$ | $l$ | $nl$ (suborbit) | $n+l$ |
|---|---|---|---|---|
| (a) | 4 | 1 | $4p$ | 5 |
| (b) | 4 | 0 | $4s$ | 4 |
| (c) | 3 | 2 | $3d$ | 5 |
| (d) | 3 | 1 | $3p$ | 4 |

Higher the value of $(n + l)$, higher is the energy. If $(n + l)$ are same, then suborbit with lower value of $n$ has lower energy. Thus,

$$3p < 4s < 3d < 4p$$
$$(4) < (2) < (3) < (1)$$

**4.** (c) Energy values are always additive.

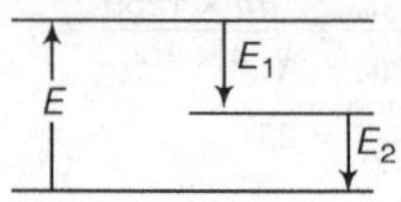

$$E_{total} = E_1 + E_2$$
$$\frac{hc}{\lambda} = \frac{hc}{\lambda_1} + \frac{hc}{\lambda_2}$$
$$\therefore \quad \frac{1}{\lambda} = \frac{1}{\lambda_1} + \frac{1}{\lambda_2}$$
$$\frac{1}{355} = \frac{1}{680} + \frac{1}{\lambda_2}$$
$$\lambda_2 = 742.77 \text{ nm} \approx 743 \text{ nm}$$

**5.** (b) $\Delta E = hu = \dfrac{2\pi^2 m Z^2 e^4 k^2}{h^2}\left[\dfrac{1}{n_1^2} - \dfrac{1}{n_2^2}\right]$

if electron falls from $n_2$-level to $n_1$- level.

$\therefore$ In $He^+$, for $n_2 = 4$ to $n_1 = 2$ transition

$$v_{He^+} = \text{constant } (4)\left[\frac{1}{2^2} - \frac{1}{4^2}\right] \qquad [\because Z_{He^+} = 2]$$
$$= \text{constant} \times 4\left[\frac{3}{16}\right]$$
$$= \frac{3}{4} \times \text{constant}$$
$$v\text{ (H)} = \text{constant } (1)^2\left[\frac{1}{n_1^2} - \frac{1}{n_2^2}\right]$$
$$= \text{constant} \times \left[\frac{1}{n_1^2} - \frac{1}{n_2^2}\right]$$

(a) For $n_2 = 3$ and $n_1 = 1$,

$$v\text{ (H)} = \text{constant}\left[\frac{1}{1} - \frac{1}{9}\right]$$
$$= \frac{8}{9} \times \text{constant} \neq \frac{3}{4} \times \text{constant}$$

(b) For $n_2 = 2$ and $n_1 = 1$,

$$v\text{ (H)} = \text{constant} \times \left[\frac{1}{1} - \frac{1}{4}\right]$$
$$= \frac{3}{4} \times \text{constant} = v(He^+)$$

(c) For $n = 3$ and $n_1 = 2$

$$v\text{ (H)} = \text{constant} \times \left[\frac{1}{2^2} - \frac{1}{3^2}\right]$$
$$= \frac{5}{36} \times \text{constant} \neq v\text{ (}He^+\text{)}$$

(d) For $n_2 = 4$ and $n_1 = 3$

$$v\text{ (H)} = \text{constant} \times \left[\frac{1}{3^2} - \frac{1}{4^2}\right]$$
$$= \frac{7}{144} \times \text{constant}$$
$$\neq v\text{ (}He^+\text{)}$$

**6.** (d) Energy required for 1 $Cl_2$ molecule $= \dfrac{242 \times 10^3}{N_A}$ J

$$E = \frac{hc}{\lambda} \text{ or } \lambda = \frac{hc}{E}$$
$$= \frac{6.626 \times 10^{-34} \times 3 \times 10^8 \times 6.02 \times 10^{23}}{242 \times 10^3}$$
$$= 494 \times 10^{-9} \text{ m}$$
$$= 494 \text{ nm}$$

**7.** (b) IE $= -E_1$

$$E_1 \text{ for } He^+ = -19.6 \times 10^{-18} \text{ J atom}^{-1}$$
$$\frac{(E_1)_{He^+}}{(E_1)_{Li^{2+}}} = \frac{(Z_{He^+})^2}{(Z_{Li^{2+}})^2}$$
$$\frac{-19.6 \times 10^{-18}}{(E_1)_{Li^{2+}}} = \frac{4}{9}$$

or
$$E_1(Li^{2+}) = \frac{-19.6 \times 9 \times 10^{-18}}{4}$$
$$= -4.41 \times 10^{-17} \text{ J atom}^{-1}$$

**8.** (c) By Heisenberg's uncertainty principle

$$\Delta x \cdot m\Delta V = \frac{h}{4\pi}$$
$$\Delta V_{acc} = 0.005\%$$

or
$$\Delta V = \frac{600 \times 0.005}{100} = 0.03 \text{ ms}^{-1}$$
$$\Delta x \times 9.1 \times 10^{-31} \times 0.03 = \frac{6.6 \times 10^{-34}}{4 \times 3.14}$$

Hence,
$$\Delta x = \frac{6.6 \times 10^{-34}}{4 \times 3.14 \times 0.03 \times 9.1 \times 10^{-31}}$$
$$= 1.92 \times 10^{-3} \text{ m}$$

**9.** (b)
$$\lambda = \frac{h}{mV}$$
$$= \frac{6.63 \times 10^{-34}}{1.67 \times 10^{-27} \times 1 \times 10^3}$$
$$= 3.97 \times 10^{-10} \text{ m}$$
$$\approx 0.40 \text{ nm}$$

**10.** (d) Ionisation enthalpy of hydrogen atom is $1.312 \times 10^6$ J mol$^{-1}$. It suggests that the energy of electron in the ground state (first orbit) is $-1.312 \times 10^6$ J mol$^{-1}$.

$$\Delta E = E_2 - E_1$$
$$= -\left(\frac{1.312 \times 10^6}{2^2}\right) - \left(\frac{-1.312 \times 10^6}{1}\right)$$
$$= 9.84 \times 10^5 \text{ J mol}^{-1}$$

**11.** (b) $n + l = 5$, maximum.

**12.** (a) ${}^{A}_{Z}X \longrightarrow {}^{A-1}_{Z}X + {}_{0}n^1$

Isotopes are species having same number of protons but different number of neutrons.

**13.** (c) $\Delta x \cdot \Delta v \geq \dfrac{h}{4\pi m}$

$$\Delta v = 300 \times 0.001 \times 10^{-2} \text{ ms}^{-1}$$
$$\Delta x = \frac{6.63 \times 10^{-34}}{4 \times 3.14 \times 9.1 \times 10^{-31} \times 300 \times 0.001 \times 10^{-2}}$$
$$= 0.01933 = 1.93 \times 10^{-2} \text{ m}$$

**14.** (d) Angular momentum of an electron

$$= mvr = \frac{nh}{2\pi} \qquad [n \text{ is orbit number}]$$

In 5th orbit $= \dfrac{5h}{2\pi} = \dfrac{2.5h}{\pi}$

**15.** (a) Hydrogen atom has $1s^1$ configuration and these $3s$, $3p$ and $3d$ orbitals will have same energy w.r.t. $1s$ orbital.

**16.** (a)
(A) 1s
(B) 2s
(C) 2p
(D) 3d
(E) 3d
In the absence of any field, 3d in (D) and (E) will be of equal energy.

**17.** (b)

| Species | Electron in central element | Electrons in other element | Charge gained | Total |
|---|---|---|---|---|
| $BO_3^{3-}$ | 5 | $3\times8=24$ | −3 | 32 |
| $CO_3^{2-}$ | 6 | $3\times8=24$ | −2 | 32 |
| $NO_3^-$ | 7 | $3\times8=24$ | −1 | 32 |
| $SO_3^{2-}$ | 16 | $3\times8=24$ | −2 | 42 |
| $CN^-$ | 6 | 7 | 1 | 14 |
| $N_2$ | 7 | 7 | 0 | 14 |
| $C_2^{2-}$ | 6 | 6 | −2 | 14 |
| $PO_4^{3-}$ | 15 | $4\times8=32$ | −3 | 50 |
| $SO_4^{2-}$ | 10 | $4\times8=32$ | −2 | 50 |
| $ClO_4^-$ | 17 | $4\times8=32$ | −1 | 50 |

Thus, (b) $SO_3^{2-}$, $CO_3^{2-}$, $NO_3^-$ are not isoelectronic.

**18.** (a) On applying Rydberg formula,

$$\frac{1}{\lambda}=\bar{\nu}_H=R_H\left[\frac{1}{n_1^2}-\frac{1}{n_2^2}\right]$$

$$=1.097\times10^7\left[\frac{1}{1^2}-\frac{1}{\infty^2}\right]$$

$$\therefore\quad \lambda=\frac{1}{1.097\times10^7}\text{ m}=9.11\times10^{-8}\text{m}$$

$$=91.1\times10^{-9}\text{ m}$$

$$=91.1\text{ nm}\qquad[\because 1\text{ nm}=10^{-9}\text{ m}]$$

**19.** (c) Any suborbit is represented as $nl$ such that $n$ is the principal quantum number (in the form of values) and $l$ is the azimuthal quantum number (its name).

Value of $l < n$

$l$ = 0 1 2 3 4
s p d f g

Value of $m : -l, -l+1 ...... 0, ...... +l$

Value of $s : +\frac{1}{2}$ or $-\frac{1}{2}$

Thus, for $4f : n=4$, $l=3$, $m$ = any value between − 3 to + 3 and $s$ = may be $\pm\frac{1}{2}$.

**20.** (b) EC of Cr ($Z=24$) is

| | $n$ | $l$ |
|---|---|---|
| $1s^2$ | 1 | 0 |
| $2s^2$ | 2 | 0 |
| $2p^6$ | 2 | 1 |
| $3s^2$ | 3 | 0 |
| $3p^6$ | 3 | 1 |
| $3d^5$ | 3 | 2 |
| $4s^1$ | 4 | 0 |

Thus, electrons with $l=1$, are 12 and with $l=2$, are 5.

**21.** (a)

| Species | Z | Electron gained or lost in the ion formation | Electrons |
|---|---|---|---|
| $K^+$ | 19 | −1 | 18 |
| $Ca^{2+}$ | 20 | −2 | 18 |
| $Sc^{3+}$ | 21 | −3 | 18 |
| $Cl^-$ | 17 | +1 | 18 |

In this set, all have 18 electrons, thus isoelectronic.

**22.** (a) Wavelength associated with the velocity of tennis ball,

$$\lambda=\frac{h}{mv}=1105\times10^{-33}\text{ m}$$

**23.** (b) The emission is in visible region [V I B G Y O R]. The visible region in H spectrum corresponds to Balmer series [$n_1=2$, $n_2=3, 4 ... \infty$]
*i.e.*, either (a) or (b). In (a), there is only one emission. Thus, the third line will be in the jump from 5 to 2.

**24.** (d) $Fe_{26}$ = [Ar] $3d^6, 4s^2$; $Fe^{2+}_{(24)}$ = [Ar] $3d^6, 4s^0$

**25.** (b) $N^{3-}=7+3=10$ electrons; $F^-=9+1=10$ electrons
$Na^+=11-1=10$ electrons

**26.** (b) For $s$-electron, $l=0$.

**27.** (a) Heisenberg uncertainty principle

$$\Delta x\cdot m\Delta v\approx\frac{h}{4\pi}$$

Uncertainty in velocity,

$$\Delta v=\frac{h}{4\times3.14\times\Delta x\times m}=\frac{6.6\times10^{-34}}{4\times3.14\times0.025\times10^{-5}}$$

$$=2.1\times10^{-28}\text{ m/s}$$

**28.** (c) Energy in nth orbit $E_n=-\frac{13.6}{n^2}$ eV

For second excited state, $n=3$

$$E_3=-\frac{13.6}{9}=-1.51\text{ eV}$$

## JEE Advanced & IIT JEE Archive

**1.** (6) $|m_l|=1$ thus, magnetic quantum number = + 1 and −1

$$(m_s)=-\frac{1}{2}$$

$n=4$ can have suborbits

| | $l$ | $m_l$ | $m_s$ |
|---|---|---|---|
| 4s | 0 | 0 | $+\frac{1}{2}-\frac{1}{2}$ |
| 4p | 1 | $\underline{-1}, 0, \underline{+1}$ | $+\frac{1}{2}-\frac{1}{2}$ |
| 4d | 2 | $-2, \underline{-1}, 0, \underline{+1}, +2$ | |
| 4f | 3 | $-3, -2, \underline{-1}, 0, \underline{+1}, +2, +3$ | |

Thus underlined electrons with ($m_l=-1, +1$) and spin $-\left(\frac{1}{2}\right)$ are six.

**2.** (5) $KE=\frac{1}{2}mv^2=\frac{3}{2}RT$

$\therefore\quad m^2v^2=2mKE$

$\therefore\quad mv=\sqrt{2mKE}$

$$\lambda\text{ (wavelength)}=\frac{h}{mv}=\frac{h}{\sqrt{2mKE}}\propto\frac{h}{\sqrt{2m(T)}}$$

where, $T$ = Temperature in Kelvin

$$\lambda(\text{He at } -73°\text{C} = 200\text{ K}) = \frac{h}{\sqrt{2 \times 4 \times 200}}$$

$$\lambda(\text{Ne at } 727°\text{C} = 1000\text{ K}) = \frac{h}{\sqrt{2 \times 20 \times 1000}}$$

$$\therefore \quad \frac{\lambda(\text{He})}{\lambda(\text{Ne})} = M = \sqrt{\frac{2 \times 20 \times 1000}{2 \times 4 \times 200}} = 5$$

Thus, $M = 5$

**3.** (c) According to Bohr's model

$$mvr = \frac{nh}{2\pi}$$

$$\Rightarrow \quad (mv)^2 = \frac{n^2h^2}{4\pi^2r^2}$$

$$\Rightarrow \quad \text{KE} = \frac{1}{2}mv^2 = \frac{n^2h^2}{8\pi^2r^2m} \quad \ldots\text{(i)}$$

Also Bohr's radius for H-atom is $r = n^2a_0$

Substituting '$r$' in Eq. (i) gives

$$\text{KE} = \frac{h^2}{8\pi^2n^2a_0^2m}$$

when $n = 2$, $\quad \text{KE} = \dfrac{h^2}{32\pi^2a_0^2m}$

**4.** (9) When $n = 3$, $l = 0, 1, 2$, i.e. there are $3s$, $3p$ and $3d$ orbitals. If all these orbitals are completely occupied as

[↿⇂] [↿⇂|↿⇂|↿⇂] [↿⇂|↿⇂|↿⇂|↿⇂|↿⇂]

Total 18 electrons, 9 electrons with $s = +\frac{1}{2}$ and 9 with $s = -\frac{1}{2}$.

**Alternatively** In any $n$th orbit, there can be a maximum of $2n^2$ electrons. Hence, when $n=3$, number of maximum electrons = 18. Out of these 18 electrons, 9 can have spin $-\frac{1}{2}$ and remaining nine with spin $= +\frac{1}{2}$.

**5.** (4) Energy of photon $= \dfrac{hc}{\lambda}\text{J} = \dfrac{hc}{e\lambda}\text{eV}$

$$= \frac{6.625 \times 10^{-34} \times 3 \times 10^8}{300 \times 10^{-9} \times 1.602 \times 10^{-19}} = 4.14\text{eV}$$

For photoelectric effect to occur, energy of incident photons must be greater than work function of metal. Hence, only Li, Na, K and Mg have work functions less than 4.14V.

**6.** (b) $S_1$ is spherically symmetrical state, i.e. it correspond to a $s$-orbital. Also it has one radial node.

Number of radial nodes $= n - l - l$

$$n - 0 - 1 = 1$$

$n = 2$, i.e. $S_1 = 2s$ orbital.

**7.** (c) Ground state energy of electron in H-atom ($E_\text{H}$)

$$E_\text{H} = \frac{kZ^2}{n^2} = k \ (Z = 1, n = 1)$$

for $S_1$ state of $Li^{2+}$, $\quad E = \dfrac{k(3)^2}{2^2} = \dfrac{9}{4}k = 2.25\,k$

**8.** (b) In $S_2$ state, $E(Li^{2+}) = k$ (given)

$$k = \frac{9k}{n^2} \Rightarrow n = 3$$

Since, $S_2$ has one radial node.

$$3 - l - 1 = 1 \Rightarrow l = 1$$

**9.** (b)

(i) Orbital angular momentum

$$(L) = \sqrt{l(l+1)}\,\frac{h}{2\pi}$$

**(i) → (q)**

i.e. $L$ depends on azimuthal quantum number only.

(ii) To describe a one electron wave function, three quantum numbers $n$, $l$ and $m$ are needed. Further to abide by Pauli exclusion principle, spin quantum number ($s$) is also needed.

**(ii) → (s)**

(iii) For shape, size and orientation, only $n$, $l$ and $m$ are needed.

**(i) → (p,q,r)**

(iv) Probability density ($\psi^2$) can be determined if $n$, $l$ and $m$ are known.

**(iv) → (p,q,r)**

**10.** (a)

(i) $$V_n = -\frac{1}{4\pi\varepsilon_0}\left(\frac{Ze^2}{r}\right)$$

$$K_n = \frac{1}{8\pi\varepsilon_0}\left(\frac{Ze^2}{r}\right)$$

$$\Rightarrow \quad \frac{V_n}{K_n} = -2\,(p)$$

(ii) $$E_n = -\frac{Ze^2}{8\pi\varepsilon_0 r} \propto r^{-1}$$

$$\Rightarrow \quad x = -1\,(q)$$

(iii) Angular momentum $= \sqrt{l(l+1)}\,\dfrac{h}{2\pi} = 0$ in 1s orbital.

(iv) $$r_n = \frac{a_0n^2}{Z}\,(p)$$

$$\Rightarrow \quad \frac{1}{r_n} \propto Z\,(s)$$

**11.** (a) The number of radial nodes is given by expression $(n - l - 1)$.

For 3s,

number of nodes $= 3 - 0 - 1 = 2$

For $2p$,

number of nodes $= 2 - 1 - 1 = 0$

**12.** (i) $$mvr = \frac{nh}{2\pi}$$

$$v = \frac{nh}{2\pi mr}$$

$$= \frac{6.625 \times 10^{-34}}{2 \times 3.14 \times 9.1 \times 10^{-31} \times 0.529 \times 10^{-10}}$$

$$= 2.18 \times 10^6 \text{ ms}^{-1}$$

(ii) $$\lambda = \frac{h}{mv}$$

$$= \frac{6.625 \times 10^{-34}}{9.1 \times 10^{-31} \times 2.18 \times 10^6}$$

$$= 0.33 \times 10^{-9}\text{ m}$$

(iii) Orbital angular momentum

$$(L) = \sqrt{l(l+1)}\,\frac{h}{2\pi}$$

$$= \sqrt{2}\left(\frac{h}{2\pi}\right) \quad [\because \text{For } p\text{-orbital, } l = 1]$$

# 2. Redox Reactions

## DPP-1 Oxidation Number

**1. Plan** To determine oxidation number of the element,

Take oxidation number of alkali metals = + 1

Oxidation number of alkaline earth elements = + 2

Oxidation number of fluorine = – 1

Oxidation number of oxygen = – 2 [except in $OF_2$ (O = + 2) in $H_2O_2$ (O = – 1)]

Oxidation number of hydrogen = + 1 (except in hydride, H = – 1)

(i) $H_4P_2O_7$

$+4 + 2x - 14 = 0 \Rightarrow x = +5$

$H_4P_2O_7$ has two units of $H_3PO_4$ joined by O-atom.

(ii) $CaO_2$,

$+2 + 2x = 0 \Rightarrow x = -1$

$CaO_2$ is calcium peroxide.

(iii) $H_2S_2O_7$,

$2 + 2x - 14 = 0 \Rightarrow x = +6$

$H_2S_2O_7$ is called **oleum**, having $H_2SO_4$ saturated with $SO_3$

$$H_2SO_4 + SO_3 \rightleftharpoons H_2S_2O_7$$

(iv) $KAl(SO_4)_2 \rightleftharpoons K^+ + Al^{3+} + 2SO_4^{2-}$

$SO_4^{2-}$

$x - 8 = -2 \Rightarrow x = +6$

(v) $Fe_3O_4$

$3x - 8 = 0$

$x = \frac{8}{3}$ per atom

**Note** $Fe_3O_4$ *is a mixed oxide of iron*

$FeO \cdot Fe_2O_3$

+2 +3

*(per atom)*

(vi) $\overset{2}{C}H_3\overset{1}{C}H_2OH$

**Oxidation number of $C_1$**

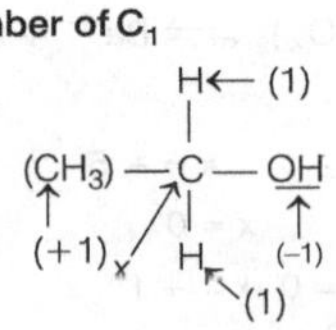

$+1 + x + 2 - 1 = 0$

$x = -2$

Thus, oxidation number of $C_1$ is –2.

**Oxidation number of $C_2$**

Charge on $CH_2OH = -2 + 2 - 2 + 1 = -1$

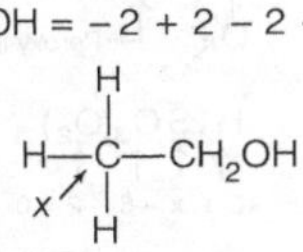

$x + 3 - 1 = 0 \Rightarrow x = -2$

**Note** *Collectively oxidation number of each carbon = – 2*

$C_2H_6O$

$2x + 6 - 2 = 0 \Rightarrow x = -2$

(vii)

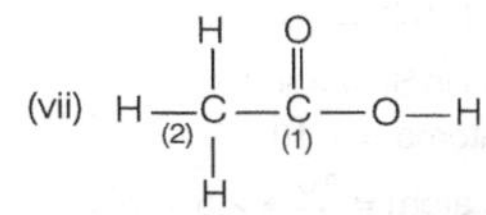

**Oxidation number of $C_1$**

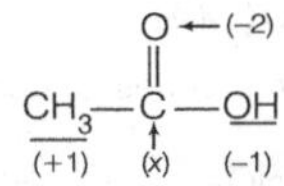

Then, $1 + x - 2 - 1 = 0 \Rightarrow x = +2$

**Oxidation number of $C_2$**

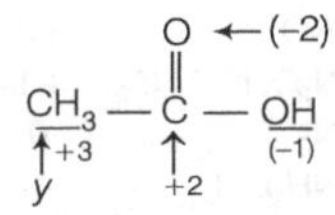

$y + 3 + 2 - 2 - 1 = 0$

$y = -2$

Collectively $x + y = 0$

$C_2H_4O_2$, $2z + 4 - 4 = 0 \Rightarrow z = 0$

**2. Plan** $H_2S_2O_3$ has two types of sulphur atoms.

One directly attached to sulphur atom.

One directly attached to three oxygen atoms.

The difference in the environmental of the two S-atoms is shown by radioactive tracer studies using ($^{35}_{16}S$) (a)radioactive isotope of sulphur $S^*SO_3^{2-}$

$$SO_3^{2-} + S^* \rightleftharpoons S^*SO_3^{2-} \rightleftharpoons S^* + SO_3^{2-}$$

$H_2S_2O_3$ decomposes into $H_2SO_3$ and S, and the entire radioactivity is associated with single S-atom.

Thus, oxidation number of terminal sulphur = – 2

and oxidation number of sulphur atom in $H_2S_2O_3$ = + 6

Net oxidation number of two S-atom = – 2 + 6 = + 4

Thus, oxidation number per sulphur atom = + 2

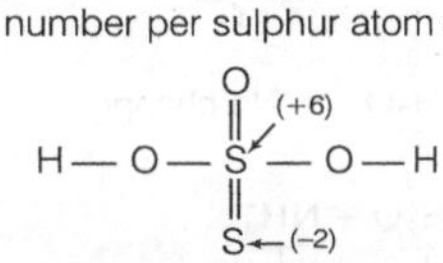

**3.** (i) **Plan** $(Ca(ClO_2)_2$ has $ClO_2^-$ ion

$$Ca(ClO_2)_2 \rightleftharpoons Ca^{2+} + 2ClO_2^-$$

In $ClO_2^-$ oxidation number of Cl – x

Oxidation number of two O-atom = – 4

Total charge = – 1

Thus, $x - 4 = -1$

$x = +3$

(ii) **Plan** $CaOCl_2$ is bleaching powder.

$$CaOCl_2 + H_2O \rightleftharpoons Ca(OH)_2 + Cl_2$$

It has two types of ions $Cl^-$ and $ClO^-$

Oxidation number of Cl in $Cl^-$ = – 1

Oxidation number of Cl in $ClO^-$ = + 1

Thus, in combined state,oxidation number of Cl = 0 (zero)

$$Ca^{2+}\begin{pmatrix} Cl^- \\ ClO^- \end{pmatrix}$$

**Note** $Cl_2$ *formed when* $CaOCl_2$ *reacts with* $H_2O$ *is called available chlorine.*

**4.** **Plan** $\underset{\text{Lewis base}}{S_2^{2-}} + \underset{\text{Lewis acid}}{2SO_3} \longrightarrow S_4O_6^{2-}$

Thus, $S_4O_6^{2-}$ is an adduct formed by reaction between Lewis base and Lewis acid.

Oxidation number of two S-atom in $S_2^{2-} = -2$

Oxidation number of two S-atoms in $SO_3 = +12$

Net oxidation number of four S-atoms = + 10

Oxidation number of sulphur per atom $= \frac{10}{4} = 2.5$

**Note** $2S_2O_3^{2-} \longrightarrow S_4O_6^{2-} + 2e^-$

**5.** (i) **Plan** Nitrogen forms series of compounds as hydrides and oxyacids/oxides

| | |
|---|---|
| $HNO_3$ | +5 |
| $NO_2, N_2O_4$ | +4 |
| $HNO_2$ | +3 |
| $NO$ | +2 |
| $N_2O, H_2N_2O_2$ | +1 |
| $N_2$ | 0 |
| $NH_2OH$ | −1 |
| $N_2H_4$ | −2 |
| $NH_3$ | −3 |

(ii) **Plan**

(a) $\underset{0}{\frac{1}{2}Cl_2} \longrightarrow \underset{-1}{Cl^-}$ Change 1

(b) $\underset{0}{\frac{1}{2}Cl_2} \longrightarrow \underset{+5}{ClO_3^-}$ 5

(c) $\underset{\uparrow\,-1}{PCl_5} \longrightarrow \underset{\uparrow\,-1}{PCl_3} + \underset{\uparrow\,0}{Cl_2}$ 1

(d) $\underset{-3}{NH_4^+} \longrightarrow \underset{0}{\frac{1}{2}N_2}$ 3

$\underset{+3}{NO_2^-} \longrightarrow \underset{0}{\frac{1}{2}N_2}$ 3

(e) $\underset{\uparrow\,+4}{CO_2} \longrightarrow \underset{\uparrow\,+4}{HCO_3^-}$ 0

**6.** (d)

I. $\underset{\uparrow\,+1}{H_2SO_4} \longrightarrow \underset{\uparrow\,0}{H_2}$ Change 1

II. $\underset{\uparrow\,+1}{H_2SO_4} \longrightarrow \underset{\uparrow\,+1}{H_2O}$ No change

III. $\underset{\uparrow\,+1}{HNO_3} \longrightarrow \underset{\uparrow\,+1}{H_2O} + \underset{\uparrow\,+1}{NH_4^+}$ No change

**7.** (b) (a) Oxidation number of V

$\underset{+4}{VO^{2+}} \longrightarrow \underset{+5}{VO_2^+}$ Change 1 unit

(b) $\underset{+5}{VO_2^+} \longrightarrow \underset{+3}{VO^{2+}}$ Change 2 units

(c) $\underset{+4}{VO^{2+}} \longrightarrow \underset{+3}{V^{3+}}$ Change 1 unit

**8.** (a) $KO_2$ (superoxide) $O_2^-$ is superoxide ion

$2x = -1$

$x = -0.5$

$OF_2$ Fluorine is most electronegative atom with oxidation number = −1

$x - 2 = 0$

$x = +2$

$H_2O_2$ $2 + 2x = 0 \Rightarrow x = -1$

**9.** (d) (I) Indicates that oxidation number of Cl is +1.

(a) $\underset{\uparrow}{Cl}O_2^-$,

$x - 4 = -1 \Rightarrow x = +3$

(b) $\underset{\uparrow}{Cl}O_3^-$,

$x - 6 = -1 \Rightarrow x = +5$

(c) $\underset{\uparrow}{Cl}O_4^-$,

$x - 8 = -1 \Rightarrow x = +7$

(d) $\underset{\uparrow}{Cl}O^-$,

$x - 2 = -1 \Rightarrow x = +1$

**10.** (a) If acid is strong, its conjugate base is weak and *vice-versa*. Greater the oxidation number of the atom (halogen) of the oxyacids, stronger the acid, and thus weaker the conjugate base.

| Acid strength | Acid | (ON of Cl) | Conjugate base | Base strength |
|---|---|---|---|---|
| ↓ | $HClO$ | +1 | $ClO^-$ | ↑ |
| | $HClO_2$ | +3 | $ClO_2^-$ | |
| | $HClO_3$ | +5 | $ClO_3^-$ | |
| | $HClO_4$ | +7 | $ClO_4^-$ | |

Thus, the order is I < II < III < IV.

**11.** (d) Fluorine is the most electronegative element. It can gain one electron

$$F + e^- \longrightarrow F^-$$

Oxidation number = − 1

Due to very high (IE), it cannot lose electron.

(a) H $H^-$ (−1), $H^+$ (+1)

(b) O −2 in oxide

−1 in peroxide

+2 in $OF_2$

(c) N −3 in $N_2O_3$, $NH_3$

+3 in $HNO_2$,

+5 in $HNO_3$,

+1 in $N_2O$

**12.** (c) $Ba(H_2PO_2)_2 \rightleftharpoons Ba^{2+} + 2H_2PO_2^-$

$2 + x - 4 = -1$

$x = +1$

$Ba(H_2PO_3)_2 \rightleftharpoons Ba^{2+} + 2H_2PO_3^-$

$+2 + x - 6 = -1$

$x = +3$

$Ba(H_2PO_4)_2 \rightleftharpoons Ba^{2+} + 2H_2PO_4^-$

$+2 + x - 8 = -1$

$x = +5$

**13.** (a) $S_8$, $8x = 0$, $x = 0$

$S_2F_2$, $2x - 2 = 0$, $x = +1$

$H_2S$, $+ x = 0$, $x = -2$

$H_2SO_5$ has one peroxy linkage.

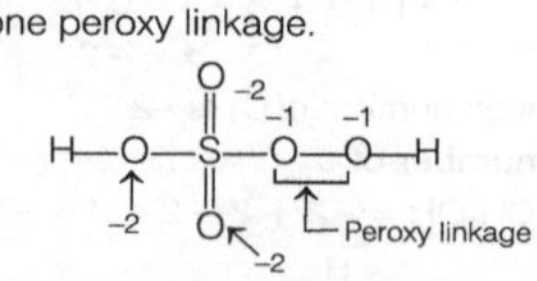


$H_2SO_3(O_2)$

$+2 + x - 6 - 2 = 0$

$x = +6$

**14.** (c) $(\underset{\uparrow\,-3}{N}H_4)_2\underset{\uparrow\,+6}{Cr_2}O_7 \longrightarrow \underset{\uparrow\,0}{N_2} + \underset{\uparrow\,+3}{Cr_2}O_3$

Oxidation number of Cr changes from +6 to +3.

**15.** (c) $K Fe^x [Fe^{II}(CN)_6]$

$+1 + x + 2 - 6 = 0$

$x = +3$

**16.** (c) $\underset{-3}{NH_4^+}\ \underset{+3}{NO_2^-} \longrightarrow \underset{0}{N_2}$

**17.** (b)

| | Element | Oxidation number |
|---|---|---|
| $S_4O_6^{2-}$ | S | 2.5 |
| $Pb_3O_4$ | Pb | 8/3 |
| $C_3O_2$ | C | 4/3 |
| $Br_3O_8$ | Br | 16/3 |

**18.** (a, b, c)

$MnO_4^{2-}$, $x - 8 = -2 \Rightarrow x = +6$

$CrO_2Cl_2$, $x - 4 - 2 = 0 \Rightarrow x = +6$

$CrO_4^{2-}$, $x - 8 = -2 \Rightarrow x = +6$

$MnO_4^-$, $x - 8 = -1 \Rightarrow x = +7$

**19.** (a, b, d)

$SO_3$

$x - 6 = 0,\ x = +6$

$CrO_3$

$x - 6 = 0,\ x = +6$

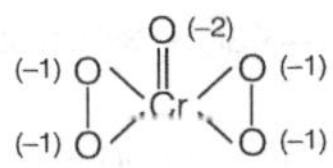


$CrO_5$ has two peroxy linkages thus,

oxidation number of four O-atom in peroxy linkages = − 4

$Cr(O_2)_2O$

$x - 4 - 2 = 0$

$x = +6$

$S_2O_8^{2-}$ has one peroxy linkage

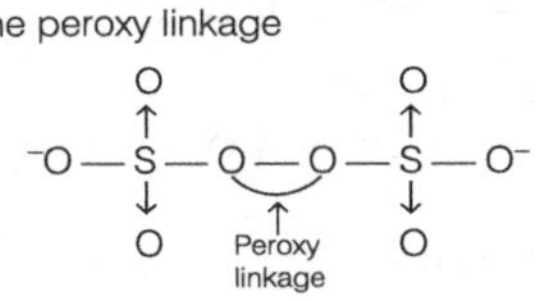


$S_2O_6(O_2)^{2-}$,

$2x - 12 - 2 = -2$

$\therefore \quad x = +6$

$SO_5^{2-}$ has one peroxy linkage

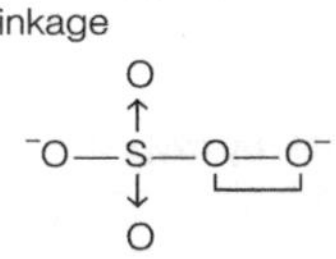

$SO_3(O_2)^{2-}$,

$x - 6 - 2 = -2 \Rightarrow x = 6$

$SO_4^{2-}$,

$x - 8 = -2 \Rightarrow x = +6$

$S_2O_3^{2-}$,

$2x - 6 = -2 \Rightarrow x = +2$

$S_4O_6^{2-}$,

$4x - 12 = -2 \Rightarrow x = 2.5$

$CrO_4^{2-}$,

$x - 8 = -2 \Rightarrow x = +6$

$MnO_4^{2-}$,

$x - 8 = -2 \Rightarrow x = +6$

**20.** (a) Let, oxidation number of Fe = $x$

$[Fe(H_2O)_5NO]^{2+}$

$x + 5 \times (0) + 1 = +2$

$\therefore \quad x = +1$

**21.** (c) $Fe^{2+}$ is added as $FeSO_4$

$Fe^+$ is formed by charge transfer from NO to $Fe^{2+}$

$NO \longrightarrow NO^+ + e^-$

$Fe^{2+} + e^- \longrightarrow Fe^+$

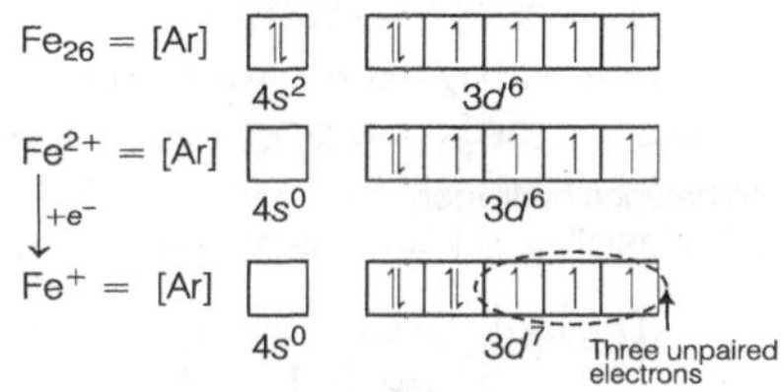


$Fe^+$ has three unpaired electrons ($N$).

Thus, magnetic moment = $\sqrt{N(N+2)}$ BM

$= \sqrt{15}$ BM

**22.** (b)

(i) $CaOCl_2$ has

$Cl^-$ ion oxidation number = − 1

$ClO^-$ ion oxidation number = + 1

**(i) → (s,t)**

(ii) $S_2O_3^{2-}$

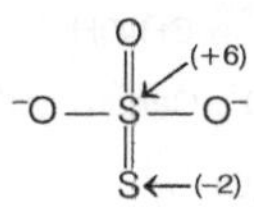


Oxidation number = + 6

Oxidation number = − 2

**(ii) → (p,u)**

(iii) $NH_4^+NO_2^-$

Oxidation number of N in $NH_4^+$ = − 3

Oxidation number of N in $NO_2^-$ = + 3

**(iii) → (q,v)**

(iv) $Fe^{III}[Fe^{II}(CN)_6]$

Oxidation number of Fe (separate) = + 3

Oxidation number of Fe (coordinated) = + 2

**(iv) → (q,r)**

(v)

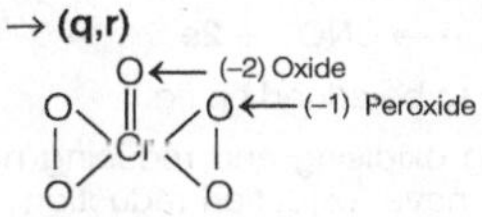


**(v) → (t,u)**

**23.** 8

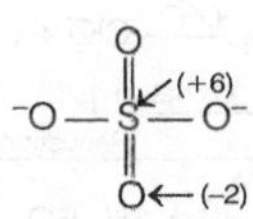


Difference in oxidation number of two S-atom = 6 − (−2) = 8

**24.** (7)

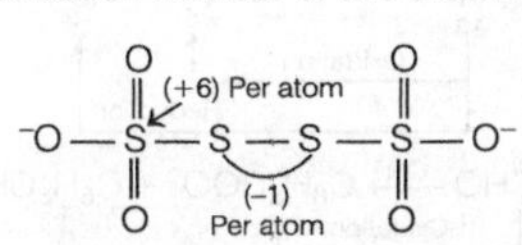


Difference in oxidation number = 6 − (−1) = 7

# DPP-2 Oxidation-Reduction and Disproportionation Reactions

**1.** (i) **Plan**

$$\underset{+12}{Cr_2O_7^{2-}} + \underset{+4}{SO_3^{2-}} \longrightarrow \underset{+6}{SO_4^{2-}} + \underset{+6}{2Cr^{3+}}$$

Oxidation ($-2e^-$); Reduction ($+6e^-$)

Oxidation number of two Cr-atoms has been taken, to balance electrons and affected atoms.

$$Cr_2O_7^{2-} + 3SO_3^{2-} \longrightarrow 3SO_4^{2-} + 2Cr^{3+}$$

To balance oxygen, add $H_2O$ on less oxygen side

$$Cr_2O_7^{2-} + 3SO_3^{2-} \longrightarrow 3SO_4^{2-} + 2Cr^{3+} + 4H_2O$$

Add $H^+$ to balance hydrogen

$$Cr_2O_7^{2-} + 3SO_3^{2-} + 8H^+ \longrightarrow 3SO_4^{2-} + 2Cr^{3+} + 4H_2O$$

(ii)

$$\frac{1}{2}Cl_2 + \frac{1}{2}Cl_2 \longrightarrow ClO_3^- + Cl^-$$

Oxidation; Reduction

$$3Cl_2 + 6OH^- \longrightarrow ClO_3^- + 5Cl^- + 3H_2O$$

**2.** **Plan** If same species is oxidised and reduced then, it is a disproportionation reaction.

(i) $(CN)_2$ is reduced to $CN^-$ as well as oxidised to $CNO^-$.

(ii) $Mn^{3+}$ is reduced to $Mn^{2+}$ as well as oxidised to $MnO_2$.

(iii) $P_4$ is reduced to $PH_3$ as well as oxidised to $H_2PO_2^-$.

(iv) $\underset{CH_2OH}{\overset{|}{CHO}}$ is reduced to $\underset{CH_2OH}{\overset{|}{CH_2OH}}$ as well

as oxidised to $\underset{CH_2OH}{\overset{|}{COO^-}}$ (Cannizzaro). Thus, all of the above reactions are disproportionation reactions.

**3.** (b) $\underset{0}{Zn} \longrightarrow \underset{+2}{Zn^{2+}} + 2e^-$

$Zn + O \longrightarrow ZnO$

Oxidation number increases, oxygen gained and loss of electrons is oxidation.

**4.** (a) Gain of hydrogen, electron and decrease in oxidation number is reduction.

**5.** (d) $CN^- \longrightarrow CNO^-$

$4 + x = -1$ $\quad$ $4 + x - 2 = -1$

$\therefore x = -5$ $\quad$ $\therefore x = -3$

Oxidation number increases. This is oxidation.

$$CN^- \longrightarrow CNO^- + 2e^-$$

Thus, two electrons are to be added on right.

**6.** (b) Compounds having oxidising and reducing nature in given reaction are said to have oxidation-reduction duality. Such compounds are said to undergo disproportionation reaction.

$$\underset{0}{Cl_2} \xrightarrow{OH^-} \underset{+1}{ClO^-} + \underset{-1}{Cl^-}$$

Oxidation; Reduction

$$4H_3\underset{+3}{P}O_3 \xrightarrow{\Delta} 3H_3\underset{+5}{P}O_4 + \underset{-3}{P}H_3$$

Oxidation; Reduction

$$C_6H_5CHO \xrightarrow{OH^-} C_6H_5COO^- + C_6H_5CH_2OH$$

Oxidation; Reduction

(This is called Cannizzaro reaction.)

$$H_2\underset{-1}{O_2} \longrightarrow H_2\underset{-2}{O} + \underset{0}{O_2}$$

Reduction; Oxidation

*Note Such compounds have O.N. of the affected atoms intermediate of oxidation part and reduction part.*

**7.** (a) If balanced, then to balance charge

$x = -1$

$O_2^-$ is superoxide ion.

**8.** (c)

$$\underset{(-1)}{I^-} + \underset{(+5)}{IO_3^-} \longrightarrow \frac{1}{2}\underset{0}{I_2} + \frac{1}{2}\underset{0}{I_2}$$

Reduction; Oxidation

To balance oxidation number, cross-multiply by change in oxidation number

$5I^- + IO_3^- \longrightarrow 3I_2 \longrightarrow 3I_2 + 3H_2O]$ balance (O) by $H_2O$

$+ 6H^+]$ balance (H) by $H^+$

Thus, $5I^- + IO_3^- + 6H^+ \longrightarrow 3I_2 + 3H_2O$

**9.** (d) $2\underset{+1}{H_2}\underset{-2}{O} \longrightarrow 2\underset{0}{H_2} + \underset{0}{O_2}$

H is reduced to $H_2$ and O is oxidised to $O_2$.

**10.** (a)

(I) $\underset{0}{Zn} + \underset{+1}{2H^+} \longrightarrow \underset{+2}{Zn^{2+}} + \underset{0}{H_2}$

$H^+$ oxidises Zn to $Zn^{2+}$.

(II) $Zn + \underset{+6}{S}O_4^{2-} \longrightarrow Zn^{2+} + \underset{+4}{S}O_2$

$SO_4^{2-}$ oxidises Zn to $Zn^{2+}$.

**11.** (b) (a) $\underset{+1}{Ag_2}O \longrightarrow Ag + \underset{0}{O_2}$

$Ag^+$ is reduced to Ag by oxygen

(b) $\underset{-3}{N}H_3 \longrightarrow \underset{0}{N_2}$

$NH_3$ is oxidised to $N_2$ by oxygen

(c) $\underset{0}{F_2} \longrightarrow H\underset{-1}{F}$

$F_2$ is reduced to $F^-$ by oxygen of $H_2O$

(d) $\underset{+1}{Ag}NO_3 \longrightarrow \underset{0}{Ag}$

$Ag^+$ is reduced to Ag by oxygen of $H_2O_2$.

Thus, except in (b), oxygen behaves as reducing agent.

**12.** (a)

$$2I^- + \underset{+12}{Cr_2O_7^{2-}} \longrightarrow I_2 + \underset{+6}{2Cr^{3+}}$$

$-2e^-$; $+6e^-$

(I) $Cr_2O_7^{2-} + 6e^- \longrightarrow 2Cr^{3+}$

(II) $2I^- \longrightarrow I_2 + 2e^-$

To balance electrons (II) is to be multiplied by (3). Thus,

$\underline{6I^-} + Cr_2O_7^{2-} + 14H^+ \longrightarrow \underline{3I_2} + 2Cr^{3+} + 7H_2O$

Thus, $I^-$ and $I_2$ are not balanced.

**13.** (c)

$$MnO_4^- + H_2O_2 + H^+ \longrightarrow Mn^{2+} + O_2 + H_2O$$

Oxidation numbers: Mn in $MnO_4^-$ = +7, O in $H_2O_2$ = −2 (2e⁻ for two atoms), $Mn^{2+}$ = +2, $O_2$ = 0; 5e⁻ gained by Mn, 2e⁻ (For two atoms) lost by O.

Cross-multiply by change in oxidation number and balance H by $H^+$ ions.

$$2MnO_4^- + 5H_2O_2 + 6H^+ \longrightarrow 2Mn^{2+} + 5O_2 + 8H_2O$$

Thus, $x = 2, y = 5$ and $z = 6$

**14.** (d) Maximum oxidation number of Cl = + 7

| $ClO^-$ | $ClO_2^-$ | $ClO_3^-$ | $ClO_4^-$ |
|---|---|---|---|
| +1 | +3 | +5 | +7 |
| Show disproportionation | | | Does not show disproportionation |

Thus, species with +7 oxidation number can be reduced but not oxidised.

**15.** (d)

(a) $$KClO_3 \longrightarrow KCl + 3O_2$$

(K +1, Cl +5, O −2 ⟶ K +1, Cl −1, O 0)

Cl is reduced and O is oxidised.

(b) $$(NH_4)_2Cr_2O_7 \longrightarrow N_2 + Cr_2O_3$$

(N −3, Cr +6 ⟶ N 0, Cr +3)

N is oxidised and Cr is reduced.

(c) $$PCl_5 \longrightarrow PCl_3 + Cl_2$$

(P +5, Cl −1 ⟶ P +3, Cl 0)

P is reduced and Cl is oxidised.

**16.** (d) Oxidising power of $F_2 > Cl_2 > Br_2 > I_2$

Thus, $2KI + Cl_2 \longrightarrow 2KCl + I_2$ (violet in $CHCl_3$)

$Cl_2$ oxidises KI to $I_2$

Thus, (I) is true.

$2KBr + Cl_2 \longrightarrow 2KCl + Br_2$ (orange in $CHCl_3$)

$Cl_2$ oxidises KBr to $Br_2$

Thus, (II) is true.

$$2KBr + Cl_2 \longrightarrow 2KCl + Br_2$$
$$2KI + Br_2 \longrightarrow 2KBr + I_2$$
$$2KI + Cl_2 \longrightarrow 2KCl + I_2$$

In experiment (III), orange colour of $Br_2$ does not appear, as it also oxidises KI to $I_2$. Hence, only violet colour appears in $CHCl_3$ layer. Thus, (III) is also true.

**17.** (b) $$XeO_6^{4-} + F^- \longrightarrow XeO_3 + F_2$$

(Xe (+8), F (−1) ⟶ Xe +6, F 0)

Since, $\Delta G° < 0$, hence it is spontaneous in forward direction.

Oxidation number of Xe decreases hence, it is an oxidising agent and oxidation number of F increases, hence it is a reducing agent.

**18.** (c)

$$Fe^{2+} + Cr_2O_7^{2-} + H^+ \longrightarrow Fe^{3+} + 2Cr^{3+} + H_2O$$

(Fe +2 ⟶ +3, +1e⁻; Cr +12 ⟶ +6, +6e⁻)

To **balance** a redox reaction, balance effected atoms

I. $Fe^{2+} \longrightarrow Fe^{3+}$ (Fe –balanced)

II. Take oxidation number of two Cr-atom

$[\underset{12}{Cr_2O_7^{2-}} \longrightarrow \underset{6}{2Cr^{3+}}]$ (Cr – balanced)

Take total (oxidation number) of each atom (oxidised, reduced) cross-multiply by change in oxidation number and add

$$I \times 6 + II \times 1$$

$$6Fe^{2+} + Cr_2O_7^{2-} \longrightarrow 6Fe^{3+} + 2Cr^{3+}$$

Balance O, by $H_2O \longrightarrow ... + 7H_2O$

Balance H, by $H^+$ ... + $14H^+$

Thus, $x = 6$

$y = 1$

$z = 14$

**19.** (b) $$Fe^{2+} + 2CrO_2^- + O_2 \longrightarrow Fe_2O_3 + CrO_4^{2-}$$

(Fe +2, Cr +3, O 0 ⟶ Fe +3, O −2, Cr +6)

$Fe^{2+}$ and $Cr^{3+}$ are oxidised

(increase in oxidation number)

$O_2$ is reduced.

**20.** (d) A species which is oxidised (deelectronation) is a reducing agent.

(i) Electronation — $H_2O_2$ is reduced

(ii) Deelectronation — $H_2O_2$ is oxidised

(iii) Electronation — $H_2O_2$ is reduced

(iv) Deelectronation — $H_2O_2$ is oxidised

Thus, (ii) and (iv) represent reducing behaviour of $H_2O_2$.

**21.** (a, c)

(a) $$2NH_4^+ + Cr_2O_7^{2-} \longrightarrow N_2 + Cr_2O_3$$

(N −3, Cr +6 ⟶ N 0, Cr +3)

N of $NH_4^+$ is oxidised to $N_2$ and Cr of $Cr_2O_7^{2-}$ is reduced to Cr.

Both elements are present in the same substance. Such reaction is called intramolecular redox reaction.

(c) $NH_4^+$ is reducing agent and reduces $Cr_2O_7^{2-}$ to $Cr^{3+}$, in turn $Cr_2O_7^{2-}$ is an oxidising agent and reduced to $Cr^{3+}$.

**22.** (a, b, c, d)

(a) Zn is oxidised to $Zn^{2+}$ in complex

$[Zn(CN)_4]^{2-}$—True

(b) $Au^+$ is reduced to Au—True

(c) Au (*s*) is displaced

(d) Since, it is a spontaneous reaction $\Delta G° < 0$

**23.** (a, b)

$$\underset{\text{Excess}}{\overset{(R)}{C}} + \overset{(O)}{\frac{1}{2}O_2} \longrightarrow CO$$

(Oxidation number of C = +2)

Carbon is reducing agent (R) and oxygen is oxidising agent (O).

CO with oxidation number of C (= + 2) is formed.

(R) is in excess hence, product of lower (oxidation number) is formed.

Thus, (a) is correct.

$$\overset{(R)}{C} + \underset{\text{Excess}}{\overset{(O)}{O_2}} \longrightarrow CO_2$$

(Oxidation number of C = +4)

(O) is in excess hence product of higher oxidation number is formed. Thus, (b) is also correct.

**24.** (b, d) $\underset{-1}{I^-} + \underset{+5}{ClO_3^-} + \underset{+6}{H_2SO_4} \longrightarrow \underset{-1}{Cl^-} + \underset{+6}{HSO_4^-} + \underset{0}{I_2}$

(Reduction: $ClO_3^- \to Cl^-$; Oxidation: $I^- \to I_2$)

To balance a redox reaction,

***Step* I** Balance effected atoms

$$\underset{-1}{I^-} \longrightarrow \frac{1}{2}\underset{0}{I_2}$$

$$\underset{+5}{ClO_3^-} \longrightarrow \underset{-1}{Cl^-}$$

***Step* II** Cross-multiply by change in oxidation number and add $H_2O$ to balance O (on RHS)

$6I^- + ClO_3^- \longrightarrow 3I_2 + Cl^- + 3H_2O$

Add $H^+$ to balance H (on LHS)

$6I^- + ClO_3^- + 6H^+ \longrightarrow 3I_2 + Cl^- + 3H_2O$

$6I^- + ClO_3^- + 3H_2SO_4 \longrightarrow 3I_2 + Cl^- + 3H_2O + 3HSO_4^- + 3H^+$

(a) Stoichiometric coefficient of $HSO_4^-$ is three and not six. (a) is incorrect.

(b) $I^-$ is oxidised to $I_2$. (b) is correct.

(c) No effect on S. (c) is incorrect.

(d) $H_2O$ is one of the products. (d) is correct.

**25.** (a, b, c, d)

(a) $\underset{+5\ -1}{PCl_5} \longrightarrow \underset{+3\ -1}{PCl_3} + \underset{0}{Cl_2}$

P is reduced and Cl is oxidised.

(b) $\underset{-3}{NH_4^+}\underset{+3}{NO_2^-} \longrightarrow \underset{+1}{N_2O} + 2H_2O$

N of $NH_4^+$ is oxidised.

N of $NO_3^-$ is reduced.

(c) $\underset{-3}{(NH_4)_2}\underset{+6}{Cr_2O_7} \longrightarrow \underset{0}{N_2} + \underset{+3}{Cr_2O_3} + H_2O$

$NH_4^+$ is oxidised.

$Cr_2O_7^{2-}$ is reduced.

(d) 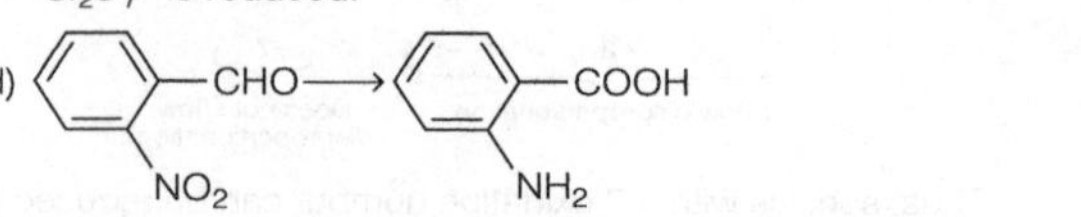


—CHO is oxidised to —COOH

—$NO_2$ is reduced to —$NH_2$

In all of the above reactions, oxidised and reduced part are in same species, hence, intramolecular redox reactions.

## DPP-3 **Equivalent in Terms of Redox Reactions** (Simple Calculations)

**1.** (c) $As_2S_3(g) + NO_3^- \longrightarrow NO + 2AsO_4^{3-} + 3SO_4^{2-}$

Oxidation number 0; (+10) (for two As); (+18) (for three S)

Increase in oxidation number of $As_2S_3 = 28 - 0 = 28$

$$\text{Equivalent weight} = \frac{M}{28}$$

**2.** (c)

$$\underset{-2}{2H_2S} + \underset{+4}{SO_2}(s) \longrightarrow \underset{0}{2S(s)} + \underset{0}{S(s)} + 2H_2O(l)$$

(Reduction: $SO_2 \to S$; Oxidation: $H_2S \to S$)

Change in oxidation number of S of $H_2S$ = 2 units

Change in oxidation number of S of $SO_2$ = 4 units

Then one mole $H_2S$ = 2 equivalents $H_2S$

One mole $SO_2$ = 4 equivalent $SO_2$

One equivalent $H_2S \equiv$ one equivalent $SO_2$ = 0.25 mole $SO_2$

**3.** (d) Oxidant is the oxidising agent (which has been reduced)

| | | | | Change in ON | Equivalent per mol |
|---|---|---|---|---|---|
| (a) | $\underset{+4}{VO^{2+}}$ | $\longrightarrow$ | $\underset{+3}{V^{3+}}$ | 1 | 1 |
| (b) | $\underset{+5}{NO_3^-}$ | $\rightarrow$ | $\underset{+4}{NO_2}$ | 1 | 1 |
| (c) | $\underset{+5}{VO_4^{3-}}$ | $\rightarrow$ | $\underset{+3}{V^{3+}}$ | 2 | 2 |
| (d) | $\underset{+5}{IO_3}$ | $\rightarrow$ | $\underset{0}{I_2}$ (in $I^-$) | 5 | 5 |

**4.** (b) (a) It is monobasic acid, hence only one $H^+$ can be neutralised

$$E = \frac{M}{1}$$

(b) $\underset{+1}{H_3PO_2} \longrightarrow \underset{+3}{H_3PO_3}$ $\quad E = \dfrac{M}{2}$

(c) $\underset{+1}{H_3PO_2} \longrightarrow \underset{+5}{H_3PO_4}$ $\quad E = \dfrac{M}{4}$

(d) $\underset{+1}{H_3PO_2} \longrightarrow \underset{-3}{PH_3}$ $\quad E = \dfrac{M}{4}$

**5.** (c) $\underset{+7}{MnO_4^-} + \underset{+4}{SO_3^{2-}} \longrightarrow \underset{+6}{SO_4^{2-}} + \underset{+6}{MnO_4^{2-}}$

($-2e^-$: $SO_3^{2-} \to SO_4^{2-}$; $+e^-$: $MnO_4^- \to MnO_4^{2-}$)

Multiply by change in oxidation number

$2MnO_4^- + SO_3^{2-} \longrightarrow SO_4^{2-} + 2MnO_4^{2-}$

Thus, $\quad 2\,MnO_4^- \equiv 1\,SO_3^{2-}$

$1\,MnO_4^- \equiv 0.5\,SO_3^{2-}$

**6.** (d) $Cr_2O_7^{2} \longrightarrow Cr^{3+}$

$\underset{+12}{Cr_2O_7^{2-}} \longrightarrow \underset{+6}{2Cr^{3+}}$ Balance Cr oxidation number of Cr-atoms on each side ($+6e^-$)

Thus, equivalent weight of

$$K_2Cr_2O_7 = \frac{\text{molar mass}}{\text{change in ON}} = \frac{M}{6}$$

**7.** (d) In a disproportionation reaction net, equivalent mass = E (oxidation part) + E (reduction part)

| | Change in ON | Equivalent weight |
|---|---|---|
| $\frac{1}{2}\underset{0}{Cl_2} \longrightarrow \underset{-1}{Cl^-}$ | 1 | $= \frac{M}{2}$ |
| $\frac{1}{2}\underset{0}{Cl_2} \longrightarrow \underset{+1}{ClO^-}$ | 1 | $= \frac{M}{2}$ |

$$\text{Net} = \frac{M}{2} + \frac{M}{2} = M = 71.0$$

**8.** (c) $\underset{1\text{ mol}}{CaOCl_2} + H_2O \longrightarrow Ca(OH)_2 + \underset{1\text{ mol}}{Cl_2}$

$Cl_2 + 2KI \longrightarrow 2KCl + \underset{1\text{ mole}}{I_2}$

$I_2 + KI \rightleftharpoons KI_3$

$\underset{1\text{ mol}}{I_2\text{ (in KI)}} + \underset{2\text{ mol}}{2Na_2S_2O_3} \longrightarrow 2NaI + Na_2S_4O_6$

**9.** (b)

| | | | ON of Mn in product | Equivalent of $MnCl_2$ |
|---|---|---|---|---|
| $MnCl_2$ | $\rightarrow$ | $MnCl_3$ | +3 | $M/1$ |
| +2 | $\rightarrow$ | $MnO_2$ | +4 | $M/2$ |
| | $\rightarrow$ | $MnO_4^{2-}$ | +6 | $M/4$ |
| | $\rightarrow$ | $MnO_4^-$ | +7 | $M/6$ |

Thus, $\frac{M}{6} < \frac{M}{4} < \frac{M}{2} < \frac{M}{1}$

IV < III < II < I

**10.** (b) Equivalent weight $= \frac{\text{molar mass}}{\text{change in ON}} = \frac{M}{2}$

Thus, change in oxidation number = 2 units

| | Change in ON | |
|---|---|---|
| (a) $\underset{+2}{MnSO_4} \longrightarrow \underset{+3}{Mn_2O_3}$ | 1 unit | $E = \frac{M}{1}$ |
| (b) $\underset{+2}{MnSO_4} \longrightarrow \underset{+4}{MnO_2}$ | 2 units | $E = \frac{M}{2}$ |
| (c) $\underset{+2}{MnSO_4} \longrightarrow \underset{+7}{MnO_4^-}$ | 5 units | $E = \frac{M}{5}$ |
| (d) $\underset{+2}{MnSO_4} \longrightarrow \underset{+6}{MnO_4^{2-}}$ | 4 units | $E = \frac{M}{4}$ |

**11.** (b) $FeSO_3 \rightleftharpoons Fe^{2+} + SO_3^{2-}$

$MnO_4^-$ oxidises $Fe^{2+}$ and $SO_3^{2-}$ both

$5Fe^{2+} + MnO_4^- + 8H^+ \longrightarrow 5Fe^{3+} + Mn^{2+} + 4H_2O$

$5SO_3^{2-} + 2MnO_4^- \longrightarrow 5SO_4^{2-} + 2Mn^{2+} + H_2O$

$5FeSO_3 + 3MnO_4^- \longrightarrow 5Fe^{3+} + 5SO_4^{2-} + 3Mn^{2+}$

$5FeSO_3 \equiv 3MnO_4^-$

$1\ FeSO_3 \equiv \frac{3}{5} MnO_4^- = 0.6$

**12.** (d) $\underset{1\,mol}{Fe_2(SO_3)_2} \rightleftharpoons 2Fe^{3+} + \underset{3\,mol}{3SO_3^{2-}}$

$Fe^{3+}$ is already in oxidised state hence $Cr_2O_7^{2-}$ is not required by $Fe^{3+}$

$SO_3^{2-}$ is oxidised to $SO_4^{2-}$ by $Cr_2O_7^{2-}$ in acidic medium

$\underset{+4}{3SO_3^{2-}} + \underset{+12}{Cr_2O_7^{2-}} + 8H^+ \longrightarrow \underset{+6}{2Cr^{3+}} + \underset{+6}{3SO_4^{2-}} + 4H_2O$

1 mole $Fe_2(SO_3)_3 \rightleftharpoons$ 3 moles $SO_3^{2-} \equiv$ 1 mole $Cr_2O_7^{2-}$

**13.** (a) Oxidation number of (six C) $\underset{24}{6CO_2} \longrightarrow \underset{0}{C_6H_{12}O_6}$

Change in oxidation number = 24 units

Hence, equivalent weight of $CO_2 = \frac{6 \times 44}{24} = 11$

and equivalent weight of $C_6H_{12}O_6 = \frac{180}{24} = 7.5$

**14.** (b)

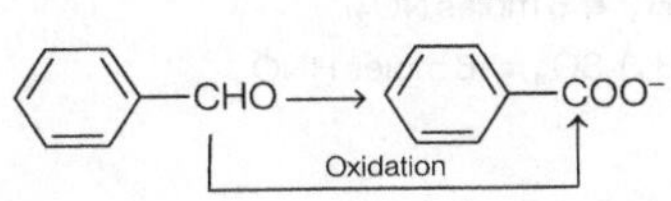


Gain of one O-atom

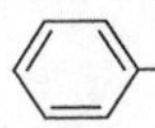 $-CHO \equiv 1(O) \equiv 16\ g \equiv 8 \times 2\ g(O)$

$\frac{M}{2} \equiv 8\ g\ (O)$

Thus, $E$ (oxidation part) $= \frac{M}{2}$

Since, $E$ (net) $= M$

Hence, $E$ (reduction part) $= \frac{M}{2}$

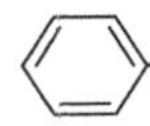 $-CHO \equiv 2H$

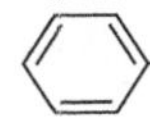 $-CHO + 2H \longrightarrow$ 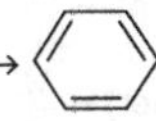 $-CH_2OH$

$\therefore$ $E$ (reduction) $= \frac{M}{2}$

Hence, $A$ is 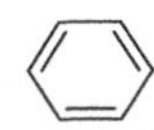 $-CH_2OH$.

**15.** (c) Equivalent weight $= \frac{\text{molar mass}}{\text{change in oxidation number}}$

| | Change in oxidation number | Equivalent weight |
|---|---|---|
| $\underset{7}{MnO_4^-} \rightarrow \underset{2}{Mn^{2+}}$ | 5 | $\frac{M}{5}$ |
| $\underset{7}{MnO_4^-} \rightarrow \underset{6}{MnO_4^{2-}}$ | 1 | $\frac{M}{1}$ |
| $\underset{7}{MnO_4^-} \rightarrow \underset{4}{MnO_2}$ | 3 | $\frac{M}{3}$ |

**16.** (a, b, c, d)

(a) Sulphate (IV) ion is $SO_3^{2-}$. It reduces $CrO_4^{2-}$ (chromate (VI)) to $Cr^{3+}$ (green)

$\underset{\substack{+4 \\ \text{Sulphate (IV)}}}{3SO_3^{2-}} + \underset{+6}{2CrO_4^{2-}} \longrightarrow \underset{+3}{2Cr^{3+}} + \underset{\substack{+6 \\ \text{Sulphate (VI)}}}{3SO_4^{2-}}$

Thus, (a) is correct.

(b) Equivalent mass of chromate $= \frac{M}{\text{change in ON}} = \frac{M}{3}$

(c) Equivalent mass of sulphate $= \frac{M}{2}$

(d) $SO_3^{2-}$ is oxidised to $SO_4^{2-}$

**17.** (a, b)

(a) $\frac{1}{2}\underset{0}{Cl_2} \longrightarrow \underset{-1}{Cl^-}$

$E$ (reduction) $= \frac{M}{2}$

$\frac{1}{2}\underset{0}{Cl_2} \longrightarrow \underset{+1}{ClO^-}$

$E$ (oxidation) $= \frac{M}{2}$

$E$ (net) $= \frac{M}{2} + \frac{M}{2} = M$

Thus, true.

(b) $H\underset{0}{C}HO \longrightarrow H\underset{2}{C}OO^-$

$E$ (oxidation) $= \frac{M}{2}$

$H\underset{0}{C}HO \longrightarrow \underset{-2}{C}H_3OH$

$E$ (reduction) $= \frac{M}{2}$

$E$ (net) $= \frac{M}{2} + \frac{M}{2} = M$

Thus, true.

(c) $\underset{+6}{MnO_4^{2-}} \longrightarrow \underset{+4}{MnO_2}$

$$E\text{(reduction)} = \frac{M}{2}$$

$$\underset{+6}{MnO_4^{2-}} \longrightarrow \underset{+7}{MnO_4^-}$$

$$E\text{(oxidation)} = \frac{M}{1}$$

$$E\text{ (net)} = \frac{M}{2} + \frac{M}{1} = 1.5\,M$$

Thus, false.

(d) $\underset{+5}{KClO_3} \longrightarrow \underset{-1}{KCl}$

$$\text{In terms of Cl} = \frac{M}{6}$$

$$\underset{(-2\times 3)}{KClO_3} \longrightarrow \frac{3}{2}\underset{0}{O_2}$$

$$\text{In terms of O} = \frac{M}{6}$$

Thus, false.

**18.** (b) $NaHC_2O_4 \rightleftharpoons Na^+ + H^+ + C_2O_4^{2-}$

It has one $H^+ \equiv NaOH$

Thus, equivalent weight when reacted with NaOH = $M$

$$C_2O_4^{2-} \longrightarrow 2CO_2 + 2e^-$$

$$\text{Equivalent weight} = \frac{M}{\text{exchange of electrons}} = \frac{M}{2}$$

**Explanations for Q. Nos. (19-20)**

$$\underset{-8}{C_3H_8} \longrightarrow \underset{+6}{3CO}$$

(oxidation number of C taking 3C-atoms together)

Change in oxidation number of C = 14 units

$$\text{Equivalent weight of } C_3H_8 = \frac{44}{14}$$

$$\underset{-8}{C_3H_8} \longrightarrow \underset{+12}{3CO_2}$$

(oxidation number of C taking 3C-atoms together)

Change in oxidation number of C = 20 units

$$\text{Equivalent weight of } C_3H_8 = \frac{44}{20}$$

$$\text{Difference} = \frac{44}{14} - \frac{44}{20} = 0.9428$$

(a) **Oxidation**

$$C_3H_8 + 6H_2O \longrightarrow 3CO_2 + 20OH^+ + 20\,e^-$$

**Reduction**

$$O_2 + 4H^+ + 4e^- \longrightarrow 2H_2O$$

Thus, (a) is correct.

(b) From (a), it is clear that (20) electrons are exchanged. Multiply reduction part by (5) and add

$$C_3H_8 + 5O_2 \longrightarrow 3CO_2 + 4H_2O$$

Thus, (c) is correct.

(c) Equivalent mass of

$$C_3H_8 = \frac{\text{molar mass}}{\text{electrons}} = \frac{44}{20} = 2.2$$

$$\text{Equivalent mass of } O_2 = \frac{32}{4} = 8$$

Thus, (c) is true.

**19.** (b) **20.** (d)

**21.** (c)

(i)

$$\underset{+3}{4H_3PO_3} \xrightarrow{\Delta} \underset{+5}{3H_3PO_4} + \underset{-3}{PH_3}$$

(Oxidation: +3 → +5; Reduction: +3 → −3)

$$E\text{ (oxidation)} = \frac{M}{2} = \frac{3M}{6}$$

$$E\text{ (reduction)} = \frac{M}{6}$$

$$E\text{ (net)} = \frac{3M}{6} + \frac{M}{6} = \frac{4M}{6} = \frac{2}{3}M$$

Thus, **(i) → (s)**

(ii) $H_3PO_3 \equiv 1\,OH^- \equiv 1\,H^+$

$$E(H_3PO_3) = M$$

Thus, **(ii) → (p)**

(iii) $H_3PO_3 \equiv 1\,Ca(OH)_2 \equiv 2OH^- \equiv 2H^+$

$$\therefore \quad 1H^+ \equiv \frac{H_3PO_3}{2}$$

$$E(H_3PO_3) = \frac{M}{2}$$

Thus, **(iii) → (q)**

(iv) $\underset{+3}{H_3PO_3} \longrightarrow \underset{+5}{H_3PO_4}$

$$E = \frac{M}{2}$$

Thus, **(iv) → (q)**

**22.** (5)

$$\underset{+2}{Mn^{2+}} \longrightarrow \underset{+7}{MnO_4^-}$$

Oxidation number of $Mn^{2+}$ changes from +2 to +7

Thus, $Mn^{2+}$ is oxidised.

$$S_2O_8^{2-} \longrightarrow SO_4^{2-}$$

In persulphate ion $S_2O_8^{2-}$, there is one peroxy linkage which is source of oxidation and reduced to oxide.

$$S_2O_8^{2-} \longrightarrow 2SO_4^{2-} + 2e^-$$

Thus, $2Mn^{2+} + 5S_2O_8^{2-} + 8\,H_2O \longrightarrow 2MnO_4^- + 10\,SO_4^{2-} + 16H+$

$SO_4^{2-}$ formed per $Mn^{2+}$ = 6

**23.** (6) $NH_4^+$ is oxidised to $N_2$ and $HNO_3$ is reduced to $NO_2$

$$\underset{-6}{2NH_4^+} + \underset{+5}{6NO_3^-} + 4H^+ \longrightarrow \underset{0}{N_2} + \underset{+4}{6NO_2} + 6H_2O$$

(for two N-atom)

2 moles $NH_4^+ \equiv$ 6 moles $NO_3^-$

1 mole $(NH_4)_2SO_4 \equiv$ 6 moles $HNO_3$

# Revisal Problems (JEE Main)

**1.** (c)

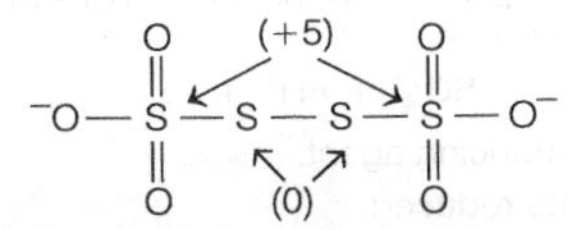


**2.** (c) Purple coloured compound $(X)$ is $MnO_4^-$

$$\underset{+4}{PbO_2} + \underset{+2}{MnO} \longrightarrow \underset{+7}{MnO_4^-} + \underset{+2}{Pb^{2+}}$$

$-5e^-$ (MnO → $MnO_4^-$); $+2e^-$ ($PbO_2$ → $Pb^{2+}$)

To balance (ON)

$$5PbO_2 + 2MnO \longrightarrow 2MnO_4^- + 5Pb^{2+}$$

Thus, $PbO_2$ and $(X)$ appear in 5 : 2 ratio.

**3.** (a) Coefficient of $SO_3^{2-}$ (product) = 6

Hence, $x = 3$ to balance S

$$\underset{ON\,=\,6}{S_2O_4^{2-}} \longrightarrow \underset{8}{2SO_3^{2-}} \quad \text{(for two S-atom)}$$

Change in ON = 2 units

Thus, to balance ON $y = 2$, $z = 2$

**4.** (d) Iron (II) oxalate is

$$\overset{II}{Fe}C_2O_4 \rightleftharpoons Fe^{2+} + C_2O_4^{2-}$$

$Fe^{2+}$ and $C_2O_4^{2-}$ both are oxidised by $MnO_4^-$ in acidic medium.

$$\underset{+2}{5Fe^{2+}} + \underset{+7}{MnO_4^-} \longrightarrow \underset{+2}{Mn^{2+}} + \underset{+3}{5Fe^{3+}}$$

$$\underset{+6\,\text{(for two C-atom)}}{5C_2O_4^{2-}} + 2MnO_4^- \longrightarrow 2Mn^{2+} + \underset{+4\,\text{(for two C-atom)}}{10CO_2}$$

Thus, $5FeC_2O_4 + 3MnO_4^- \longrightarrow 3Mn^{2+} + 5Fe^{3+} + 10CO_2$

3 moles $MnO_4^-$ oxidise = 5 moles $FeC_2O_4$

1 mole $MnO_4^-$ oxidises $= \frac{5}{3}$ moles $FeC_2O_4$

**5.** (c)

| | Change in ON | Equivalent per mol |
|---|---|---|
| (a) $\underset{+7}{MnO_4^-} \longrightarrow \underset{+4}{MnO_2}$ | 3 | 3 |
| (b) $\underset{+7}{MnO_4^-} \longrightarrow \underset{+2}{Mn^{2+}}$ | 5 | 5 |
| (c) $\underset{+7}{MnO_4^-} \longrightarrow \underset{+6}{MnO_4^{2-}}$ | 1 | 1 |
| (d) $\underset{+7}{MnO_4^-} \longrightarrow \underset{+3}{Mn_2O_3}$ | 4 | 4 |

**6.** (c) $Pb_3O_4$ is a double oxide of Pb

$$2PbO + PbO_2$$
$$2 : 1$$

PbO is also a basic oxide thus, acid-base reaction converts PbO into $PbCl_2$

$$\underset{Base}{PbO} + \underset{Acid}{2HCl} \longrightarrow PbCl_2 + H_2O$$

Out of $Pb^{4+}$ and $Pb^{2+}$, $Pb^{2+}$ is more stable than $Pb^{4+}$ (inert-pair effect) hence

$Pb^{4+} + 2e^- \longrightarrow Pb^{2+}$ is spontaneous. Thus, $PbO_2$ oxidises HCl to $Cl_2$.

$PbO_2 + 4HCl \longrightarrow PbCl_2 + Cl_2 + 2H_2O$

Thus, (a) is correct.

When $HNO_3$ reacts with $Pb_3O_4$, there is normal acid-base reaction

$$\underset{Base}{PbO} + \underset{Acid}{2HNO_3} \longrightarrow Pb(NO_3)_2 + H_2O$$

$HNO_3$ is also the oxidising agent, hence being competitive with $HNO_3$, $PbO_2$ separates without any change.

Thus, (b) is also correct.

It is the passive nature of $PbO_2$ against $HNO_3$ that makes the reaction different from the one that follows with HCl.

**7.** (b) $\underset{+5}{HNO_3} \xrightarrow{+4e^-} \underset{+1}{(X)}$

Species $(X)$ has O.N. = + 1

(a) $N_2$, ON = 0

(b) $N_2O$, ON = + 1

(c) $NO_2$, ON = + 4

(d) $NH_3$, ON = − 3

Thus, 0.5 mole $N_2O$ is formed.

**8.** (b) If $\Delta G° < 0$, that reaction is thermodynamically feasible (spontaneous in forward side)

Based on reaction I

$$\Delta G° \,(CO) = -\frac{460}{2}$$
$$= -230 \text{ kJ mol}^{-1}$$

Based on reaction II

$$\Delta G° \,(ZnO) = -\frac{360}{2}$$
$$= -180 \text{ kJ mol}^{-1}$$

$\Delta G°$ (element) = zero

Thus, $\Delta G°$ of $ZnO + C \longrightarrow Zn + CO$ is

$$\Delta G° = \Delta G° \,(CO) - \Delta G° \,(ZnO)$$
$$= -230 - (-180)$$
$$= -50 \text{ kJ mol}^{-1}$$

Reduced by carbon.

**9.** (c) Ba Xe O

Atoms $n$ 1 6

$$2n + 8 - 12 = 0$$
$$2n = 4$$
$$n = 2$$

Thus, compound is $Ba_2XeO_6$.

**10.** (d

I. $(NH_3)_3Cr^{IV}O_x$

$0 + 4 - x = 0$

$x = 4$

(in peroxide, O has ON = − 1)

II. $K_3\overset{V}{Cr}O_y$

$3 + 5 - y = 0$

$y = 8$

I is $(NH_3)_3Cr(O_2)_2$

II is $K_3Cr(O_2)_4$

**11.** (c) (a) $\underset{Excess}{NH_3} + 3Cl_2 \longrightarrow \underset{+3}{NCl_3} + 3HCl$, **True**

(b) $\underset{Excess}{8NH_3} + 3Cl_2 \longrightarrow N_2 + 6NH_4Cl$, **True**

**12.** (b) Compound is neutral species, hence sum of oxidation number of all the elements based on formula should be zero.

$$A_3(BC_4)_2$$
$$+6 + [5 + (-8)] \times 2$$
$$6 + (-3) \times 2 = 0$$

13. (c) $[\overset{III}{Cr}(O_2)(NH_3)_4(H_2O)]Cl_2$

$$\underset{+3}{Cr} + \underset{x}{(O_2)} + \underset{0}{(NH_3)_4} + \underset{0}{(H_2O)} + \underset{-2}{Cl_2} = 0$$

$$x = -1$$

Thus, $O_2$ carries (−1) change.
It is $O_2^-$ (superoxide ion) $-2e^-$

14. (a) $FeCr_2O_4$ is $\underset{+2}{FeO}.\underset{+3}{Cr_2O_3}$

15. (a) $2\underset{-1}{HI} + H_2\underset{+6}{S}O_4 \longrightarrow \underset{+4}{S}O_2 + \underset{0}{I_2}$ (Reduction: +6 → +4; oxidation: −1 → 0)

$H_2SO_4$ is reduced to $SO_2$ and $I^-$ is oxidised to $I_2$.

(b) $\underset{\text{Base}}{Ca(OH)_2} + \underset{\text{Acid}}{H_2SO_4} \longrightarrow CaSO_4 + 2H_2O$

It is neutralisation reaction.

(c) Decomposition reaction

(d) Decomposition reaction

16. (a) $S_8$ is represented as, each sulphur is joined to two sulphur atoms

Thus, covalency = 2

Oxidation number = 0

17. (d) A species which is oxidised (de-electronation) is a reducing agent.

(1) Electronation — $H_2O_2$ is reduced
(2) De-electronation — $H_2O_2$ is oxidised
(3) Electronation — $H_2O_2$ is reduced
(4) De-electronation — $H_2O_2$ is oxidised

Thus, (2) and (4) represent reducing behaviour of $H_2O_2$.

18. (b) ON of S in $SO_2$ = + 4

It is the intermediate of maximum ON (+6) and minimum ON (−2). Thus, $SO_2$ can be oxidised.

$SO_2 + 2H_2O \longrightarrow SO_4^{2-} + 4H^+ + 2e^-$

and thus is a reducing agent.

and $SO_2$ can be reduced.

$SO_2 + 6H^+ + 6e^- \longrightarrow H_2S + 2H_2O$

Thus, it is also the oxidising agent.

Similarly, $H_2O_2$ can be oxidised and reduced and thus, has dual nature.

$$H_2O_2 \longrightarrow O_2 + 2H^+ + 2e^-$$

$$H_2O_2 + 2H^+ + 2e^- \longrightarrow 2H_2O$$

Oxidation number of oxygen is zero in $O_3$.

It can be reduced to −1, − 2 but cannot be increased beyond zero. Thus, $O_3$ is the oxidising agent.

In $HNO_3$, ON of N is +5 (maximum). Thus, it can be decreased but cannot be increased.

Thus, it is the oxidising agent only.

Thus, Statement I is correct but Statement II is not the correct explanation of Statement I.

19. (a) By isotopic labelling, it is found that all the $O_2$ is from $H_2O$. Hence, to get six $O_2$, coefficients of $H_2O$ is taken as 12 and not as 6.

Thus, Statements I and II both are correct and Statement II is the correct explanation of Statement I.

20. (a) $Pb_3O_4$ is writen as $2PbO.PbO_2$

Thus, it is a double oxide of lead.

Thus, Statement I is correct

$$Pb_3O_4 + 4HNO_3 \longrightarrow PbO_2 + 2Pb(NO_3)_2 + 2H_2O$$

$PbO_2(+4)$ and $Pb^{2+}$ (+2) appear on reaction with $HNO_3$.

Thus, Statement II is correct and is the correct explanation of Statement I.

## Revisal Problems (JEE Advanced)

1. (a) $\underset{-3}{N}H_4^+ + \underset{+5}{N}O_3^- \longrightarrow \underset{0}{N_2}$

Oxidation number of N in $NH_4^+$ changes from −3 to 0 and that of N in $NO_3^-$ changes from +5 to 0.

Thus, different N-atoms are involved. Hence, N does not undergo disproportionation.

(b) $\underset{+1}{Cu_2O} \longrightarrow \underset{-2}{Cu} + \underset{+2}{CuO}$

(c) $\underset{-1}{H_2O_2} \longrightarrow \underset{-2}{H_2O} + \underset{0}{O_2}$

(d) $\underset{0}{S} \longrightarrow \underset{+4}{S}O_3^{2-} + \underset{-2}{S^{2-}}$

2. (b) $\underset{0}{Zn}(s) \longrightarrow \underset{+2}{Zn^{+2}}(aq) + 2e^-$

Zn (s) has been oxidised and causes reduction of $NO_3^-$. Thus, it is reductant

$\underset{+5}{N}O_3^-(aq) + 8e^- \longrightarrow \underset{-3}{N}H_4^+(aq)$

$NO_3^-$ has been reduced and causes oxidation of Zn.

Thus, it is oxidant.

3. (c) Molar mass of $M(Br_2) = 160$

$$\frac{1}{2}\underset{0}{Br_2} \longrightarrow \underset{-1}{Br^-}$$

$$E\text{ (reduction)} = \frac{M}{2}$$

$$\frac{1}{2}\underset{0}{Br_2} \longrightarrow \underset{y}{(X)}$$

(ON of Br in $X$) = $y$

$$E\text{ (oxidation)} = \frac{M}{2y}$$

$$E\text{ (net)} = \frac{M}{2} + \frac{M}{2y} = 96$$

$$\frac{160}{2} + \frac{160}{2y} = 96$$

$$\frac{80}{y} = 16$$

$$y = 5$$

Thus, ON of Br in oxidised part $X$ = 5

Thus, $X$ is $BrO_3^-$.

**Note.** *To determine equivalent weight* = $\dfrac{\textit{formula weight}}{\textit{change in ON}}$

$$\frac{1}{2}\underset{\substack{\uparrow 0\\ \text{Half formula weight}}}{Br_2} \longrightarrow \underset{-1}{Br^-} \quad \text{Balance affected atom}$$

Thus, we take $(M/2)$, $E = \frac{M/2}{1}$

or $$\underset{0}{Br_2} \longrightarrow \underset{-2}{2Br^-}$$

$$0 \qquad -2 \quad \text{(for two Br atoms)}$$

$$E = \frac{M}{2}$$

**4.** (a) $\underset{+3\ -2}{As_2S_3} + 10\underset{+5}{NO_3^-} + 4H^+ \longrightarrow 2\underset{+5}{AsO_4^{3-}} + 3\underset{0}{S} + 10\underset{+4}{NO_2} + 2H_2O$

Thus, $As^{3+}$ is oxidised to $AsO_4^{3-}$.

$S^{2-}$ is oxidised to S.

$NO_3^-$ is reduced to $NO_2$.

**5.** (c) $\underset{\substack{-4\\ \text{ON of two N-atom in } N_2H_4}}{N_2H_4} \longrightarrow \underset{\substack{+6\\ \text{ON of two N-atomic } Y}}{Y} + 10e^-$

$N_2H_4$ loses 10 electrons, hence oxidation number of two N-atoms is $Y = +6$

Thus, oxidation state of N in $Y = +3$

**Note** *Oxidation number per atom is called oxidation state.*

**6.** (c) This reaction takes place during metallurgical extraction of Cu from sulphide ore.

$$2Cu_2S + 3O_2 \xrightarrow{\Delta} 2Cu_2O + 2SO_2$$
$$2Cu_2O + Cu_2S \xrightarrow{\Delta} 6Cu(l) + SO_2$$
$$3Cu_2S + 3O_2 \longrightarrow 6Cu(l) + 3SO_2$$

or $$Cu_2S + O_2 \longrightarrow 2Cu(l) + SO_2$$

**7.** (d) 1 mole NaOH ≡ 3 moles $H^+$

Pure salt has $3H^+$ per mole salt

Thus, 4.62 moles NaOH ≡ $\frac{4.62}{3}$ moles salt

= 1.54 moles salt = 3.08 moles oxalate

(Salt has two $C_2O_4^{2-}$ per mole)

$$2MnO_4^- + 5C_2O_4^{2-} + H^+ \longrightarrow 2Mn^{2+} + 10CO_2$$

5 moles $C_2O_4^{2-}$ ≡ 2 moles $MnO_4^-$

$\therefore$ 3.08 moles $C_2O_4^{2-} = \frac{2}{5} \times 3.08$

= 1.232 moles $MnO_4^-$

**8.** (a, d) $2H^+ + 2e^- \longrightarrow H_2$

ON $\quad +1 \qquad\qquad 0$

ON decreases, electrons are gained thus, reduction.

**9.** (b, d)

(b) $\underset{+1}{Cu_2O} \longrightarrow \underset{0}{Cu} + \underset{+2}{Cu^{2+}}$ (Reduction: +1 → 0; Oxidadtion: +1 → +2)

(d) $H\underset{+1}{Cu}Cl_2 \longrightarrow \underset{0}{Cu} + \underset{+2}{Cu^{2+}}$

A reaction in which same atom is oxidised as well as reduced is a disproportionation reaction (auto oxidation-auto reduction).

**10.** (a,b,d) Oxidising power of $F_2 > Cl_2 > Br_2 > I_2$

Thus,

(a) Weak oxidising agent $I_2$ oxidises $S_2O_3^{2-}$ (ON of S = + 2) to $S_4O_6^{2-}$ (ON of S = 2.5)

(b) and better oxidising agent $Br_2$ oxidises $S_2O_3^{2-}$ to $SO_4^{2-}$ (ON of S = + 6)

Thus, true.

(c) $I_2$ cannot oxidise $Br^-$ to $Br_2$ being least oxidising agent. Thus, it is not feasible.

(d) $Cl_2$ is better oxidising agent thus, $I^-$ is oxidised. Thus, feasible.

**11.** (a, b, c)

(a) $\underset{0}{Cl_2} + 2KOH \xrightarrow{\text{Cold}} K\underset{-1}{Cl} + K\underset{+1}{ClO} + H_2O$

$Cl_2$ is reduced to $Cl^-$ thus, $Cl_2$ is an oxidising agent.

$Cl_2$ is oxidised to $ClO^-$ thus, $Cl_2$ is a reducing agent.

(b) $H_2O_2 \longrightarrow O_2 + 2H^+ + 2e^-$

$H_2O_2$ is oxidised to $O_2$ thus, it is a reducing agent.

$$H_2O_2 + 2H^+ + 2e^- \longrightarrow 2H_2O$$

$H_2O_2$ is reduced to $H_2O$ thus, it is an oxidising agent.

(c) $\underset{+3}{NO_2^-} + H_2O \longrightarrow \underset{+5}{NO_3^-} + 2H^+ + 2e^-$

$NO_2^-$ is oxidised to $NO_3^-$ thus, $NO_2^-$ is a reducing agent

$$\underset{+3}{NO_2^-} + 2H^+ + 2e^- \longrightarrow \underset{+2}{NO} + H_2O$$

$NO_2^-$ is reduced to NO thus, it is an oxidising agent.

(d) Maximum oxidation state of S is 6 as in $H_2SO_4$ thus, $H_2SO_4$ can be reduced and cannot be oxidised.

Thus, it is only the oxidising agent.

**12.** (a, b, c, d) The corrosion reaction can be written as

$$2Cu + H_2O + CO_2 + O_2 \longrightarrow Cu_2(OH)_2CO_3$$

It involves oxidation of Cu to $Cu^{2+}$ (as CuO) reduction of $O_2$ to $OH^-$.

acid base reaction, $\underset{\text{Base}}{CuO} + \underset{\text{Acid}}{H_2CO_3}$ ($CO_2$ in $H_2O$) $\longrightarrow CuCO_3$

precipitation of basic carbonate

$$Cu(OH)_2 + CuCO_3 \longrightarrow Cu_2(OH)_2CO_3$$

**13.** (a, c) $CN^-$ is a reducing agent. When added to $Cu^{2+}$, $Cu^+$ is formed.

$$CuCl_2 + 2KCN \longrightarrow Cu(CN)_2 + 2KCl$$

$$2Cu(CN)_2 \longrightarrow 2CuCN + \underset{\text{Cyanogen}}{(CN)_2}$$

CuCN changes to soluble $[Cu(CN)_4]^{3-}$ by complex formation.

$$CuCN + 3KCN \longrightarrow K_3[Cu(CN)_4]$$

Thus, $CN^-$ is also a complexing agent.

**14.** (a, d)

(a) $2Br^-(aq) + Cl_2(g) \longrightarrow 2Cl^-(aq) + Br_2(aq)$

$Br_2(aq)$ is soluble in $CS_2$ thus, extracted into it.

(d) $I^-(aq) + 3Cl_2(g) + 3H_2O \longrightarrow 6Cl^-(aq) + IO_3^-(aq) + 6H^+(aq)$

**15.** (c)

$$\underset{+4}{BrO_2} \longrightarrow \underset{+1}{Br_2} + A \quad \text{(Reduction)}$$

Change in ON when $Br_2O$ is formed from $BrO_2$ = 3 units

Hence, change in ON when $A$ is formed from $BrO_2$ by oxidation = 3 units

Thus, $Br_2O_7$

**16.** (c) $Br^-$ is oxidised to $Br_2$ and

$BrO_3^-$ is reduced to $Br_2$

Thus, (c) is correct.

Thus, (a) and (b) are incorrect.

$$\underset{-1}{Br^-} \longrightarrow \frac{1}{2}\underset{0}{Br_2} \qquad E_{ox} = \frac{M}{2} = \frac{5M}{10}$$

$$\underset{+5}{BrO_3^-} \longrightarrow \frac{1}{2}\underset{0}{Br_2} \qquad E_{ox} = \frac{M}{10}$$

$$E_{net} = \frac{5M}{10} + \frac{M}{10} = 0.6\,m$$

Thus, (d) is incorrect.

**17.** (a,b,c)

(a) $NO \xleftarrow{1} HNO_2 \xrightarrow{2} HNO_3$ Yes
$\uparrow$ +2, $\uparrow$ +3, $\uparrow$ +5

(b) $HNO_3 \xleftarrow{3} NO_2 \xrightarrow{4} NO_3^-$ Yes
+3, +4, +5

(c) $NO^+ \xleftarrow{8} N_2O_4 \xrightarrow{9} NO_3^-$ Yes
+3, +4, +5

(d) $NO^+ \xleftarrow{8} N_2O_4 \xrightarrow{10} NO_2^-$ No
+3, +4, +3

**18.** (a,d)

$Br^- + 4OH^- \longrightarrow BrO_2^- + 2H_2O + 4e^-$
$\uparrow$ −1, $\uparrow$ +3

Equivalent mass of $Br^-$ $= \dfrac{\text{atomic weight}}{\text{change in ON}} = \dfrac{80}{4} = 20$

Thus, (a) and (d)

**Explanations for Q. Nos. (19-20)**

$Cr_2O_7^{2-}$ (orange) is reduced to $Cr^{3+}$ (green) by Zn in acidic medium

$\underset{\text{Orange}}{Cr_2O_7^{2-}} + 3Zn + 14H^+ \longrightarrow \underset{\text{Green}}{2Cr^{3+}} + 7H_2O + 3Zn^{2+}$

$\underset{\text{Green}}{2Cr^{3+}} + Zn \longrightarrow Zn^{2+} + \underset{\text{Blue}}{2Cr^{2+}}$

$\underset{\text{Blue}}{Cr^{2+}} \xrightarrow{\text{Air}} \underset{\text{Green}}{Cr^{3+}} + e^-$

**19.** (c) **20.** (a)

**21.** (d)

(i) $N_2$ is oxidised and $O_2$ is reduced a combination redox
(ii) Decomposition redox
(iii) $H_2$ is formed by displacement — displacement redox
(iv) Disproportionation of $NO_2$ to $NO_3^-$ (by oxidation) and to $NO_2^-$ (by reduction).
(v) Disproportionation of $S_8$ to $S_2O_3^{2-}$ (by oxidation) and to $S^{2-}$ (by reduction)

(i) → (t), (ii) → (s),
(iii) → (r), (iv) → (q),
(v) → (q)

**22.** (a)

(i) Iron (III) oxalate is ferric oxalate

$Fe_2(C_2O_4)_3 \rightleftharpoons 2Fe^{3+} + 3C_2O_4^{2-}$

$Fe^{3+}$ is already in oxidised state and cannot be oxidised further. $KMnO_4$ is required by $C_2O_4^{2-}$ only.

$2MnO_4^- + 5C_2O_4^{2-} \longrightarrow 2Mn^{2+} + 10CO_2$

2 moles $MnO_4^-$ oxidise = 5 moles $C_2O_4^{2-}$

1 mole $MnO_4^-$ oxidises = 2.5 moles $C_2O_4^{2-}$

$= \dfrac{2.5}{3}$ moles of $Fe_2(C_2O_4)_3 = 0.83$ mole

Thus, **(i) → (r)**

(ii) Iron (II) sulphate (IV) is ferrous sulphite $FeSO_3$ (Fe +2, S +4)

$FeSO_3 \rightleftharpoons Fe^{2+} + SO_3^{2-}$

Both ions are oxidised by $KMnO_4$ in acidic medium

$5Fe^{2+} + MnO_4^- \longrightarrow Mn^{2+} + 5Fe^{3+}$

$5SO_3^{2-} + 2MnO_4^- \longrightarrow Mn^{2+} + SO_4^{2-}$

$5FeSO_3 \equiv 3MnO_4^-$

1 mole $MnO_4^- \equiv \dfrac{5}{3}$ mole $FeSO_3 = 1.67$ moles

Thus, **(ii) → (s)**

(iii) Iron (II) oxalate is ferrous oxalate $FeC_2O_4$

$FeC_2O_4 \rightleftharpoons Fe^{2+} + C_2O_4^{2-}$

Both ions are oxidised by $MnO_4^-$ in acidic medium.

$2MnO_4^- + 5C_2O_4^{2-} \longrightarrow$

$MnO_4^- + 5Fe^{2+} \longrightarrow 3MnO_4^- \equiv 5FeC_2O_4$

$1\ MnO_4^- \equiv \dfrac{5}{3} FeC_2O_4 = 1.67$ moles

Thus, **(iii) → (s)**

(iv) Iron (III) sulphate (VI) is ferric sulphate

$Fe_2(SO_4)_3 \rightleftharpoons 2Fe^{3+} + 3SO_4^{2-}$

Both ions are in oxidised state and no further reaction with $MnO_4^-$ is possible. Hence, $KMnO_4$ is not used.

Thus, **(iv) → (q)**

(v) Iron (III) sulphate (IV) is ferric sulphite $Fe_2(SO_3)_3$

$Fe_2(SO_3)_3 \rightleftharpoons 2Fe^{3+} + 3SO_3^{2-}$

Only $SO_3^{2-}$ is oxidised.

$5SO_3^{2-} + 2MnO_4^- \longrightarrow 2Mn^{2+} + 5SO_4^{2-}$

$2MnO_4^-$ oxidise = 5 moles $SO_3^{2-} = \dfrac{5}{3}$ moles of $Fe_2(SO_3)_3$

$1\ MnO_4^-$ oxidises $= \dfrac{5}{6}$ mole of $Fe_2(SO_3)_3 = 0.83$ mole

Thus, **(v) → (s)**

**23.** (b)

(i) $\underset{\text{Peroxy disulphuric acid}}{H_2S_2O_8} + H_2O \longrightarrow H_2O_2 + 2H_2SO_4$

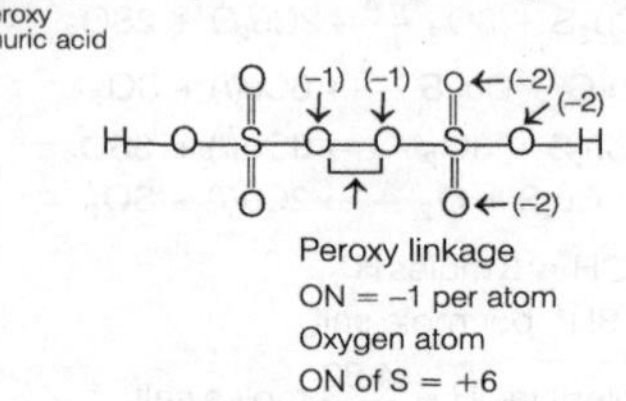


$H_2S_2O_8$ is called Marshall's acid.

Thus, **(i) → (p,q,r,s)**

(ii)

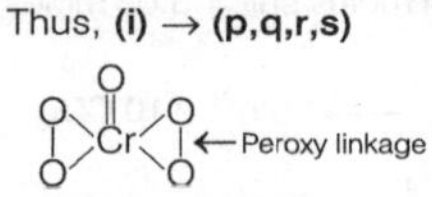


ON of Cr = + 6

Thus, **(ii) → (p,q)**

(iii) $H_2SO_5$

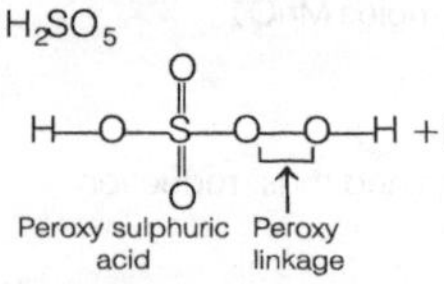


$+ H_2O \longrightarrow H_2SO_4 + H_2O_2$

ON of S = + 6

It is called Caro's acid.

Then, **(iii) → (p,q,r,s)**

(iv) $(NH_4)_2Cr_2O_7 \rightleftharpoons 2NH_4^+ + Cr_2O_7^{2-}$

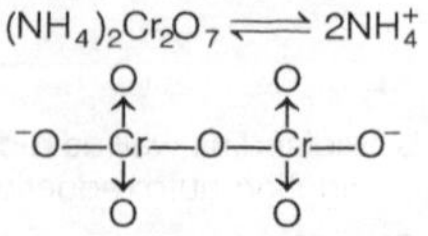

ON of Cr = + 6

Thus, **(iv) → (q)**

**24.** (2) $\underset{+n}{A^{n+}} + \underset{+7}{MnO_4^-} \longrightarrow \underset{+2}{Mn^{2+}} + \underset{+5}{AO_3^-}$

ON — Reduction (+7 → +2); Oxidation (+n → +5)

Increase in ON = (5 − n) units

To balance (ON), cross-multiply by change in ON

$$5A^{n+} + (5-n)\,MnO_4^- \longrightarrow (5-n)\,Mn^{2+} + 5AO_3^-$$

$$\frac{\text{Moles of } A^{n+}}{\text{Moles of } MnO_4^-} = \frac{5}{(5-n)} = \frac{2.68 \times 10^{-3}}{1.61 \times 10^{-3}}$$

$$(5-n) = \frac{5 \times 1.61 \times 10^{-3}}{2.68 \times 10^{-3}} = 3$$

Thus, $n = 2$

**25.** (0) $CrO_3$ ON = + 6

$CrO_5$ has two peroxy linkages

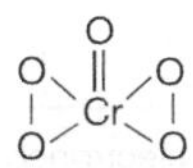

Thus,

$$\underset{x}{Cr}\ \underset{-2}{O}\ \underset{-4}{(O_2)_2}$$

$x - 2 - 4 = 0$

$x = 6$

Thus, no change.

**26.** (6)

$$\underset{+1}{ClO^-} + \underset{+3}{Cr(OH)_4^-} \longrightarrow \underset{+6}{CrO_4^{2-}} + \underset{-1}{Cl^-}$$

(Cr: +3 → +6 Oxidation; Cl: +1 → −1 Reduction)

To balance ON

$$3ClO^- + 2Cr(OH)_4^- \longrightarrow 2CrO_4^{2-} + 3Cl^-$$

$$2Cr(OH)_4^- \equiv 3ClO^-$$

$\therefore$ $4Cr(OH)_4^- \equiv 6ClO^-$

**27.** (7)

$$\underset{+5}{NO_3^-} \xrightarrow{+7e^-} \frac{1}{2}\underset{-2}{N_2H_4}$$

Thus, seven electrons are required to reduce one mole of $NO_3^-$ to $N_2H_4$.

**28.** (1) $Cl_2 \longrightarrow Cl^- + A$

$$\frac{1}{2}\underset{0}{Cl_2} \xrightarrow{\text{Reduce}} \underset{-1}{Cl}$$

Equivalent mass of reduction part $= M/2 = \frac{71}{2}$

$$\frac{1}{2}\underset{0}{Cl_2} \xrightarrow{\text{Oxidation}} A^{n+} + n$$

Equivalent mass of oxidation part $= M/2n = \frac{71}{2n}$

$$\frac{71}{2} + \frac{71}{2n} = 71 \qquad \text{(given)}$$

$$\frac{1}{2} + \frac{1}{2n} - 1$$

$$n = 1$$

## JEE Main & AIEEE Archive

**1.** (d) A species which is oxidised (de-electronation) is a reducing agent.

(I) Electronation — $H_2O_2$ is reduced

(II) De-electronation — $H_2O_2$ is oxidised

(III) Electronation — $H_2O_2$ is reduced

(IV) De-electronation — $H_2O_2$ is oxidised

Thus, (II) and (IV) represent reducing behaviour of $H_2O_2$.

**2.** (c) The half equations of the reaction are

$$MnO_4^- \longrightarrow Mn^{2+}$$

$$C_2O_4^{2-} \longrightarrow CO_2$$

The balanced half equations are

$$(MnO_4^- + 8H^+ + 5e^- \longrightarrow Mn^{2+} + 4H_2O) \times 2$$

$$(C_2O_4^{2-} \longrightarrow 2CO_2 + 2e^-) \times 5$$

Equating number of electrons, we get

$$2MnO_4^- + 16H^+ + 10e^- \longrightarrow 2Mn^{2+} + 8H_2O$$

$$5C_2O_4^{2-} \longrightarrow 10CO_2 + 10e^-$$

On adding both the equations, we get

$$2MnO_4^- + 5C_2O_4^- + 16H^+ \longrightarrow 2Mn^{2+} + 2 \times 5\,CO_2 + \frac{16}{2}H_2O$$

Thus, *x*, *y* and *z* are 2, 5 and 16, respectively.

**3.** (a) $2\overset{-1}{H}I + H_2\overset{+6}{S}O_4 \longrightarrow \overset{0}{I_2} + \overset{+4}{S}O_2 + 2H_2O$

$H_2SO_4$ — Reduced to $\longrightarrow SO_2$

Oxidising agent

**4.** (d) $[Cr(NH_3)_4Cl_2]^+$

Let the oxidation state of Cr = *x*

$NH_3 = 0$

$Cl = -1$

Net charge = + 1

$\therefore$ $[Cr(NH_3)_4Cl_2]^+$

$$x + 4 \times 0 + 2(-1) = +1$$

$\therefore$ $x = 3$

**5.** (a) $Cr_2O_7^{2-} + 14H^+ + 6I^- \longrightarrow 2Cr^{3+} + 7H_2O + 3I_2$

$Cr_2O_7^{2-}$ is reduced to $Cr^{3+}$.

Thus, final state of Cr is + 3.

**6.** (d) $CaOCl_2$ is written basically as Ca(OCl)Cl.

$OCl^- \longrightarrow$ Cl has + 1 oxidation state

$Cl^- \longrightarrow$ Cl has − 1 oxidation state

**7.** (c) $\overset{+7}{Mn}O_4^- \longrightarrow Mn^{2+}$ change 5

$\longrightarrow \overset{+6}{Mn}O_4^{2-}$ change 1 $\longrightarrow \overset{+4}{Mn}O_2$ change 3

$\longrightarrow \overset{+3}{Mn_2}O_3$ change 4

**8.** (d)

(a) $\overset{+1}{Na}\overset{-1}{Cl} + \overset{+1}{K}\overset{-1}{NO_3} \longrightarrow \overset{+1}{Na}\overset{-1}{NO_3} + \overset{+1}{K}\overset{-1}{Cl}$

(b) $\overset{+2}{Ca}\overset{-2}{C_2O_4} + 2\overset{+1}{H}\overset{-1}{Cl} \longrightarrow \overset{+2}{Ca}\overset{-1}{Cl_2} + \overset{+1}{H_2}\overset{-2}{C_2O_4}$

(c) $\overset{+2}{Ca}\overset{-1}{(OH)_2} + 2\overset{-3}{N}\overset{+1}{H_4}\overset{-1}{Cl} \longrightarrow \overset{+2}{Ca}\overset{-1}{Cl_2} + 2\overset{-3}{N}\overset{+1}{H_3} + 2\overset{+1}{H_2}\overset{-2}{O}$

In all these cases during reaction, there is no change in oxidation state of ion or molecule or constituent atom. These are simply ionic reactions.

(d) $2K[Ag(CN)_2] + Zn \longrightarrow 2Ag + K_2[Zn(CN)_4]$

$Ag^+ \longrightarrow Ag$ gain of $e^-$, reduction

$Zn \longrightarrow Zn^{2+}$ loss of $e^-$, oxidation

**9.** (c) Oxidation takes place at anode. Hence, (a), (b) is not feasible as these are reduction reactions. But $Cr^{3+}$ is oxidised to $Cr^{6+}$ ($Cr_2O_7^{2-}$) under given condition.

# 3. Chemical Bonding

## DPP-1 Ionic Bonds and Fajans' Rule

**1. Plan** In ionic compounds, there is exchange of electrons to complete duplet or octet.

Charge on cation = electrons lost

Charge on anion = electrons gained

Cations and anions are joined by electrostatic force of attraction ($F$).

$$F = k\frac{z_1 z_2 e^2}{Dr^2}$$

$k$ = constant, $z_1 e$ = charge on cation, $z_2 e$ = charge on anion, $r$ = internuclear distance, $D$ = dielectric constant of medium

(i) $Na[1s^2 2s^2 2p^6 3s^1] \longrightarrow Na^+[1s^2 2s^2 2p^6] + e^-$

One electron extra than octet, it can lose electron — Octet complete

$F[1s^2 2s^2 2p^5] + e^- \longrightarrow F^-[1s^2 2s^2 2p^6]$

One electron short of octet, it can gain electron — Octet complete

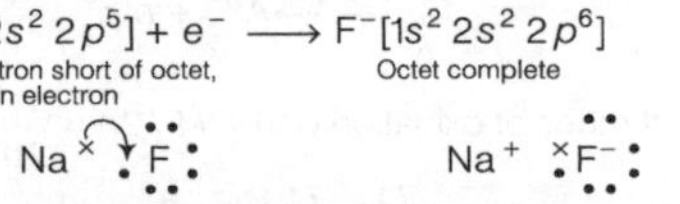

Na loses one electron hence, gains + ve charge and F gains one electron hence, gains –ve charge.

(ii) $Mg[1s^2 2s^2 2p^6 3s^2] \longrightarrow Mg^{2+}[1s^2 2s^2 2p^6] + 2e^-$

Two electrons extra than octet, it can lose two electrons — Octet complete

$Cl[1s^2 2s^2 2p^6 3s^2 3p^5] + e^- \quad Cl^-[3s^2 3p^6]$

One electron short of octet, it can gain one electron

Electrons lost by Mg are gained by two Cl atoms (one each).

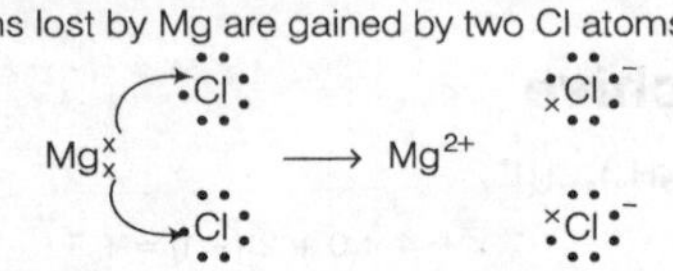

(iii) $Li[1s^2 2s^1] \longrightarrow Li^+[1s^2] + e^-$

One electron extra than duplet, it can lose one electron — Dupeet complete

$O[1s^2 2s^2 2p^4] + 2e^- \longrightarrow O^{2-}[1s^2 2s^2 2p^6]$

Two electrons short of octet, it can gain two electrons — Octet complete

O gains two electrons lost by two Li atoms.

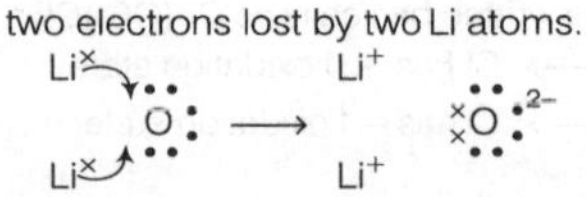

(iv) $K[\cdots 3s^2 3p^6 4s^1] \longrightarrow K^+[\cdots 3s^2 3p^6] + e^-$

One electron extra than octet, it can lose one electron — Octet complete

$Br[\ldots 4s^4 4p^5] + e^- \longrightarrow Br^-[\ldots 4s^4 4p^6]$

One electron short of octet, it can gain one electron

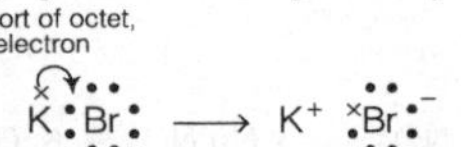

**2. Plan** Steps given (i) to (v) involve formation of lattice $M^+X^-(s)$ from $M(s)$ and $X_2(g)$ – a part of **Born-Haber cycle**.

| | | Process | $\Delta H$ |
|---|---|---|---|
| (i) | $M(s) \longrightarrow M(g)$ | Sublimation | $S$ (heat of sublimation energy) |
| (ii) | $M(g) \longrightarrow M^+(g) + e^-$ | Ionisation | $I$ (ionisation energy) |
| (iii) | $X_2(g) \longrightarrow 2X(g)$ | Dissociation | $D$ (bond energy) |
| (iv) | $X(g) + e^- \longrightarrow X^-(g)$ | Electron gain | $-E$ (electron gain energy) |
| (v) | $M^+(g) + X^-(g) \longrightarrow M^+X^-(s)$ | Lattice formation | $-U$ (lattice energy) |

$$\Delta H(\text{total}) = \left(S + I + \frac{D}{2} - E - U\right)$$

Processes (iv) and (v) are exothermic.

**3. Plan** If atomic numbers are known, then based on valence electrons, type of bonding can be decided.

| Element | Z | Group | |
|---|---|---|---|
| A | 8 | VI A (16) | Oxygen |
| B | 9 | VII A (17) | Fluorine |
| C | 10 | VIII (18) | Neon |
| D | 11 | IA (1) | Sodium |

(i) Oxygen is more electronegative and sodium is more electropositive – ionic

(ii) As in (a) – ionic

(iii) Similar elements – covalent

(iv) Similar elements – covalent

**4. Plan** Ionic bond strength is the force of attraction between cation and anion given by

$$F = k\frac{z_1 z_2 e^2}{Dr^2}$$

$z_1 = z_2$ = charge on cation and anion (+ 1 on cation and – 1 on anion)

$e$ = electronic charge

$D$ = dielectric constant of medium

$r$ = bond length

$$r_{NaCl} = r_{Na^+} + r_{Cl^-} = 0.097 + 0.181 = 0.278 \text{ nm}$$

$$r_{KF} = r_{K^+} + r_{F^-} = 0.133 + 0.133 = 0.266 \text{ nm}$$

$$F(NaCl) = \frac{ke^2}{D(0.278)^2}$$

$$F(KF) = \frac{ke^2}{D(0.266)^2}$$

$$\therefore \quad \frac{F(NaCl)}{F(KF)} = \frac{(0.266)^2}{(0.278)^2} = 0.9155$$

Thus, ionic bond strength of NaCl is 91.55% that of KF.

**5.** (c) $CsI_3 \rightleftharpoons Cs^+ + I_3^-$

$I^-$ dissolves $I_2$ forming $I_3^-$.

Cs is electropositive element hence, it forms $Cs^+$ ion. Hence, no covalent bonding.

$$\underset{[Xe]6s^2}{Cs} \longrightarrow \underset{[Xe]}{Cs^+} + e^-$$

$Cs^+$ has stable inert gas configuration thus, $Cs^{3+}$ is not formed.

**6.** (a) $Mg \longrightarrow Mg^{2+} + 2e^-; \quad \Delta H^\circ = IE$

$O + 2e^- \longrightarrow O^{2-}; \quad \Delta H^\circ = EA$

Smaller the IE of Mg, easier the formation of $Mg^{2+}$.

Greater the EA of O, easier the formation of $O^{2-}$.

**7.** (c) KCl is an ionic compound. Force of attraction ($F$) between $K^+$ and $Cl^-$ is

$$F = k\frac{z_1 z_2 e^2}{D}$$

$z_1, z_2$ = charge (± 1 for KCl)

$D$ = dielectric constant of solvent (maximum for $H_2O$)

$$F \propto \frac{1}{D}$$

Greater the value of $D$, smaller the value of $F$, hence larger the solubility.

**Note** *D for $CH_3OCH_3$ is least out of given solvents, thus solubility is minimum.*

**8.** (b) Mg($1s^2\,2s^2\,2p^6 3s^2$) is an electropositive element and Cl($1s^2\,2s^2\,2p^6\,3s^2\,3p^5$) is an electronegative element.

$$:\text{Mg}_{[\text{Ne}]3s^2} \longrightarrow \text{Mg}^{2+}_{[\text{Ne}]} + 2e^-$$

$Mg^{2+}$ attains stable inert gas configuration.

$$2\text{Cl}_{[\text{Ne}]\,3s^2\,3p^5} + 2e^- \longrightarrow 2\,[\overset{\times}{:}\text{Cl}:]^-_{[\text{Ar}]}$$

$Cl^-$ also attains stable inert gas configuration.

$$\cdot\text{Al}_{[\text{Ne}]\,3s^2 3p^1} \longrightarrow \text{Al}^{3+}_{[\text{Ne}]} + 3e^-$$

$Al^{3+}$ attains stable inert gas configuration.

$$:\text{O}:_{[1s^2 2s^2 2p^4]} + 2e^- \longrightarrow [:\text{O}:]^{2-}_{[\text{Ne}]}$$

$O^{2-}$ attains stable inert gas configuration.

**9.** (c) $\cdot\dot{M}\cdot$ + 3 (Valency)

(Three electrons extra than octet, it can donate three electrons)

$\cdot\ddot{Z}\cdot$ − 3 (Valency)

(Three electrons short of octet, it can gain three electrons)

Thus, $M^{3+}\ Z^{3-}$

Compound is $MZ$.

**10.** (d) $CuSO_4 \cdot 5H_2O$ is actually $[Cu(H_2O)_4]SO_4 \cdot H_2O$.

I. σ-bonds in $H_2O$ molecules $H \overset{\sigma}{—} O \overset{\sigma}{—} H$

II. π-bonds in $SO_4^{2-}$

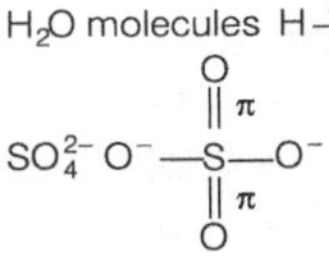

III. Coordinate bond in complex

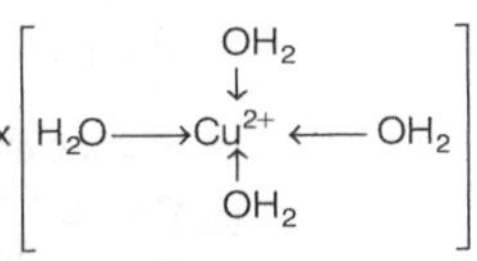

IV. Electrostatic force of attraction between $[Cu(H_2O)_4]^{2+}$ and $SO_4^{2-}$.

V. $H_2O$ molecules joined together by dipole-dipole interaction.

VI. Outer $H_2O$ molecule joined to $SO_4^{2-}$ by ion-dipole interaction.

**11.** (b) I. $K_4[Fe(CN)_6] \rightleftharpoons \underbrace{4K^+ + [Fe(CN)_6]^{4-}}_{\text{Ionic bond between cation and anion}}$

$[Fe(CN)_6]^{4-}$ has ($^-C \equiv N$) covalent bond. $CN^-$ is coordinated to $Fe^{2+}$.

II. $NH_4^+, Cl^-$ ionic,

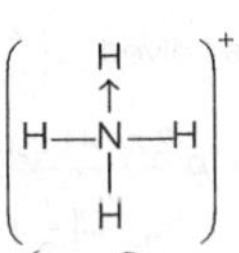

Covalent and coordinate bond.

III. $H^+, SO_4^{2-}$ ionic,

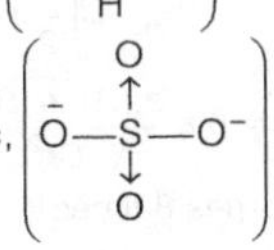

Covalent and coordinate bond.

**12.** (d) By Fajans rule,

Smaller the size of cation, larger the size of anion. Larger the charge on cation or anion, larger the polarising power and thus, greater the covalent nature.

| | |
|---|---|
| Size of cations | $Na^+ > Mg^{2+} > Al^{3+} > Si^{4+}$ |
| Charge on cations | $Na^+ < Mg^{2+} < Al^{3+} < Si^{4+}$ |
| Polarising power of cations | $Na^+ < Mg^{2+} < Al^{3+} < Si^{4+}$ |
| Size of anions | $F^- < O^{2-} < N^{3-} < C^{4-}$ |
| Charge on anions | $F^- < O^{2-} < N^{3-} < C^{4-}$ |
| Polarising power of anions | $F^- < O^{2-} < N^{3-} < C^{4-}$ |

Thus, by Fajans' rule, covalent nature is

$NaF < MgO < AlN < SiC$ (maximum)

**13.** (a) Like dissolves like. Polar solvents (as $H_2O$) dissolve polar compounds (ionic compounds) and non-polar solvents (as ether) dissolve non-polar compounds (covalent compounds).

By Fajans' rule smaller the size of cation, larger the size of anion. Larger the charge on cation or anion then larger the polarising power and thus, larger the covalent nature and thus, solubility in non-polar solvents.

In NaCl, $MgCl_2$, $AlCl_3$, $SiCl_4$. Anion is common ($Cl^-$).

Size of cations

$Na^+ > Mg^{2+} > Al^{3+} > Si^{4+}$ (smallest cation)

Charge on cations

$Na^+ < Mg^{2+} < Al^{3+} < Si^{4+}$ (largest charge)

Polarising power of cations

$Na^+ < Mg^{2+} < Al^{3+} < Si^{4+}$

Covalent nature and solubility in ether

$NaCl < MgCl_2 < AlCl_3 < SiCl_4$

**14.** (b) By Fajans' rule,

Smaller the size of cation, larger the size of anion. Larger the charge on cation or anion, larger the polarising power and thus, greater the covalent nature.

NaF, $Na_2O$, $Na_3N$, $Na^+$ is same.

Size of anion ($F^- < O^{2-} < N^{3-}$)

Charge on anion

$F^- < O^{2-} < N^{3-}$

Polarising power to polarise cation

$F^- < O^{2-} < N^{3-}$

Covalent nature

$NaF < Na_2O < Na_3N$

**15.** (a) By Fajans' rule,

Smaller the size of cation, larger the size of anion.

Larger the charge or $d$-electrons present, then larger the polarising power of cation and larger the covalent nature and smaller the solubility in water.

$S^{2-}$ (common)

| | $Na^+$ | $Zn^{2+}$ | $Cu^{2+}$ |
|---|---|---|---|
| **Charge** | +1 | <+2 | =+2 |
| **Size** | $Na^+$ | $>Cu^{2+}$ | $>Zn^{2+}$ |
| **$d$-electrons** | 0 | $d^9$ | $d^{10}$ |

$Cu^{2+}$ due to shielding effect causes more screening than $Zn^{2+}$.

Thus, polarising power of $Na^+ < Zn^{2+} < Cu^{2+}$ hence, ionic nature and solubility in water is in order

$CuS < ZnS < Na_2S$

**16.** (d) Based on Fajans' rule,

Smaller the size of cation, larger the size of anion,

larger the charge,

$d$-$f$-electrons present then larger the polarising power, larger the covalent nature and larger the lattice energy.

(a) $K^+ > Mg^{2+}$ (size)

$+1 < +2$ (charge)

$Cl^- > O^{2-}$ (size)

$-1 < -2$ (charge)

MgO > KCl (lattice energy) thus, correct.

(b) $Al^{3+} < Mg^{2+}$ (size)

$+3 > +2$ (charge)

$N^{3-} > O^{2-}$ (size)

$-3 > -2$ (charge)

AlN > MgO (lattice energy) thus, correct.

(c) $Be^{2+} < Mg^{2+}$ (size)

$CO_3^{2-}$ same, Charge same

$BeCO_3 > MgCO_3$ (lattice energy) thus, correct.

(d) Thus, incorrect.

**17.** (a) By Fajans' rule,

Smaller the size of cation, larger the size of anion, larger the charge, then larger the polarising power, larger the covalent nature, means low melting point.

Smaller the ionic nature, means low melting point

$$Li^+ \text{ (same)}$$

$$F^- < Cl^- < Br^- < I^-$$

$\longrightarrow$ Size

$\longrightarrow$ Polarising power

$\longleftarrow$ Ionic nature LiF (maximum)

$\longleftarrow$ Melting point LiF (maximum)

Hence, most stable is LiF.

**18.** (d) By Fajans' rule, smaller the size of cation, larger the size of anion, larger the charge on cation or anion,

*d*- and *f*-electrons present on the ion then greater the polarising power of the ion to polarise the other ion, thus larger the covalent nature.

$Na^+ \; Mg^{2+} \; Al^{3+} \; Si^{4+}$

Size $\xleftarrow{\text{max}}$

Charge $\xrightarrow{\text{max}}$

Polarising power $\xrightarrow{\text{max}}$

$F^- \; O^{2-} \; N^{3-} \; C^{4-}$

Size $\xrightarrow{\quad}$

Charge $\xrightarrow{\quad}$

Polarising power $\xrightarrow{\text{max}}$

Thus, compound with $Si^{4+}$ (smallest cation) and $C^{4-}$ (largest anion) will be most covalent.

**19.** (d) Polarising power of $Li^+ > Na^+$ and $Al^{3+} > Na^+$

Thus, LiCl and $AlCl_3$ have covalent nature and extracted into ether.

CuCl due to *d*-electrons has covalent nature. (Polarising power of $Cu^+ > Na^+$).

Thus, only NaCl (ionic) remains in water.

**20.** (d) $Cu^+ = [Ar]\, 3d^{10}$, $Cu^+$ has inner 18-electrons shell. Shielding effect is decreased. Thus, electronegativity is increased. This results in greater polarising power of $Cu^+$. Hence, CuCl is covalent.

**21.** (a, b, d) Born Haber cycle involves

| Process | Energy |
|---|---|
| Sublimation | $S$ |
| Ionisation | $I$ |
| Electron affinity | $-E$ |
| Lattice energy | $-U$ |

$$A(s) \longrightarrow A(g), \quad \Delta H = S$$
$$A(g) \longrightarrow A^+(g) + e^-, \quad \Delta H = I$$
$$B(g) + e^- \longrightarrow B^-(g), \quad \Delta H = -E$$
$$A^+(g) + B^-(g) \longrightarrow AB(s), \quad \Delta H = -U$$
$$A(s) + B(g) \longrightarrow A^+B^-(s), \quad \Delta H = S + I - E - U$$

If $\Delta_f H^\circ (AB) = Q$

Then $\quad Q = S + I - E - U$

Out of these, one parameter, *I*, *E*, *U* can be calculated.

**22.** (a, d)

(a) In $SbF_5$ (expansion of octet from 5 to 10 and further to 12).

(b) $NH_3$, $BF_3$ both have complete octet. $NH_3$ is electron rich, donates lone pair to $BF_3$.

(c) $AlCl_3$ is electron deficient, accepts lone pair from $Cl^-$.

(d) In $SF_4$, expansion of octet from 8 to 10 and further to 12.

**23.** (a, b) $A \longrightarrow A^+ + e^-$ is favourable since, IE of *A* is low.

$B + e^- \longrightarrow B^-$ is favourable since, EA of *B* is high.

Thus, $A^+B^-$ is favoured by low IE of *A* and high EA of *B*.

**24.** (b) 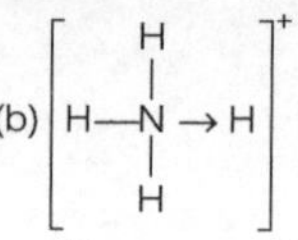 $Cl^-$ $\pi$-bond absent.

$Na^+$ 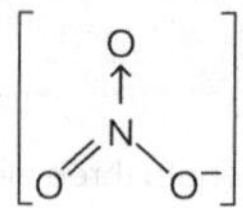 All bonds are present.

$H^+$ 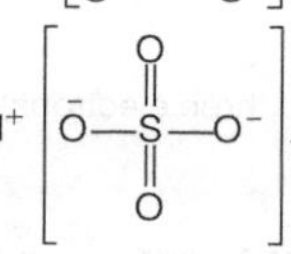 All bonds except coordinate ($\rightarrow$) bond.

$Na^+ \; {}^-N{=}N{\equiv}N$ ($\rightarrow$) All bonds except coordinate ($\rightarrow$) bond.

**Explanations for Q. Nos. (25-26)**

$$A(g) \longrightarrow A^+(g) + e^-; \qquad \Delta H = 5.5 \text{ eV}$$
$$B(g) + e^- \longrightarrow B^-(g) \quad ; \qquad \Delta H = -1.5 \text{ eV}$$
$$A(g) + B(g) \longrightarrow A^+B^-; \qquad \Delta H = 4.0 \text{ eV}$$
$$F = 1.44 \frac{Z_1 Z_2}{r^2} = 1.44 \frac{Z_1 Z_2}{(r_{AB})^2}$$
$$F = 4.0 \text{ eV}$$
$$\therefore \quad (r_{AB})^2 = \frac{1.44 \times 1 \times 1}{4.0} = 0.36$$
$$\therefore \quad r_{AB} = \sqrt{0.36} = 0.6 \text{ nm}$$
$$r_{AB} = r_{A^+} + r_{B^-} = 0.6$$
$$\therefore \quad r_{A^+} = 0.6 - 0.2 = 0.4 \text{ nm}$$

**25.** (a) **26.** (a)

**27.** (8)

$$F \text{ (strength of ionic bond)} = k \frac{z_1 z_2 e^2}{Dr^2} = k \frac{z_1(A)\, z_2(B) e^2}{Dr^2}$$

$z_1$ is doubled to $2z_1$ and $r$ is halved to $\left(\frac{r}{2}\right)$

Then $$F' = p \frac{2z_1(A)\, z_2(B) e^2}{D\left(\frac{r}{2}\right)^2}$$

$$= 8k \frac{z_1(A)\, z_2(B) e^2}{Dr^2} = 8 F$$

Thus, ionic strength becomes 8 times.

## DPP-2 Covalent and Coordinate Bonding

**1. Plan** Coordinate bonds are formed between electron deficient species (Lewis acid, lone pair acceptor) and electron rich species (Lewis-base, lone pair donor).

(i) $N{\equiv}N{\rightarrow}O$ (ii) 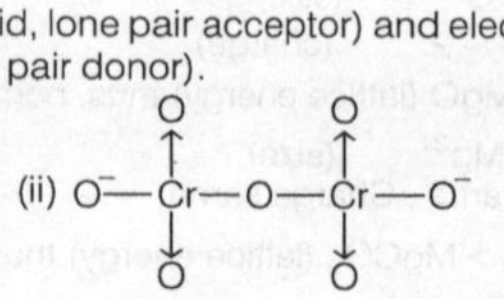 (iii) 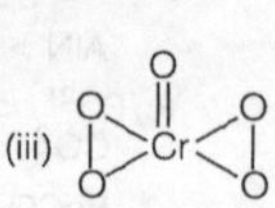

(iv) 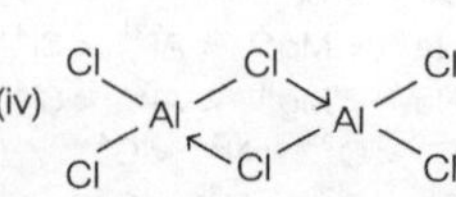 (v) 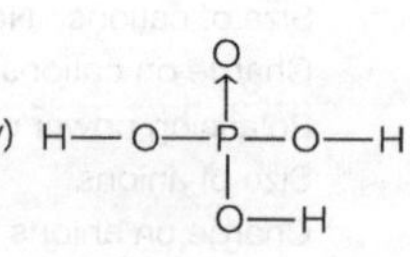

(vi) 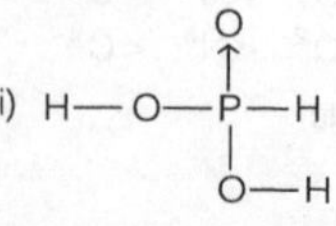 (vii) 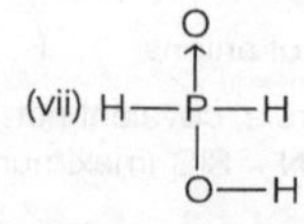 (viii) 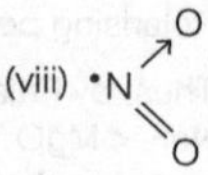

**2.** **Plan** Additional electrons (lost or gained) to form ion for complete octet decide valency of the ion.

$\cdot\dot{\underset{\cdot}{X}}$ three electrons in valence shell hence it can form $X^{3+}$ ion to complete octet.

$$X \longrightarrow X^{3+} + 3e^-$$

$:\ddot{\underset{\cdot\cdot}{Y}}$ has six electrons in valence shell. Thus, it gains two electrons to complete its octet.

$$Y + 2e^- \longrightarrow Y^{2-}$$

Thus, compound is $X_2Y_3$.

**3.** **Plan** Lewis acid is electron-deficient species thus, lone pair acceptor (electrophile).

Lewis base is electron rich species thus, lone pair donor (nucleophile).

**Lewis bases**

(vi) $:\bar{C}\equiv N:$

**Lewis acids**

(iii) $H^+$ (duplet incomplete)

(iv) $BF_3$ (octet incomplete)

(v) $AlH_3$ (octet incomplete)

(vii) $Fe^{2+}$ (acts as Lewis acid in complex formation)

(viii) $\overset{\oplus}{C}H_3$ (octet incomplete)

**4.** **Plan** In covalent compounds, bond length is the sum of the covalent radii of bonding atoms.

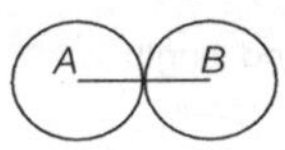


$r_{AB} = r_A + r_B$

In $BF_3$, B—F bond has also got some double bond character and due to delocalisation of double bond, all B—F bonds are equivalents.

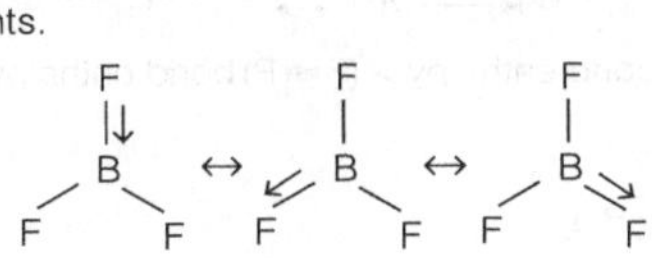


(B—F) bond length is smaller than the sum of covalent radii of B and F

B(80 pm) F(72 pm)

B—F = 152 pm

In $BF_3$ = 130 pm < 152 pm

Also, B-atom in $BF_3$ is $sp^2$ hybridised.

In $BF_4^-$, B-atom is $sp^3$ hybridised and all (B—F) bonds are single bond and (B—F) bond length in $BF_4^-$ > (B—F) bond length in $BF_3$.

**5.** (i) **Plan** The molecular shape of diborane is shown below :

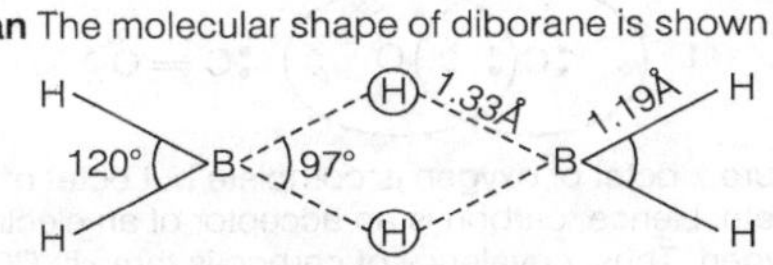


The two boron atoms and the four terminal hydrogen atoms lie in one plane; the two bridging hydrogen atoms (encircled) lie symmetrically above and below the plane. There are total 12 valence electrons —6 in two B and 6 in six H-atoms and they form in all eight B—H bonds. Evidently, there are not enough electrons to fill all the available orbitals to form eight normal —two-centre, two-electron (2c-2e) or electron pair bonds.

The two bridging hydrogen atoms lie symmetrically above and below this plane (3c-2e bonding).

This has actually multicentre bonds or odd-electron bonds and B—H bond lengths (of terminal sides) is less than that of central B—H bonds.

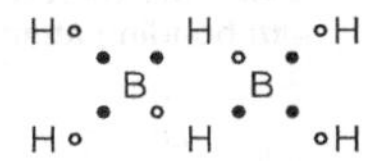


• Boron electron
∘ Hydrogen electron

(Bridging groups are electron deficient)

2c – 2e B — H terminal bond : B — H

3c – 2e B — H — B bridging bond : B—H—B (H bridging the two B atoms)

(ii) **Plan** In case of $(CH_3)_3N$, with three —$CH_3$ groups and one lone pair on N-atom, it has $sp^3$ hybridised structure, but (*lp-bp*) repulsion causes distortion and gives pyramidal structure.

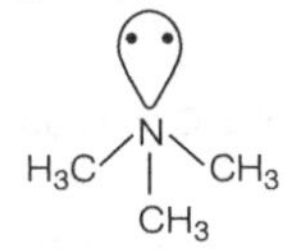


In case of $N(SiH_3)_3$, lone pair on N-atom is transferred to the empty *d*-orbital of silicon ($p\pi$-$d\pi$ overlap) thereby, causing planarity of unit.

(iii) **Plan**

Total valence electrons = $\overset{O}{\overline{3 \times 6}} + \overset{H}{\overline{2 \times 1}} = 20$

H—O—O—O—H

Duplet of H and O are complete.

**6.** (d) $N(7) = 1s^2 2s^2 2p^3$ $\overset{\times}{_\times}N\overset{\times}{_\times}$,

$O(8) = 1s^2 2s^2 2p^4$ $:\dot{\underset{\cdot}{O}}:$

To complete octet, N shares 2 electrons and oxygen shares 3 electrons.

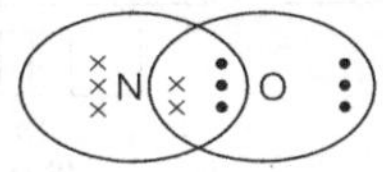


Thus, (a) is correct.

Electrons shared = 5,

Bond order = $\frac{5}{2} = 2.5$

Thus, (b) is correct.

Unpaired electron ↓

$N \overset{\cdot}{=} O$ (paramagnetic)

Thus, (c) is correct.

**7.** (a)

| Element | EC | Valence electrons | Required $e^-$ |
|---|---|---|---|
| H | $1s^1$ | 1 | One short of duplet |
| O | $1s^2 2s^2 2p^4$ | 6 | Two short of octet |
| N | $1s^2 2s^2 2p^3$ | 5 | Three short of octet |
| Cl | $1s^2 2s^2 2p^6\, 3s^2 3p^5$ | 7 | One short of octet |

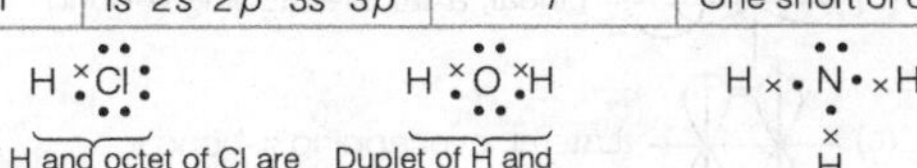


Duplet of H and octet of Cl are completed, bond is covalent in gaseous, state

Duplet of H and octet of O are completed

Duplet of H and octet of N are completed

**8.** (d) *s*-and *p*-electrons have generally low energy than *d*-and *f*-electrons thus take part in chemical bonding.

Maximum covalency = (*s* + *p*) valence electrons.

covalency of C = 4, valence electrons = 4

S = 6 valence electrons = 6

**9.** (b)

| Species | Valence electrons | Electrons shared with bonding atom | Total bonding electrons |
|---|---|---|---|
| $NH_3$ | 5 | 3 | 8 |
| $BCl_3$ | 3 | 3 | 6 |
| $CCl_4$ | 4 | 4 | 8 |
| $PCl_3$ | 5 | 3 | 8 |

Thus, $BCl_3$ has incomplete octet (two electrons short).

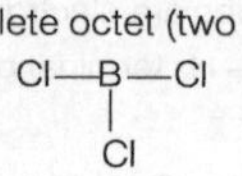

**10.** (a) $N(7) = 1s^2 2s^2 2p^3$

Nitrogen does not have (2*d*) orbitals. Thus, (more than 8) electrons cannot be accomodated in second orbit.

**11.** (c)

(a) 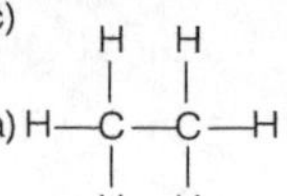 Octet of C and duplet of H complete.

(b) 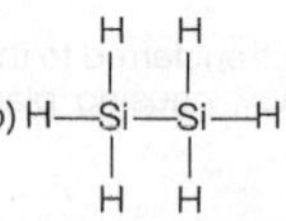 Octet of Si and duplet of H complete.

(c) 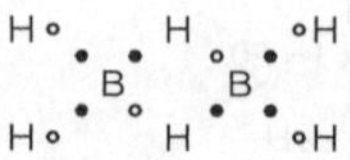 Octet of B is incomplete bridging H-atoms are also electron deficient.

• Boron electron
∘ Hydrogen electron

(d) 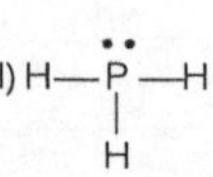 Octet of P and duplet of H complete.

**12.** (a)

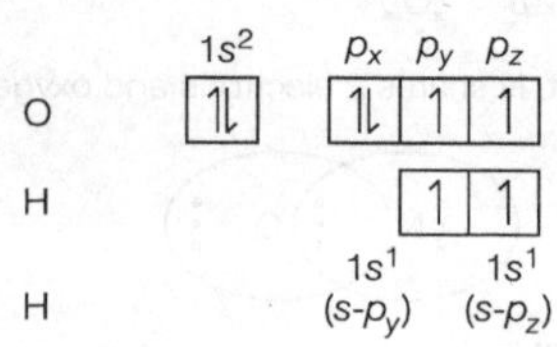


Overlapping between H and O-atom.

**13.** (d) Based on structure :

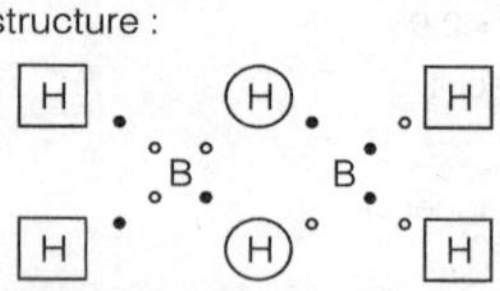

B Ⓗ bond has one electron; bond order $y = 0.5$.

B [H] bond has two electrons; bond order $x = 1.0$.

**14.** (c)

(a) 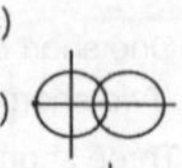 Linear, axial overlapping σ-bond

(b) 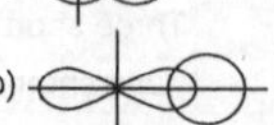 Linear, axial overlapping σ-bond

(c) 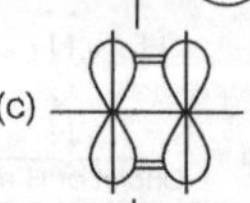 Lateral, overlapping π-bond

(d) 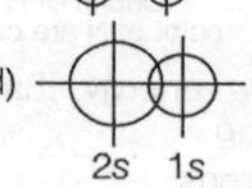 Linear, axial overlapping σ-bond



**15.** (b) Bond length $\propto \dfrac{1}{\text{number of bonds between two atom}}$

(a) 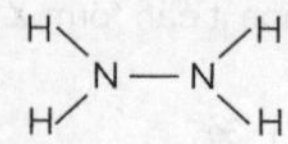 number of bonds between N-atoms = 1

(b) $N\equiv N$ = 3

(c) 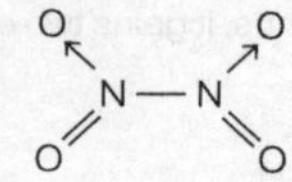 = 1

(d) $N\equiv N{=}O$ = 2 and 3

Due to resonance in $N_2O$ (N to N) bond is between 2 and 3.

**16.** (a)

| S.No. | Species | Valence electrons on central atom | Bonding pair | Non- bonding pair |
|---|---|---|---|---|
| I. | $SF_4$ | 6 | 4 | 1 |
| II. | $ICl_3$ | 7 | 3 | 2 |
| III. | $SO_2$ | 6 | 2 | 1 |
| IV. | $NH_3$ | 5 | 3 | 1 |
| V. | $SiF_4$ | 4 | 4 | 0 |

**17.** (d)

| | Lone pair | σ-bonds | π-bond |
|---|---|---|---|
| $H{-}C\equiv N\colon$ | 1 | 2 | 2 |
| $H{-}\overset{+}{N}\equiv\overset{-}{C}\colon$ | 1 | 2 | 2 |

Thus (d) is correct.

**18.** (b) When two atoms approach to form a bond, there is decrease in potential energy (PE). After bond formation, at a minimum distance (called bond length). PE further increases.

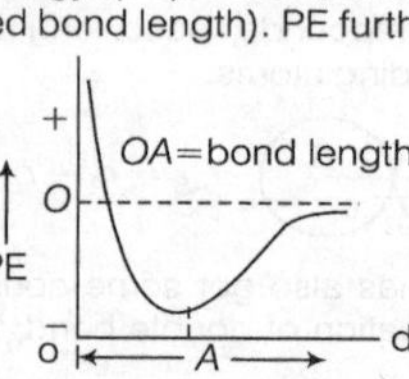


**19.** (a) $(N\equiv N)$ bond enthalpy > $(P\equiv P)$ bond enthalpy hence, $(N\equiv N)$ is stable.

($P_4$) 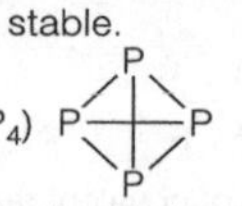

Bond enthalpy of six (P—P) bond = 6 × 215
= 1290 kJ $mol^{-1}$

($N_4$) 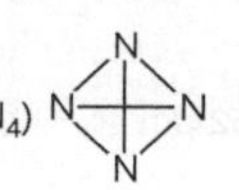

Bond enthalpy of six (N—N) bond = 6 × 160
= 960 kJ $mol^{-1}$

Thus, $P_4$ is more stable than $N_4$.

**20.** (b)

I. 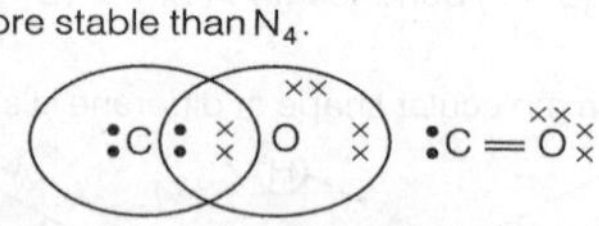

In structure I, octet of oxygen is complete but octet of carbon is incomplete. Hence, carbon is an acceptor of an electron pair from oxygen. Thus, covalency of carbon is three in CO.

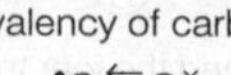

**21.** (a, b, c, d)

(a) Octet of Cl expanded to (9).
(b) Octet Xe expanded to (10).
(c) Octet of P expanded to (10).
(d) Octet of S expanded to (12).

**22.** (a, b, c, d)

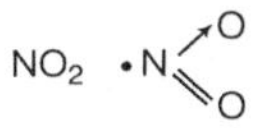

Due to unpaired electron paramagnetic.

N-atom in $NO_2$ is electron deficient thus, to complete octet, dimer is formed.

$$\underset{\text{Brown}}{2NO_2} \rightleftharpoons \underset{\text{Very light brown}}{N_2O_4}$$

In $N_2O_4$ formation, each N-atom gets ($\delta+$) charge. Dipole-dipole repulsion increases (N—N) bond length which is larger than (N—N) covalent bond.

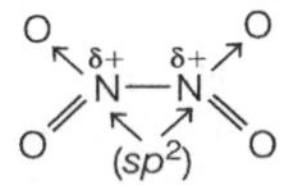

Thus, (a), (b), (c) and (d) are correct.

**23.** (a, c, d) Incorrect (b)

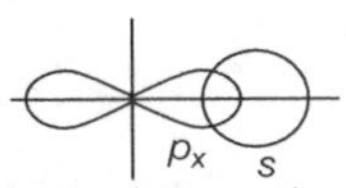

two *sp*-hybrid orbitals are along *x*-axis.

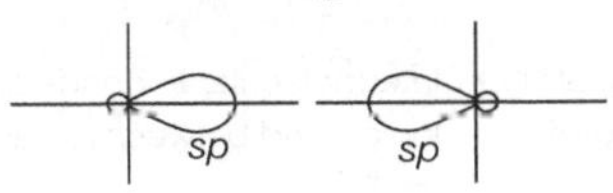

**24.** (b, c) $AlCl_3$–Octet of Al is incomplete; thus electrophile.

$SO_3$–It has positive charge on sulphur due to resonance.

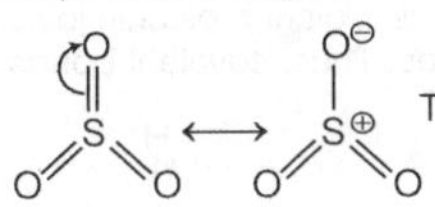

Thus, electrophile.

$\overset{\oplus}{N}O_2$–Octet of N incomplete.

Thus, electrophile. $\overset{\oplus}{C}H_3$– as $\overset{\oplus}{N}O_2$, $CH_3\overset{\oplus}{C}O$– as $\overset{\oplus}{C}H_3$

**25.** (b, d)

(a)

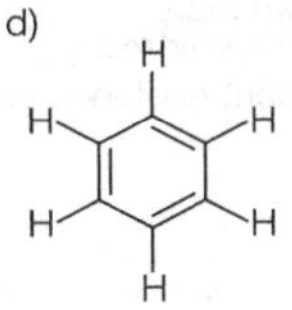

C—H $\sigma$-bonds = 6,

C—C $\sigma$-bonds = 6,

C—C $\pi$-bonds = 3

Ratio of $\sigma$ and $\pi$-bonds = $\frac{12}{3}$ = 4

(b)

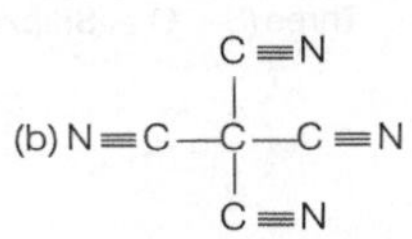

C—C $\sigma$-bonds = 8,

C—N $\pi$-bonds = 8,

Ratio = 1

(c)

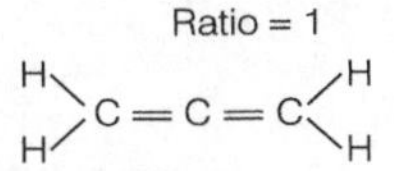

C—H $\sigma$ bonds = 4,

C—C $\sigma$ bonds = 2

C—C $\pi$ bonds = 2,

Ratio = 3

(d) O=C=O

C—O $\sigma$ bonds = 2,

C—O $\pi$ bonds = 2,

Ratio = 1

**Explanations for Q. Nos. (26-27)**

EN difference $\Delta$(EN) $\propto$ polarity.

| | | | $\Delta$(EN) |
|---|---|---|---|
| (a) | H | Br | 0.7 |
| (b) | N | H | 0.9 |
| (c) | N | O | 0.5 |
| (d) | P | H | 0.0 |

Thus, (b) is correct.

A molecule (H—X) is 50% ionic when $\Delta$(EN) = 1.7

**26.** (b) **27.** (d)

**28.** (a) Greater the electronegativity of carbon atom

$$(sp^3 < sp^2 < sp)$$

Greater the attraction for (C—H) bonding pair hence, shorter the (C—H) bond.

**29.** (c) Bond length $\propto \frac{1}{\text{bond order}}$

$$C—C > C=C > C\equiv C$$

**30.** (c) ($C \overset{\sigma}{—} C$) bond

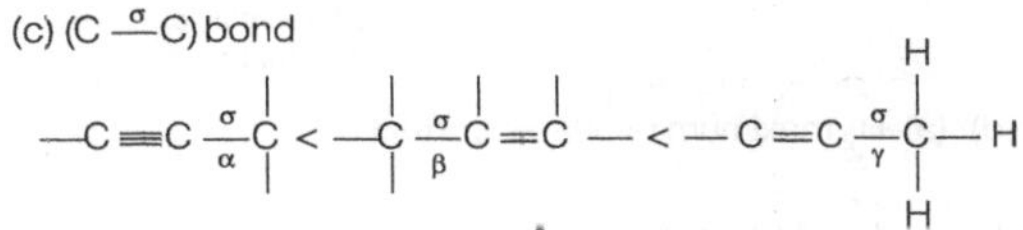

$$(C \overset{\alpha}{—} C) < C \overset{\beta}{—} C < C \overset{\gamma}{—} C$$

**31.** (a) (i)

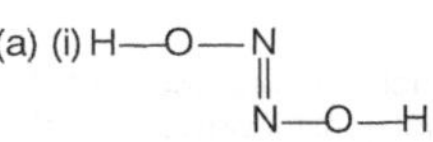

$\sigma$ and $\pi$ bonds are present.

*cis*-and *trans*-isomer due to (N=N) bond.

Thus, **(i) → (p,q)**

(ii) $2\dot{N}O_2 \rightleftharpoons N_2O_4$

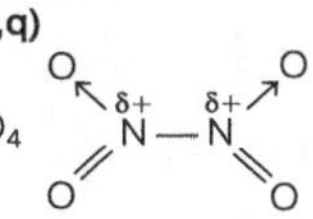

$\sigma$ and $\pi$-bonds are present

due to dipole-dipole repulsion when (N—N) bond is formed (N—N) bond length is increased.

Thus, **(ii) → (p,r)**

(iii)

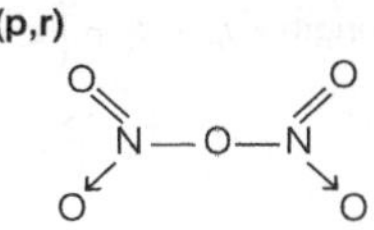

**(iii) → (p)**

(iv) H—O—N=O

$\sigma$ and $\pi$ bonds exist.

Oxidation number of N = + 3

It can be increased to +5 (maximum)

and decreased to –3 (minimum)

$HNO_2$ is also a ligand.

Thus, $HNO_2$ is the reducing agent.

$$\underset{+3}{NO_2^-} + H_2O \longrightarrow \underset{+5}{NO_3^-} + 2H^+ + 2e^-$$

$HNO_2$ is also the oxidising agent

$$\underset{+3}{2NO_2^-} + 8H^+ + 6e^- \longrightarrow \underset{0}{N_2} + 4H_2O$$

$HNO_2$ is also the complexing agent

Thus, **(iv) → (p,s)**

**32.** (5)

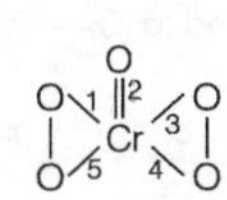

**33.** (4)

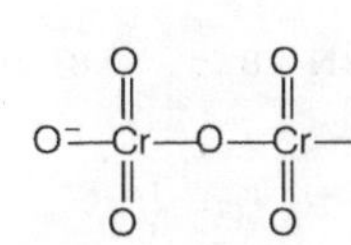

# DPP-3 Covalent Bonding

**1.** (b)

| Species | Element | Valence electrons | Bonding electrons | Total |
|---|---|---|---|---|
| $BF_3$ | B | 3 | 3 | 6 |
| $BeCl_3$ | Be | 2 | 2 | 4 |
| $NO_2$ | N | 5 | 2 | 7 |

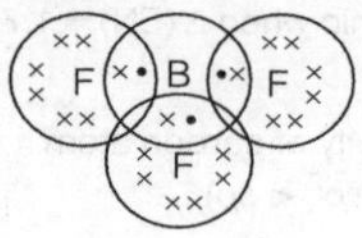

Octet of B is incomplete

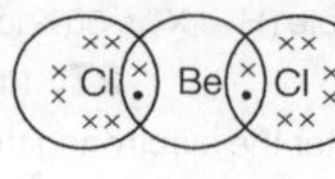

Octet of Be is incomplete

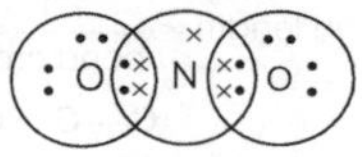

Octet of N is incomplete

**2.** (d) (a) $H_2$ maximum overlap – correct.

(b) 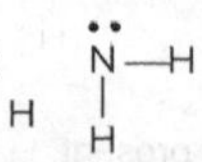 localised electron – correct.

(c) Based on valence bond approach,–no prediction of change of structure due to (*lp-lp*), (*lp-bp*) repulsions– correct.

**3.** (d) (*A*—*A*) bond length = *x*

$$\therefore \quad r_A = \frac{x}{2}$$

$$(C—C) \text{ bond length} = y \Rightarrow r_C = \frac{y}{2}$$

$$(A—B) \text{ bond length} = z$$

$$\text{Then} \quad r_B = z - r_A = \left(z - \frac{x}{2}\right)$$

$$\text{Thus, (B—C) bond length} = r_B + r_C = \left(z - \frac{x}{2}\right) + \frac{y}{2}$$

$$= z + \frac{y}{2} - \frac{x}{2} = z + \left(\frac{y - x}{2}\right)$$

**4.** (b) $CN^-$ (14*e*), $\overset{+}{N}O$ (14*e*) and $N_2$ (14*e*) are isoelectronic.

In each bond order = 3

Electrons in $N_2$ are uniformly distributed

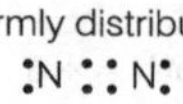

Thus, there is no polarity on N-atoms. Thus, chemically inert.

**5.** (b) H is smallest in size thus, (C—H) bond length is least.

$$(C=C) \text{ bond order} = 2$$

$$(C—C) \text{ bond order} = 1$$

$$\text{Bond length} \propto \frac{1}{\text{bond order}}$$

Thus, $\quad (C=C) < (C—C)$

Also (EN) of O > C

Hence, bonding pair in (C—O) is nearer to nucleus decreasing bond length as compared to (C—C).

Thus, I < IV < II < III.

**6.** (d)

| | σ | π | Ratio |
|---|---|---|---|
| (a) 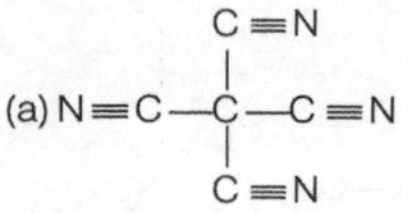 | 8 | 8 | 1.0 |
| (b) 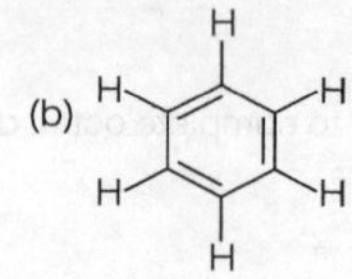 | 12 | 3 | 4.0 |
| (c) 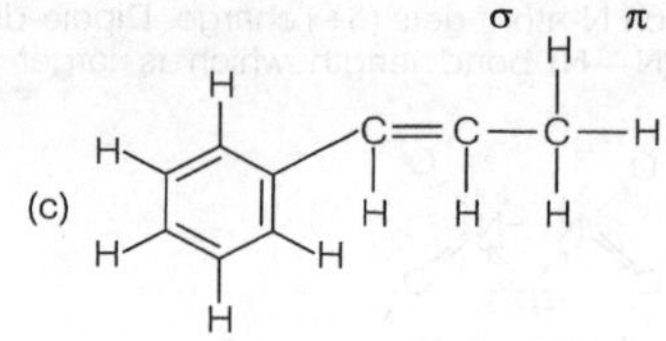 | 19 | 4 | 4.75 |
| (d) 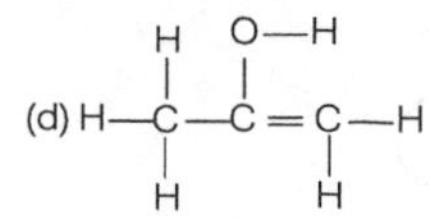 | 9 | 1 | 9.0 |

**7.** (d)

| | | |
|---|---|---|
| (a) | Boiling of $H_2O$ | Intermolecular H-bonds are cleaved. |
| (b) | Melting of KCN | Ionic bond between $K^+$ and $CN^-$ are cleaved. |
| (c) | Boiling of $CF_4$ | Forces of attraction between $CF_4$ molecules (molecular forces are affected) |
| (d) | Melting of $SiO_2$ | $SiO_2$ is a giant molecule joined by covalent bonds. Thus, covalent bonds are cleaved. |

**8.** (d) 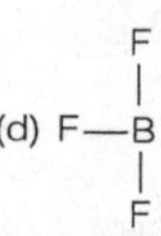

Octet of B is incomplete; Lewis acid, accepts lone pair from $NH_3$ (Lewis acid)

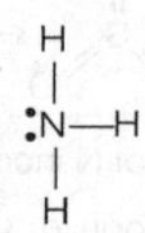

N has extra lone pair (Lewis base) can be donated to $BF_3$

Thus, in ($NH_3$,$BF_3$), both have independent existence and thus, it can be separated. ($H_3N \longrightarrow BF_3$)

**9.** (d) I. 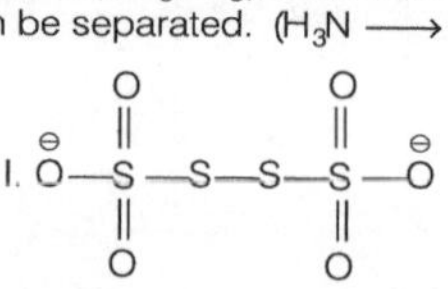 No (S—O—S) linkage

II. 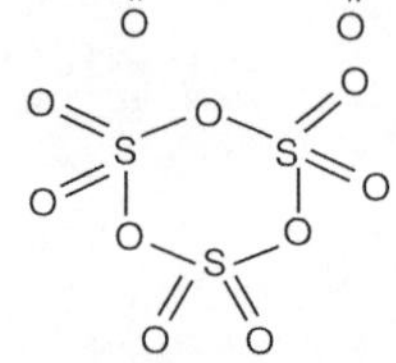 Three (S—O—S) linkages.

III. 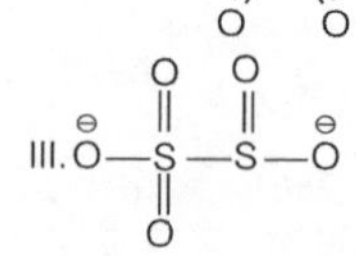 No (S—O—S) linkage.

IV. 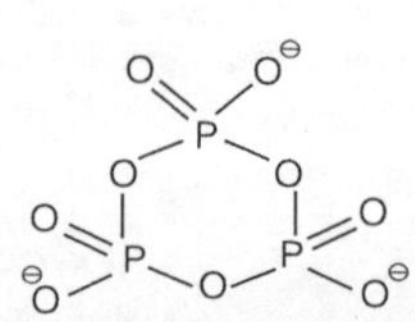 Three (P—O—P) linkages.

**10.** (c) I, II and III are axial overlapping thus, extent is larger than IV (lateral overlapping).

Axial overlapping (*p-p*) < (*s-p*) < (*s-s*)

Thus, I is the strongest and IV is the weakest.

**11.** (d) There are three lone pairs on $Y$ and four $\sigma$ -bonds between $X$ and $Y$.

Thus, octets of $X$ and $Y$ are complete.

$X$ has four electrons in valence shell.

$Y$ has seven electrons in valence shell.

Thus, $X$ = Si (group 14), $Y$ = Cl (group 17)

**12.** (b) Hyperconjugation takes place in allylic compounds in which alkyl group (having at least one H) is attached to (C = C) bond.

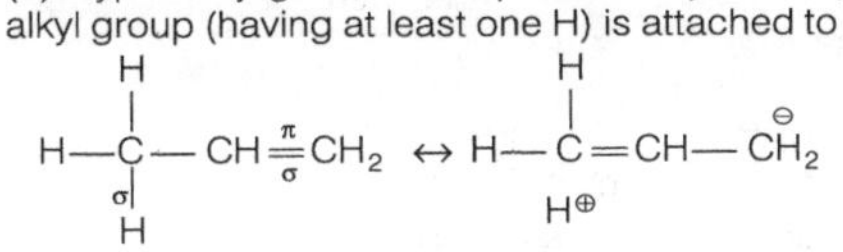

$\sigma$-bond of (— $CH_3$) and $\pi$ -bond of (ene) are involved. $\pi$-bond is formed due to overlapping of $p$-orbitals. Thus, **$\sigma$-orbital** and ***p*-orbital** are involved.

**13.** (b)

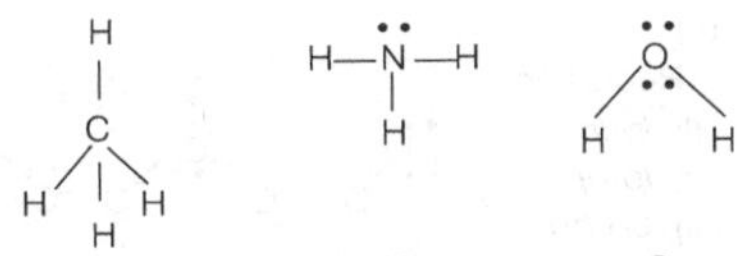

| | | | |
|---|---|---|---|
| $A$ | C | N | O |
| $X$ | 4 | 3 | 2 |
| $E$ | 0 | 1 | 2 |
| Thus, | $AX_4$ | $AX_3E$ | $AX_2E_2$ |

**14.** (b) (a) If conducting – ionic, (b) gas – covalent

(c) ionic, (d) ionic

**15.** (b)

(a) 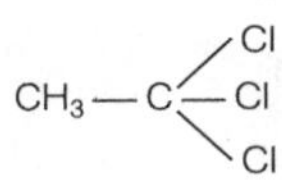 (EN) of $CCl_3 > CH_3$

(b) $PCl_5$ triangular bipyramidal, symmetrical, no lone pair.

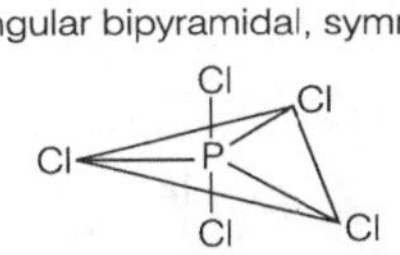

$\mu = 0$

(c) $CH_2Cl_2$ 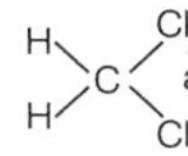 as in (b) (d) 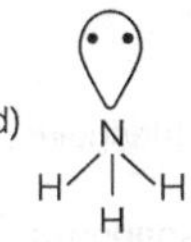 Due to lone pair

**16.** (d) ICl, NO are diatomic molecules with an electronegativity difference between the bonded atoms.

$SO_2$ is a bent molecule with an electronegativity difference between S and O.

$Cl_2$ (EN) difference zero being identical atoms.

$BF_3$ no lone pair on B-atom.

**17.** (c)

(i) ($d$-$d$) $\sigma$-bonding (ii) ($p$-$d\pi$ bonding)

(iii) ($p$-$d\pi$ anti-bonding) (iv) ($d$-$d\sigma$ anti-bonding)

(i) → (q), (ii) → (r), (iii) → (p), (iv) → (s)

**18.** (a, b, c, d)

(a) When the two H-atoms are away from points $D$, potential energy = 0. Thus, true.

(b) As two H-atoms approach in the formation of $H_2$ molecule, (PE) falls. Thus, true.

(c) Stability is said to be maximum at minimum energy. Thus, true.

(d) When atom approach more closely ($C$ to $B$) PE increases. Thus, true.

**19.** (c, d) Orthophosphorous acid ($H_3PO_3$).

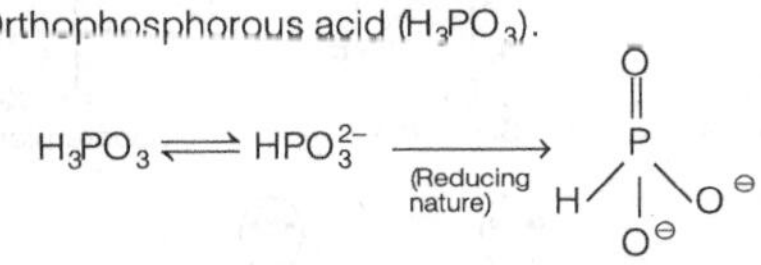

(a) It can accept $H^+$ but cannot ionise to form $H^+$. Thus, (a) is incorrect.

(b) P=O, P—$O^{\ominus}$ are of different length.

(c) Bond order = $\frac{4}{3}$ = 1.33 for (P—O) bond.

(d) It is $H^+$ ion acceptor hence, Lowry-Bronsted base.

## DPP-4 Chemical Bonding (VSEPR Model) and Hybridisation

**1. Plan** Based on VSEPR model, F-atoms and electron pairs are placed so, as to minimise repulsion.

(i) If VSEPR is not considered, it has triangular bipyramidal geometry.

(ii) To minimise repulsion, it has linear geometry having three lone pairs on equatorial positions.

(iii) $I_3^-$

**2.** (d)

| | Bond pair | Lone pair | Structure |
|---|---|---|---|
| $[ClF_4]^-$ | 4 | 2 | Square planar |
| $[ClF_2]^-$ | 2 | 3 | Linear |

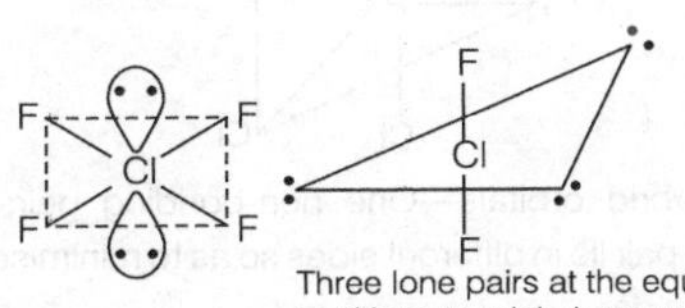

Three lone pairs at the equatorial positions to minimise repulsion.

**3.** (b)

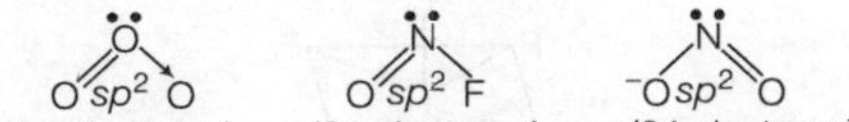

(24 electrons) (24 electrons) (24 electrons)

$\theta = 117°$ $\theta = 110°$ $\theta = 115°$

In each central atom is $sp^2$-hybridised, each having one lone pair. Difference in the extent of repulsion changes value of $\theta$.

**4.** (b) $XeF_2$-$sp^3d$ hybridised Xe

Three lone pairs on equatorial positions to minimise repulsion; two F-atom at axial position. Thus, it is linear.

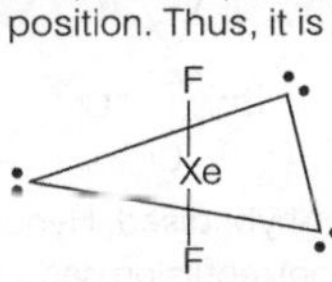

(a) $TeF_2$, $Te_{50}$ - Six electrons in valence shell

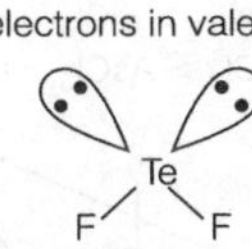

$sp^3$-hybridised Te with two lone pairs thus, V-shaped.

(b) $ICl_2^-$ –$sp^3d$ hybridised I-atom with three lone pairs at equatorial positions to minimise repulsion thus, linear.

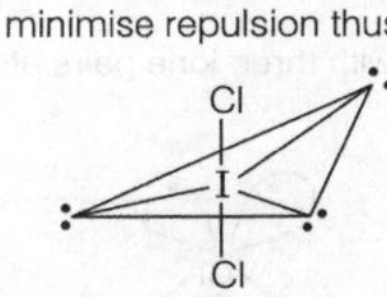

(c) $SbCl_3$ – $sp^3$-hybridised pyramidal Sb–one lone pair on Sb.

(d) $BaCl_2$ – ionic.

**5.** (d)

(a) 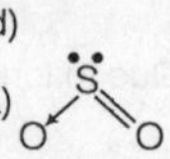 Triangular planar ($sp^2$)

(b) $BrF_3$ – T-shaped ($sp^3d^2$)

(c) $SiO_3^{2-}$ – Triangular planar ($sp^2$)

(d) 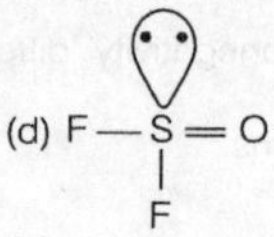 – Pyramidal ($sp^3$)

**6.** (c)

| Species | Central atom | Valence electrons | Bonds | Lone pair | Structure |
|---|---|---|---|---|---|
| $CF_4$ | C | 4 | 4 | 0 | Tetrahedral |
| $SF_4$ | S | 6 | 4 | 1 | See-saw |
| $XeF_4$ | Xe | 8 | 4 | 2 | Square planar |

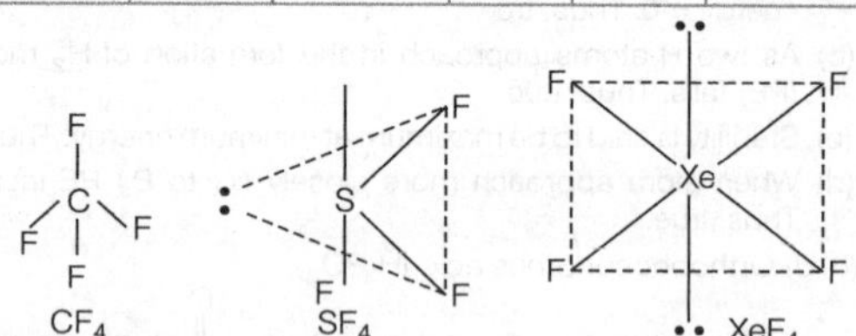

$CF_4$ Tetrahedral; $SF_4$ See – saw; $XeF_4$ Square planar

**7.** (b)

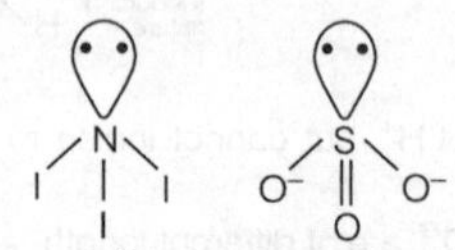

($sp^3$) ($sp^3$)

Both have one lone pair – trigonal pyramidal.

$I_3^-$ – Linear ($sp^3d$ - hybridised I)

$NO_3^-$ – Trigonal ($sp^2d$- hybridised N)

**8.** (a)

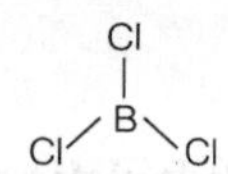

Has no lone pair thus, bond angle is 120°. B-atom is $sp^2$-hybridised.

In $PCl_3$, $AsCl_3$, $BiCl_3$

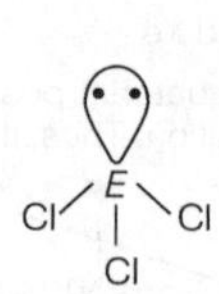

Central atom ***E*** is $sp^3$-hybridised. Hence, angle < 120°. As we go down the group, (*lp-bp*) repulsion decreases. Hence, angle (Cl—*E*—Cl) $PCl_3 > AsCl_3 > BiCl_3$.

Thus, order is $BCl_3 > PCl_3 > AsCl_3 > BiCl_3$.

**9.** (a) $F_2Cl^-$

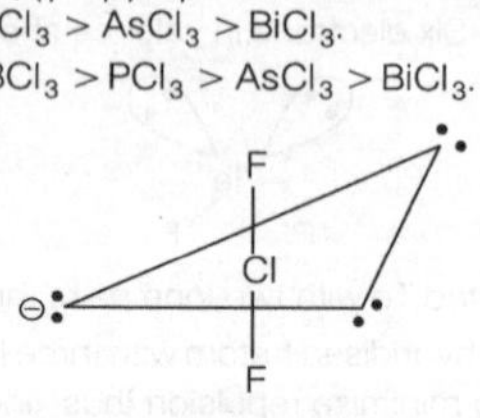

$sp^3d$-hybridised Cl with three lone pairs at equatorial positions to minimise repulsion.

$F_2Cl^+$

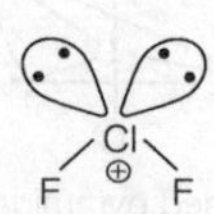

$sp^3$-hybridised Cl with two lone pairs on Cl-atom.

**10.** (b)

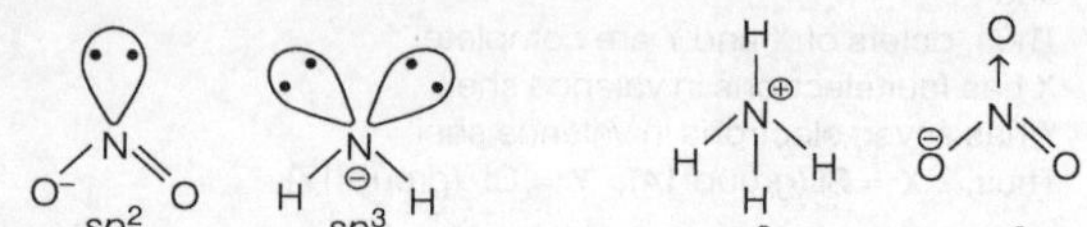

$sp^2$ $sp^3$ $sp^3$ $sp^2$

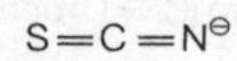

*sp*

(a) $NO_2^\ominus$ ↑ $sp^2$, $NH_2^\ominus$ ↑ $sp^3$  (c) $NH_4^\oplus$ ↑ $sp^3$, $NO_3^\ominus$ ↑ $sp^2$

(b) $NO_2^\ominus$ ↑ $sp^2$, $NO_3^\ominus$ ↑ $sp^2$  (d) $SCN^\ominus$ ↑ *sp*, $NH_2^\ominus$ ↑ $sp^3$

**11.** (a)

Repulsion

(I) *lp-lp*

(II) *lp-bp*

(III) *bp-bp*

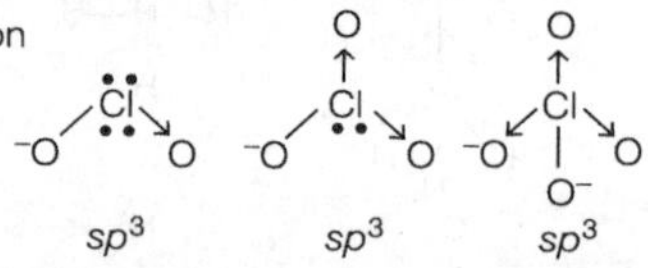

$sp^3$ (I), (II), (III)  $sp^3$ (II), (III)  $sp^3$ (III) only

Bond angle ⟶ Maximum

**12.** (c) Sn(50) = [Kr]$4d^{10}5s^25p^2$ **lone pair**

(a) 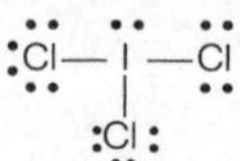 11

(b) 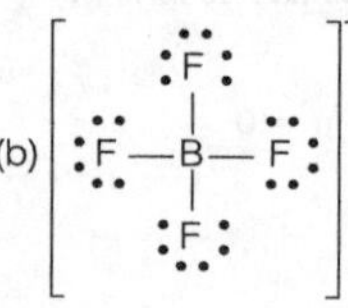 12

(c) 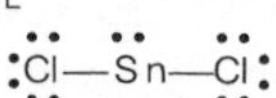 7

(d) 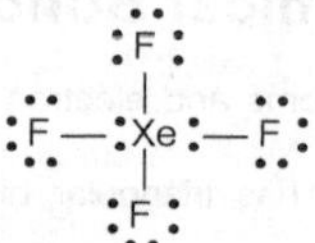 14

**13.** (a) $PCl_5$

$sp^3d$-hybrid orbitals—No non-bonding pair trigonal bipyramidal structure.

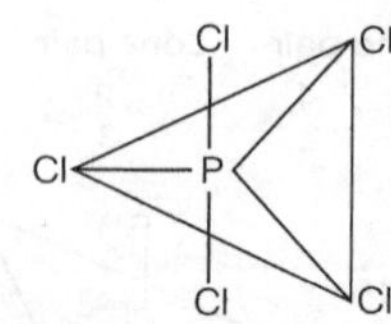

$IF_5$ $sp^3d^2$-hybrid orbitals—One non-bonding pair; one F and non-bonding pair lie in different sides so as to minimise repulsion. Square pyramidal.

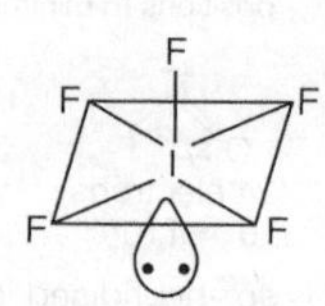

**14.** (a)

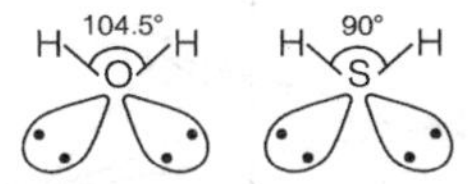


There is (i) $lp$ - $lp$ (ii) $lp$ - $bp$ (iii) $bp$ – $bp$ repulsion

Thus, bond angle is less than 109.5°. In $H_2O$, 104.5° > 90° in $H_2S$. It is due to larger size of S-atom which minimises the repulsion and allows the bonds in $H_2S$ to be purely $p$-type.

**15.** (c)

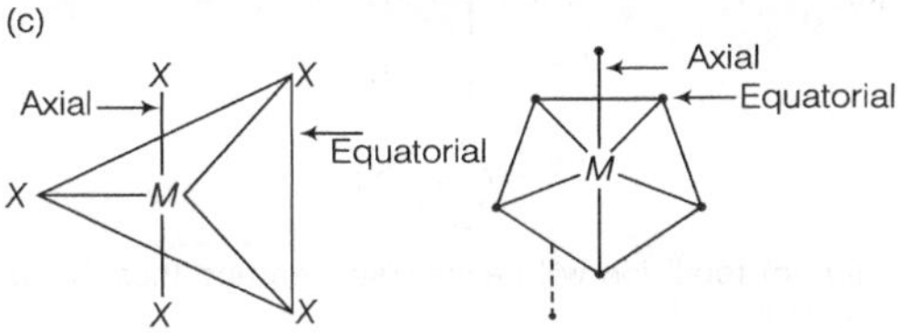


Trigonal bipyramidal Pentagonal bipyramidal

**16.** (d)

| | Species | Hybridisation of central atom | Shape |
|---|---|---|---|
| (a) | $H_2O$ | O is $sp^3$-hybridised | Bent |
| | $H_2S$ | S is $sp^3$-hybridised | Bent |
| (b) | $XeO_3$ | Xe is $sp^3$-hybridised. | Trigonal pyramidal |
| | $BrF_3$ | Br is $sp^3d$-hybridised | T-shaped |
| (c) | $[Ni(CN)_4)]^{2-}$ | Ni is $dsp^2$-hybridised | Square planar |
| | $[Ni(CO)_4]$ | N is $sp^3$-hybridised | Tetrahedral |
| (d) | $BeCl_2$ | $sp$-hybridised Be | Linear |
| | $I_3^-$ | $sp^3d$ -hybridised I | Linear |

**17.** (d) (a) $\underset{107°}{NH_3} < \underset{102°}{NF_3}$ (incorrect)

The high electronegativity of F pulls the bonding electrons farther away from N than in $NH_3$. Thus, repulsion between bond pairs is less in $NF_3$ than in $NH_3$. Hence, the lone pair in N causes a greater distortion than $NH_3$.

(b) as in (a)

(c) $\underset{93.8°}{PH_3} < \underset{91.3°}{SbH_3}$ (incorrect)

Going down the group in periodic table, as size of central atom increases, repulsion increases.

**18.** (a, b, c)

O-atom is $sp^3$-hybridised and thus, bent due to presence of lone pairs on O-atom molecule is polar and planar.

**19.** (a, b)

(a) $ICl_4^-$ $sp^3d^2$ -hybridisation atom of I. Square planar structure due to two lone pairs lying above and below the plane in which I and four Cl lie.

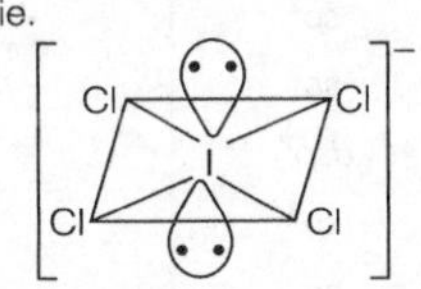


(b) $XeF_4$, $sp^3d^2$-hybridisation of Xe. As above square planar structure.

(c) $POCl_3$, $sp^3$-hybridised P no lone pair.

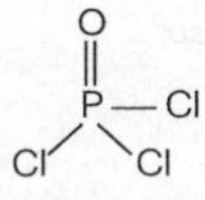


(d) $XeO_2F_2$ has $sp^3d^2$– hybridised Xe with two $\pi$ -bonds and one lone pair ($AX_4E$ type)-trigonal pyramidal structure.

**20.** (a, d)

| | | |
|---|---|---|
| $NH_3$ | N atom $sp^3$ -hybridised | – Bipyramidal |
| $H_2O$ | O atom $sp^3$ -hybridised | – V-shaped |
| $H_2S$ | S atom is $sp^3$ -hybridised | – Bent |
| $SO_2$ | S atom is $sp^2$ -hybridised | – Bent |
| $Ni(CO)_4$ | Ni atom is $sp^3$-hybridised | – Tetrahedral |
| $[NiCl_4]^{2-}$ | Ni atom is $sp^3$-hybridised | –Tetrahedral. |
| $SF_4$ | S atom is $sp^3d$ -hybridised | – See-saw |
| $PCl_5$ | P atom is $sp^3$ -hybridised | – Bipyramidal |

**21.** (a, b, c, d)

(a)

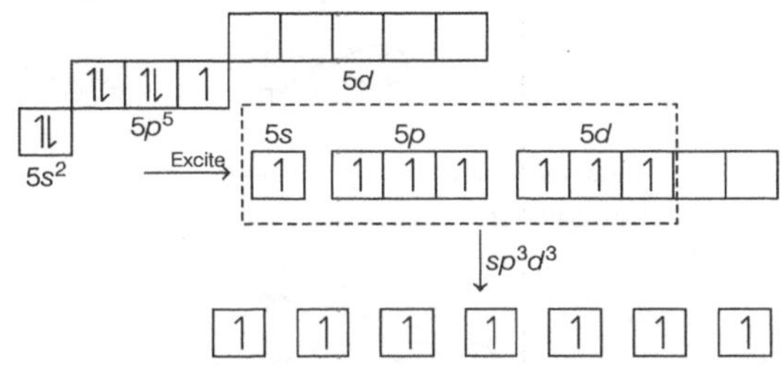


Hybridisation of I–True

(b) Oxidation state of I = +7 (highest) – I rue.

(c) It has two types of bonds two axial (shortest) and five equatorial (longest) – True.

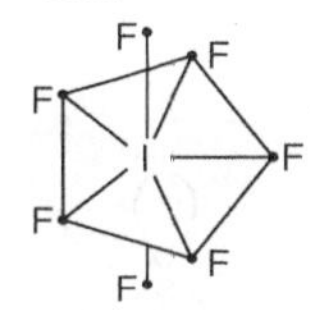


(d) Pentagonal bipyramidal structure – True.

## Explanations for Q. Nos. (22-23)

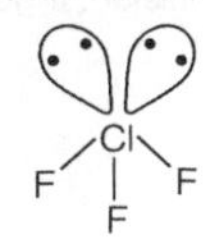


$sp^3d$ -hybridised Cl-atom – (T-shaped structure)

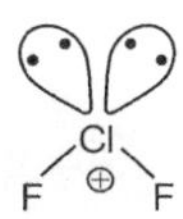


$sp^3$ -hybridised. Cl-atom (Bent)

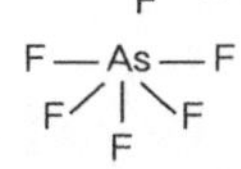


$sp^3d$ -hybridised As-atom (Trigonal bipyramidal).

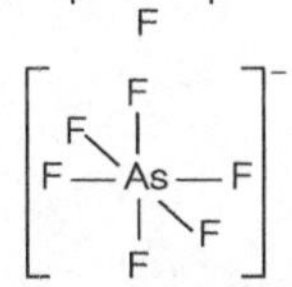


$sp^3d^2$ -hybridised As-atom (Octahedral).

**22.** (b) **23.** (b)

**24.** (d)

(i) $ClF_3$

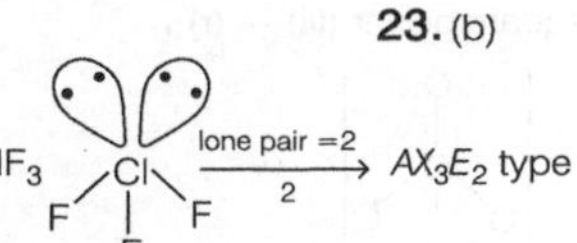


Based on VSEPR model, it is (T-shaped).

Thus, **(i)** → **(r)**

(ii) $ICl_4^-$,

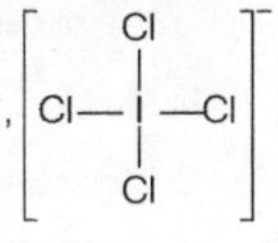


, lone pair = 2,

$AX_4E_2$ (Square planar)

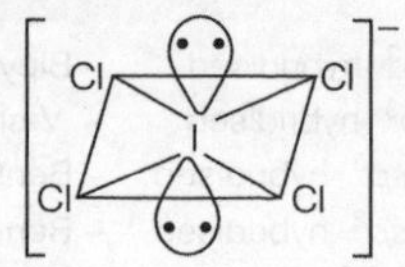

Thus, **(ii) → (s)**

(iii) $F—\underset{F}{\overset{F}{S}}—F$ lone pair = 1, $AX_4E$ (See-saw)

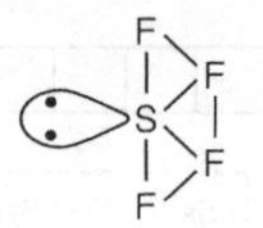

Thus, **(iii) → (q)** $I_3^-$ $(I—I—I)^-$

lone pair = 3

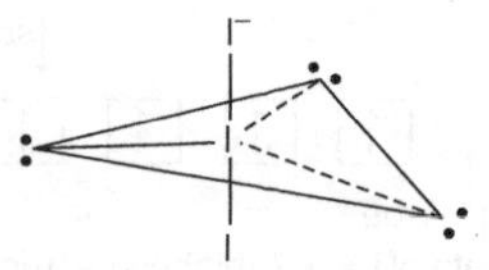

Linear structure

Three lone pairs are at equatorial position to minimise repulsion.

Thus, **(iv) → (p)**

**25.** (a) (i) $OSF_2$

$sp^3$-hybridised S -atom (*lp* - *bp*) repulsion, trigonal pyramidal

**(i) → (s)**

(ii) $O_2SF_2$

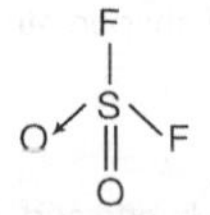

$sp^3$-hybridised s-atom tetrahedral

**(ii) → (p)**

(iii) $XeF_4$

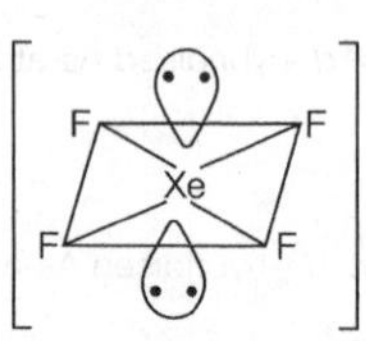

$sp^3d^2$-hybridised Xe-atom.

Two lone pairs, square planar **(iii) → (r)**

(iv) $ClO_4^-$

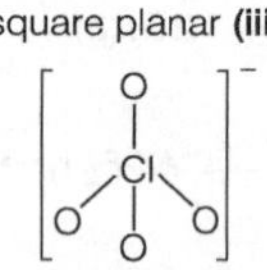

$sp^3$-hybridised Cl atom, tetrahedral.

**(iv) → (p)**

(v) $I_3^-$

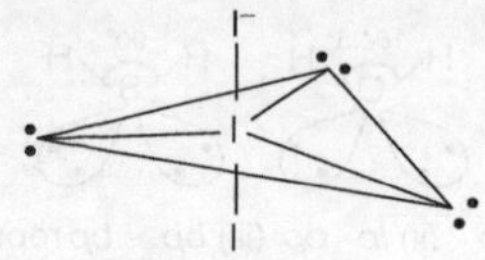

$sp^3d$-hybridised I, three lone pairs at equatorial position linear.

**(v) → (q)**

**26.** (i) (0)

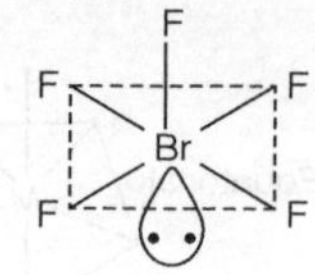

(*lp-bp*) repulsion will cause distortion and thus, bond angle < 90° in each.

(ii) (3)

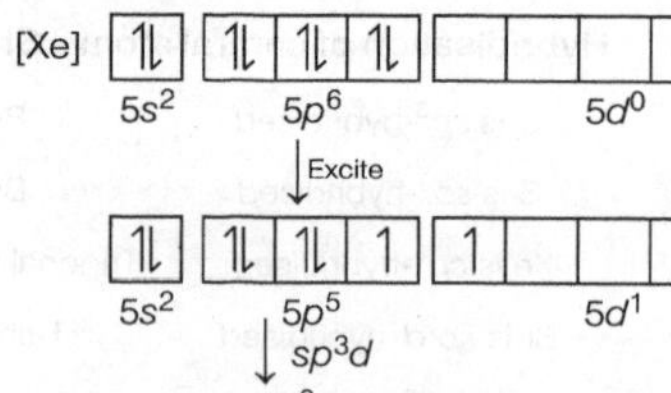


Thus, three lone pairs at equatorial position to minimise repulsion.

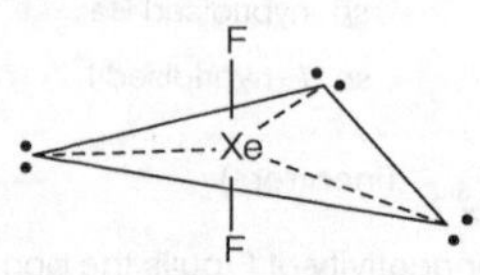

Linear structure

(iii) (6) Melamine is heterocyclic basic compound (six N-atoms) each N having one lone pair.

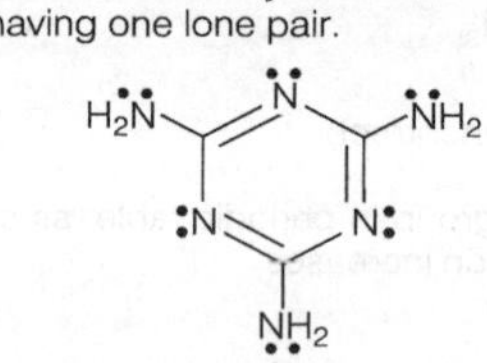

**27.** (5)

| Species | Hybridisation of central atom | Lone pairs | Geometry |
|---|---|---|---|
| $XeF_4$ | $sp^3d^2$ | Two | Square planar |
| $SF_4$ | $sp^3d$ | One | See-saw |
| $SiF_4$ | $sp^3$ | - | Tetrahedral |
| $BF_4^-$ | $sp^3$ | - | Tetrahedral |
| $[Cu(NH_3)_4]^{2+}$ | $dsp^2$ | - | Square planar |
| $[FeCl_4]^{2-}$ | $sp^3$ | - | Tetrahedral |
| $[CoCl_4]^{2-}$ | $sp^3$ | - | Tetrahedral |
| $[IF_4]^-$ | $sp^3d^2$ | Two | Square planar |
| $[BrF_4]^-$ | $sp^3d^2$ | Two | Square planar |
| $[PtCl_4]^{2-}$ | $dsp^2$ | - | Square planar |

# DPP-5 Hybridisation

**1. Plan**

Each C-atom has **three σ-bonds** and one π-bond. Only σ-bonds (lone pair and coordinate bonds if any) are involved in hybridisation.

(i) Thus, each C-atom is $sp^2$-hybridised.

(ii) (a) $C^{\oplus}$ -$sp^2$ (no change)

(b) $C^{\ominus}$ -$sp^3$ (one lone pair extra)

**2. Plan** Maximum valency is based on the unpaired electrons obtained due to excitation of paired electrons to empty orbital.

(i) One of the 3s paired electrons is taken to 3p -orbital.

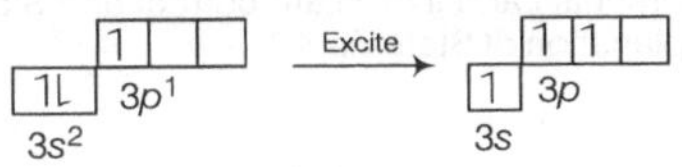


Three unpaired electrons explain trivalent nature of Al

(ii) Al is $sp^2$-hybridised, having three orbitals. There is no electron pair.

(iii)

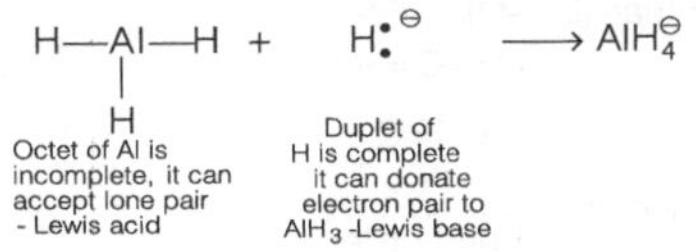


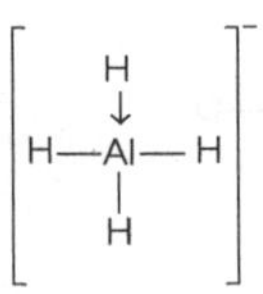

Al is $sp^3$-hybridised in $AlH_4^-$.

(iv) $SiH_4$ has $sp^3$-hybridised Si-atom but is not formed by Lewis acid Lewis base reaction.

**3. Plan**

(i) number of σ-bonds + lone pair + single electron =

2 $sp$

3 $sp^2$

4 $sp^3$ (or $dsp^2$, since, there is no $d$-electron in C-atom hence, only $sp^3$)

5 $sp^3d$

6 $sp^3d^2$

(a)

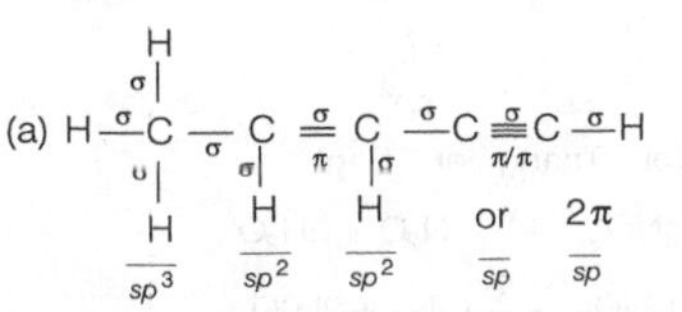


(b) $sp^3$ – 25%; $sp^2$ – 33.33%; $sp$ – 50% s-character.

(c) (EN) of C-atom $sp^3 < sp^2 < sp$ greater the s-character, greater the electronegativity.

(d) Bond ($p$) is formed by ($sp^3$-$sp^2$) overlapping.

($q$) is formed by ($sp^2$-$sp^2$) overlapping.

($r$) is formed by ($sp^2$-$sp$) overlapping.

($s$) is formed by ($sp$-$sp$) overlapping.

Thus, bond length

$$s < q < r < p$$

and bond energy $p < r < q < s$

(double bond is shorter than single bond)

(ii) $\underset{sp^2}{CH_2} = \underset{sp}{C} = \underset{sp^2}{CH_2}$ (allene)

**4.** (a)

| Species | Electrons | Hybridisation of central atom |
|---|---|---|
| $NO_3^-$ | 7 + 24 + 1 = 32, | $sp^2$-hybridised N |
| $CO_3^{2-}$ | 6 + 24 + 2 = 32, | $sp^2$-hybridised C |
| $ClO_3^-$ | 17 + 24 + 1 = 42, | $sp^3$-hybridised Cl |
| $SO_3$ | 16 + 24 = 40, | $sp^2$-hybridised S |

$NO_3^-$ and $CO_3^{2-}$ are isoelectronic, central atom is $sp^2$-hybridised, triangular planar structure.

Thus, $NO_3^-$ and $CO_3^{2-}$ are isoelectronic and isostructural.

**5.** (c) $NO_2$

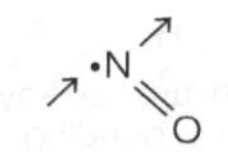

Unpaired electron

$sp^2$-hybridised nitrogen hence, paramagnetic.

It is coloured due to unpaired electron.

When dimerised, unpaired electron takes part in bonding and $N_2O_4$ is colourless

$$2NO_2 \longrightarrow N_2O_4.$$

Due to formation of $N_2O_4$, paramagnetism is lost.

Thus, incorrect statement is (c).

**6.** (c) Molecule has zero dipole moment, thus it does not have **lone pair**. Hence, non-polar.

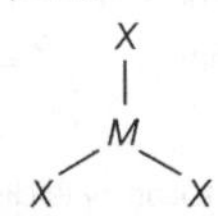

Thus, ($M$—$X$) bonds indicate $sp^2$-hybridisation.

**7.** (d) Vinyl acetylene is

$$\underset{sp}{\overset{4}{H—C}} \equiv \underset{sp}{\overset{3}{C}} \overset{\sigma}{—} \underset{sp^2}{\overset{2}{CH}} = \underset{sp}{\overset{1}{CH_2}}$$

When (C$=$C) bond and (C$\equiv$C) bond are at equal position, numbering of C-chains based on IUPAC nomenclature is done from (C$=$C) side.

Thus, ($C_2 \overset{\sigma}{—} C_3$) bond is formed by ($sp^2$-$sp$) overlapping.

**8.** (d)

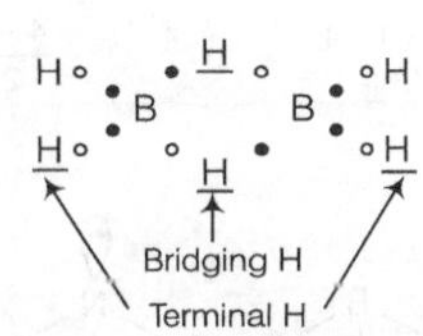


I. $sp^3$-hybridised (B is bonded to four H-atom) – True

II. B—H—B angle ≠ 180° – False.

III. Two types of (B—H) bonds – True.

IV. 12 bonding electrons – True.

**9.** (c) Molecules with (C$\equiv$C) have bond angle 180°. In (d) there is also (C$=$C) and is not symmetrical.

(c) is symmetrical w.r.t. (C$\equiv$C), thus, linear (θ = 180°).

**10.** (d)

| Species | Central atom | Bonding pairs | Lone pairs | Hybridisation |
|---|---|---|---|---|
| $PCl_5$ | P | 5 | - | $sp^3d$ |
| $SF_4$ | S | 4 | 1 | $sp^3d$ |
| $I_3^-$ | I | 2 | 3 | $sp^3d$ |
| $SbCl_6^-$ | Sb | 5 | - | $sp^3d^2$ |

**11.** (a)

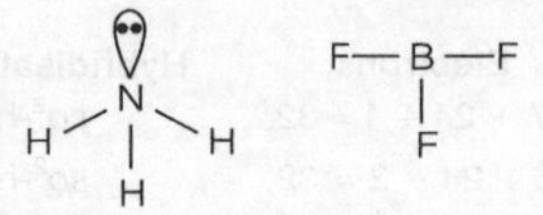

Initial state

After 1 : 1 complex is formed in which $NH_3$ is electron donor and $BF_3$ is electron acceptor

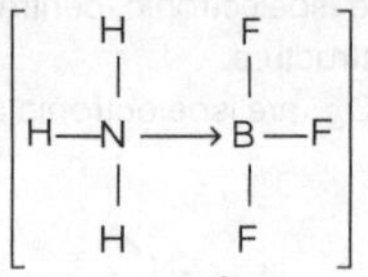

Each atom (N, B) both are $sp^3$-hybridised. Since, lone pair of N-atom is also involved in bonding, hence, tetrahedral.

**12.** (b) Three electron pairs are inclined at 120°.

Two electron-pairs are at 90° with the first three.

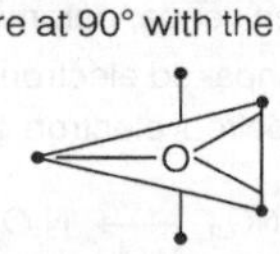

Structure is trigonal bipyramidal.

**13.** (b) :Br: Valence electrons = 7

Bonding atoms = 5

$$\text{Lone pair} = \frac{7-5}{2} = 1$$

Thus, it is $AX_5E$.

Due to one lone pair, geometry is distorted to square pyramidal (instead of trigonal bipyramidal).

**14.** (a) $A$ = Central atom, $X$ = Bonding atoms, $E$ = Lone pair

| Central atom ($A$) | Valence electrons | Bonding atoms ($X$) | Lone pairs ($E$) | Representation |
|---|---|---|---|---|
| S | 6 | 4 | $\frac{6-4}{2}=1$ | $AX_4E$ |
| Xe | 8 | 4 | $\frac{8-4}{2}=2$ | $AX_4E_2$ |
| Si | 4 | 4 | $\frac{4-4}{2}=0$ | $AX_4$ |
| $B^-$ | 3 + 1 = 4 | 4 | $\frac{4-4}{2}=0$ | $AX_4$ |

**15.** (b)

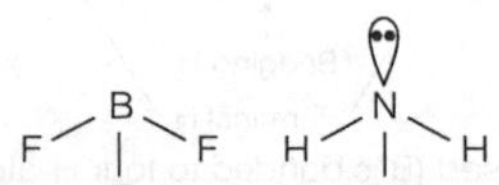

$sp^2$-hybridised B-atom $\quad$ $sp^3$-hybridised N-atom

(*lp-bp*) repulsion make it pyramidal.

Thus, Statement I and II both are correct but Statement II is not the correct explanation of Statement I.

**16.** (d) $B_2H_6$ has following types of bonding

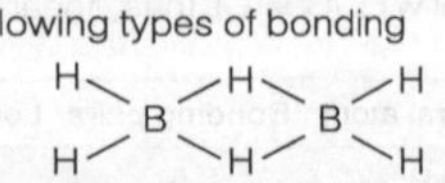

Bridging H-atoms and B-atoms are electron deficient. Each B-atom is however, joined to four H-atoms thus $sp^3$-hybridised.

There is no lone pair or unpaired electron.

Thus, Statement I is incorrect but Statement II is correct.

**17.** (a)

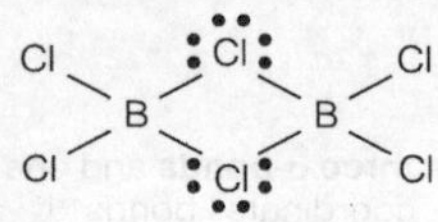

Size of B is too small to form stable 4-membered ring system. (Also delocalised $\pi$-bonding is disturbed). Hence, dimer of $BCl_3$ is not formed.

Dimer $Al_2Cl_6$ is stable due to stable ring.

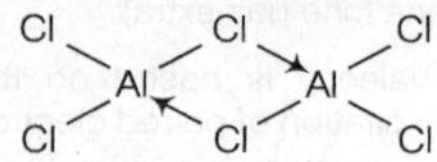

Thus, Statements I and II both are correct and Statement II is the correct explanation of Statement I.

**18.** (b, c, d)

(a)

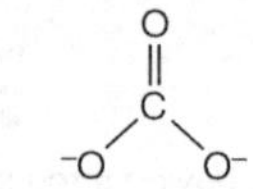

Due to resonance (C—O) bond length is larger than (C=O) but smaller than (C—O).

$$O=C=O$$

$$C \overset{\leftarrow}{\equiv} O$$

Also, Bond length $\propto \frac{1}{\text{bond order}}$

Thus, $CO < CO_2 < CO_3^{2-}$

(b) $\quad O=O \quad (O—O)^- \quad {}^-O—O^-$

Bond-order $\quad$ 2 $\quad$ 1.5 $\quad$ 1

Thus, $\quad O_2 < O_2^- < O_2^{2-}$

(c) As in (a), $NO < NO_2 < NO_3^-$

(d)

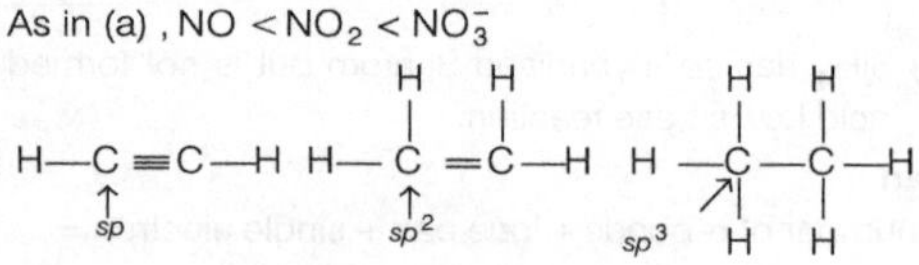

Electronegativity of C – atom

$$sp > sp^2 > sp^3$$

Thus, (C—H) bond length decreases as (EN) increases.

Thus, (C—H) is bond length.

$$CH \equiv CH < CH_2 = CH_2 < CH_3 — CH_3$$

**19.** (b, c, d) $\quad \overset{+}{N}H_4 \quad NO_3^- \quad NO_2^-$

$\uparrow \quad \uparrow \quad \uparrow$

$sp^3 \quad sp^2 \quad sp^2$

Tetrahedral $\quad$ Triangular $\quad$ Bent

$$NH_4NO_3 \xrightarrow{\Delta} N_2O + 2H_2O$$

$$NH_4NO_2 \xrightarrow{\Delta} N_2 + 2H_2O$$

(a) Both N in anion are $sp^2$-hybridised.

(b) $NH_4^+$ is $sp^3$ and $NO_2^-$ and $NO_3^-$ are $sp^2$-hybridised.

(c) Decomposition products are different.

(d) Structured $NO_2^-$ and $NO_3^-$ are different.

**20.** (a, b, c, d)

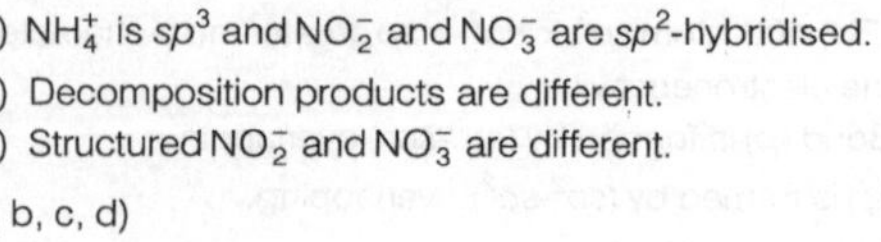

Bond angle 120° $\quad$ Bond angle 109° 28′ (on each C-atom)

(C—H) bond length < (C—H) bond length

Nodal, plane of $\pi$-bond is located in molecular plane.

21. (a) (a) $\overset{I}{P}Cl_5 \rightleftharpoons \overset{II}{P}Cl_4^+ + \overset{III}{P}Cl_6^-$ (oxidation states: +5, +5, +5)

No, change.

(b) $PCl_5 \longrightarrow PCl_4^{\oplus} + PCl_6^-$

$sp^3d$ — $sp^3$ — $sp^3d^2$

Triangular planar — Tetrahedral — Octahedral

Thus, (c) is correct.

22. (b) Lone pair on each phosphorus atom = 0

23. (c) (i) $H_2S$

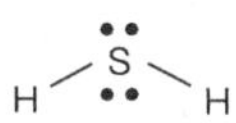

Bent (angular) (p)

Thus, **(i) → (q)**

(ii) $N_2O_4$

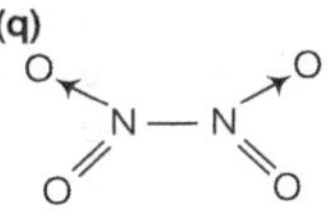

(Triangular) planar (r)

Thus, **(ii) → (r)**

(iii) H—CN — H—C≡N

Linear (p)

Thus, **(iii) → (p)**

(iv) $SbCl_6$

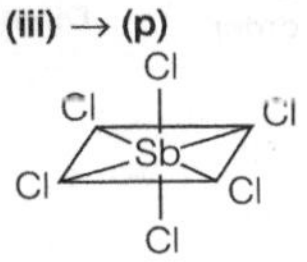

Octahedral (t)

Thus, **(iv) → (t)**

(v) $BF_4^-$

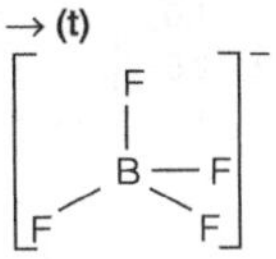

Tetrahedral (s)

Thus, **(v) → (s)**

24. (b) (i) $\underset{sp^2}{AlH_3} \longrightarrow \underset{sp^3}{AlH_4^-}$

**(i) → (p)**

(ii) $CH_3—\underset{sp^2}{C}(=O)OH \longrightarrow CH_3\underset{sp^3}{C}H_2OH$

**(ii) → (p)**

(iii) $CH_3\underset{sp^3}{C}H(OH)—CH_3 \longrightarrow CH_3\underset{sp^2}{C}H{=}CH_2$

**(iii) → (q)**

(iv) $CH_3\overset{sp^3}{C}H(Cl)CH_2(Cl) \longrightarrow CH_3—C{\equiv}\underset{sp}{C}H$

**(iv) → (r)**

(v) $CH_3—\overset{sp}{C}{\equiv}\overset{sp}{C}H \longrightarrow CH_3—\underset{sp^2}{C}(=O)—\underset{sp^3}{C}H_3$

**(v) → (s,t)**

25. (1)

Electronegativity of

$sp^3 < sp^2 < sp$-hybridised C-atom

(Greater the s-character, greater the electronegativity).

(C—H) electron pair is taken out by most (EN) atom most easily, making it is acidic.

$---- C \equiv C—H \longrightarrow ---- C \equiv C^{\ominus} + H^{\oplus}$

## DPP-6 **Molecular Orbital Theory**

1. (i) Molecular orbitals are formed when there is linear combination of atomic orbitals (LCAO) method.

$$\psi_{MO} = \phi_1 + \phi_2$$

$\phi_1$ and $\phi_2$ are atomic orbitals

| | | **Bonding** | **Anti-bonding** |
|---|---|---|---|
| 1s + 1s | $\xrightarrow{\text{Form}}$ | $\sigma 1s$ | $\sigma^* 1s$ |
| 2s + 2s | $\xrightarrow{\text{Form}}$ | $\sigma 2s$ | $\sigma^* 2s$ |
| $2p_x + 2p_x$ | $\xrightarrow{\text{Form}}$ | $\sigma 2p_x$ | $\sigma^* 2p_x$ |
| $2p_y + 2p_y$ | $\xrightarrow{\text{Form}}$ | $\pi 2p_y$ | $\pi^* 2p_y$ |
| $2p_z + 2p_z$ | $\xrightarrow{\text{Form}}$ | $\pi 2p_z$ | $\pi^* 2p_z$ |

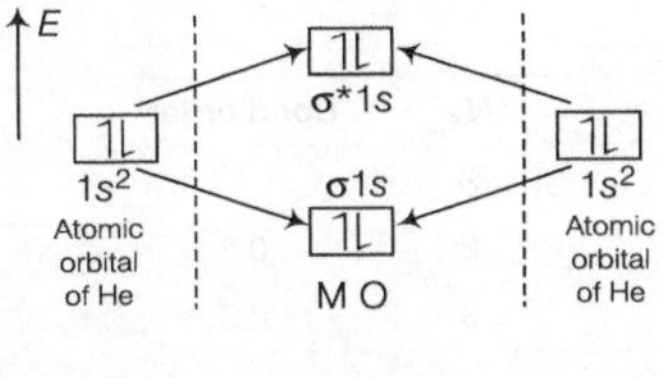


MO of higher energy is called anti-bonding MO indicated by (*).

Bond order = number of bonds between two atoms

$$= \frac{N_B - N_A}{2}$$

where, $N_B$ = electrons in bonding MO

$N_A$ = electrons in anti-bonding MO

Species is paramagnetic if there are unpaired electrons, otherwise diamagnetic.

(ii) $\sigma 1s < \sigma^* 1s < \sigma 2s < \sigma^* 2s < \underbrace{\begin{pmatrix}\pi 2p_y \\ \pi 2p_z\end{pmatrix}}_{\text{Degenerate}} < \sigma 2p_x$

(iii) $\sigma 1s < \sigma^* 1s < \sigma 2s < \sigma^* 2s < \sigma 2p_x < \begin{pmatrix}\pi 2p_y \\ \pi 2p_z\end{pmatrix} < \begin{pmatrix}\pi^* 2p_y \\ \pi^* 2p_z\end{pmatrix} \sigma^* 2p_x$

$\begin{pmatrix}\pi 2p_y \\ \pi 2p_z\end{pmatrix}$ and $\begin{pmatrix}\pi^* 2p_y \\ \pi^* 2p_z\end{pmatrix}$ are degenerate levels.

M O are filled using (I) Hund's rule

(II) Aufbau's rule

**2.** All species have electrons ≤ 14

| Species | Electrons | Molecular orbital E C | | | | | | | Electrons in | | Bond order $= \frac{N_B - N_A}{2}$ | Paramagnetic $(P)$ or Diamagnetic $(D)$ |
|---|---|---|---|---|---|---|---|---|---|---|---|---|
| | | $\sigma 1s$ | $\sigma * 1s$ | $\sigma 2s$ | $\sigma * 2s$ | $\pi 2p_y$ | $\pi 2p_z$ | $\sigma 2p_x$ | Bonding MO $(N_B)$ | Anti-bonding M O $(N_A)$ | | |
| $C_2$ | 12 | 2 | 2 | 2 | 2 | 2 | 2 | - | 8 | 4 | 2.0 | $D$ |
| $C_2^+$ | 11 | 2 | 2 | 2 | 2 | 2 | 1 | - | 7 | 4 | 1.5 | $P$ |
| $C_2^{2-}$ | 14 | 2 | 2 | 2 | 2 | 2 | 2 | 2 | 10 | 4 | 3.0 | $D$ |
| CO | 14 | 2 | 2 | 2 | 2 | 2 | 2 | 2 | 10 | 4 | 3.0 | $D$ |
| $CN^-$ | 14 | 2 | 2 | 2 | 2 | 2 | 2 | 2 | 10 | 4 | 3.0 | $D$ |
| CN | 13 | 2 | 2 | 2 | 2 | 2 | 2 | 1 | 9 | 4 | 2.5 | $P$ |
| $CN^+$ | 12 | 2 | 2 | 2 | 2 | 2 | 2 | - | 8 | 4 | 2.0 | $D$ |
| $B_2$ | 10 | 2 | 2 | 2 | 2 | 1 | 1 | - | 6 | 4 | 1.0 | $P$ |
| $Li_2$ | 6 | 2 | 2 | 2 | - | - | - | - | 4 | 2 | 1.0 | $D$ |

$\pi 2p_y$ and $\pi 2p_z$ are degenerate and follow Hund's rule.

**3.** All species have electrons > 14, Bond order $= \frac{N_B - N_A}{2}$

| Species | Electrons | Molecular orbital E C | | | | | | | | | | $N_B$ | $N_A$ | Bond order | Paramagnetic $(P)$ or Diamagnetic $(D)$ |
|---|---|---|---|---|---|---|---|---|---|---|---|---|---|---|---|
| | | $\sigma 1s$ | $\sigma * 1\sigma$ | $\sigma 2s$ | $\sigma * 2s$ | $\sigma 2p_x$ | $\pi 2p_y$ | $\pi 2\pi_z$ | $\pi * 2p_y$ | $\pi * 2p_z$ | $\sigma * 2p_x$ | | | | |
| $O_2$ | 16 | 2 | 2 | 2 | 2 | 2 | 2 | 2 | 1 | 1 | - | 10 | 6 | 2 | $P$ |
| $O_2^-$ | 17 | 2 | 2 | 2 | 2 | 2 | 2 | 2 | 2 | 1 | - | 10 | 7 | 1.5 | $P$ |
| $O_2^{2-}$ | 18 | 2 | 2 | 2 | 2 | 2 | 2 | 2 | 2 | 2 | - | 10 | 8 | 1.0 | $D$ |
| $O_2^+$ | 15 | 2 | 2 | 2 | 2 | 2 | 2 | 2 | 1 | - | - | 10 | 5 | 2.5 | $P$ |
| $F_2$ | 18 | 2 | 2 | 2 | 2 | 2 | 2 | 2 | 2 | 2 | - | 10 | 8 | 1.0 | $D$ |
| NO | 15 | 2 | 2 | 2 | 2 | 2 | 2 | 2 | 1 | - | - | 10 | 5 | 2.5 | $P$ |
| OF | 17 | 2 | 2 | 2 | 2 | 2 | 2 | 2 | 2 | 1 | - | 10 | 7 | 1.5 | $P$ |
| $Ne_2$ | 20 | 2 | 2 | 2 | 2 | 2 | 2 | 2 | 2 | 2 | 2 | 10 | 10 | 0.0 | $D$ |

$\pi 2p_y, \pi 2p_z, \pi * 2p_y, \pi * 2p_z$ are degenerate levels, follow Hund's rule.

**4.** (a) $B_2$ (10 electrons)

MO electronic configuration is $(\sigma 1s)^2 (\sigma * 1s)^2 (\sigma 2s)^2 (\sigma * 2s)^2 (\pi 2p_y)^2 (\pi 2p_z)^0$ Hund's rule is violated.

Bonding electrons = 6

Anti-bonding electrons = 4

Bond order $= \frac{6-4}{2} = 1$

No unpaired electron-diamagnetic.

**5.** (c) Species with (zero) bond order will not exist.

**Electrons in orbital**

| Species | Electrons in orbital | | Bond order $= \left(\frac{N_B - N_A}{2}\right)$ |
|---|---|---|---|
| | Bonding, $N_B$ | Anti-bonding, $N_A$ | |
| $H_2^+$ (1e) | 1 | 0 | 0.5 |
| $He_2^{2+}$ (2) | 2 | 0 | 1.0 |
| $H_2^-$(3) | 2 | 1 | 0.5 |
| $H_2^{2+}$ (0) | 0 | 0 | 0.0 |
| $He_2$(4) | 2 | 2 | 0.0 |

Thus, $H_2^{2+}$ and $He_2$ will not exist.

**6.** (b)

| Electronic configuration | $N_B$ | $N_A$ | Bond order |
|---|---|---|---|
| $Li_2$ (Six electrons) $(\sigma 1s)^2 (\sigma * 1s)^2 (\sigma 2s)^2$ | 4 | 2 | 1 |
| $Li_2^+$ (Five electrons) $(\sigma 1s)^2 (\sigma * 1s)^2 (\sigma 2s)^1$ | 3 | 2 | 0.5 |
| $Li_2^-$ (Seven electrons) $(\sigma 1s)^2 (\sigma * 1s)^2 (\sigma 2s)^2 (\sigma * 2s)^1$ | 4 | 3 | 0.5 |

$Li_2$ with maximum bond order is most stable $Li_2^+$ and $Li_2^-$ have same bond order but $Li_2^-$ is less stable having one electron in anti-bonding $(\sigma^* 2s)$ orbital.

Then, $Li_2^- < Li_2^+ < Li_2$

**7.** (c) If ($2s$-$2p$) mixing is not operative, then

| | $\sigma 1s$ | $\sigma^* 1s$ | $\sigma 2s$ | $\sigma^* 2s$ | $\sigma 2p_x$ | $\pi 2p_y$ $\pi 2p_z$ | $\pi^* 2p_y$ $\pi^* 2p_z$ | $\sigma^* 2p_x$ |
|---|---|---|---|---|---|---|---|---|
| $Be_2$(8) | 2 | 2 | 2 | 2 | | | | Diamagnetic |
| $B_2$ (10) | 2 | 2 | 2 | 2 | 2 | | | Diamagnetic |
| $C_2$(12) | 2 | 2 | 2 | 2 | 2 | 1 | – | Paramagnetic |
| $N_2$(14) | 2 | 2 | 2 | 2 | 2 | 2 | – | Diamagnetic |

**8.** (a) $O_2^+ (15e) = (\sigma 1s)^2 (\sigma^* 1s)^2 (\sigma 2s)^2 (\sigma^* 2s)^2 (\sigma 2p_x)^2 (\pi 2p_y)^2 (\pi 2p_z)^2 (\pi^* 2p_y)^1$

$O_2 (16e) = (\sigma 1s)^2 (\sigma^* 1s)^2 (\sigma 2s)^2 (\sigma^* 2s)^2 (\sigma 2p_x)^2 (\pi 2p_y)^2 (\pi 2p_z)^2 (\pi^* 2p_y)^1 (\pi^* 2p_z)^1$

$O_2^+$ Paramagnetic Bond order = 2.5

$O_2$ Paramagnetic Bond order = 2.0

**9.** (b)

| | **Electrons** |
|---|---|
| $CN^-$ | 6 + 7 + 1 = 14 |
| CO | 6 + 8 = 14 |
| $NO^+$ | 7 + 8 − 1 = 14 |
| $N_2$ | 7 + 7 = 14 |

Thus, all these are isoelectronic. MO electronic configuration is $(\sigma 1s)^2 (\sigma^* 1s)^2 (\sigma 2s)^2 (\sigma^* 2s)^2 (\pi 2p_x)^2 (\pi 2p_y)^2 (\sigma 2p_x)^2$

No electron unpaired – Diamagnetic

Bonding electrons = 10

Anti-bonding electrons = 4

Thus, bond- order = $\dfrac{10-4}{2} = 3$

**10.** (b)

| **Species** | **Electrons** | **Bond-order** |
|---|---|---|
| $N_2$ | 14 | 3 ↑ |
| $N_2^-$ | 15 | 2.5 |
| $N_2^{2-}$ | 16 | 2.0 |

$N_2^{2-} < N_2^- < N_2$

**11.** (c)

| Species | Electrons | Bond order | Unpaired electron | |
|---|---|---|---|---|
| OF | 8 + 7 = 15 | 1.5 | 1 | Paramagnetic |
| $F_2$ | 9 + 9 = 18 | 1.0 | 0 | Diamagnetic |

Bond energy of O—F (1.5) > F—F (1)

Thus, OF is more stable than $F_2$ thus, (c).

**12.** (d) A molecule is stabilised if bond-order increases, an anti-bonding electron is lost

$$X_2 \longrightarrow X_2^+ + e^-$$

| Species ($X_2$) | Electrons | Electronic configuration | Bond order in $X_2$ | Electron lost to form $X_2$ | Bond-order in $X_2^+$ |
|---|---|---|---|---|---|
| $C_2$ | 12 | $(\sigma 1s)^2 (\sigma^* 1s)^2 (\sigma 2s)^2 (\sigma^* 2s)^2 (\pi 2p_y)^2 (\pi 2p_z)^2$ | 2 | Bonding | 1.5 |
| CN | 13 | $(\sigma 2p_x)^1$ | 2.5 | Bonding | 2.0 |
| $N_2$ | 14 | $(\sigma 2p_x)^2$ | 3.0 | Bonding | 2.5 |
| $O_2$ | 16 | $(\sigma 1s)^2 (\sigma^* 1s)^2 (\sigma 2s)^2 (\sigma^* 2s)^2 (\sigma 2p_x)^2 (\pi 2p_y)^2 (\pi 2p_z)^2 (\pi^* 2p_y)^1 (\pi^* 2p_z)^1$ | 2.0 | Anti-bonding | 2.5 |

**13.** (b) Electron-charge density in a bonding molecular orbital is high in the internuclear region as shown in II. In an anti-bonding molecular orbital, it is high in parts of the molecule away from the internuclear region.

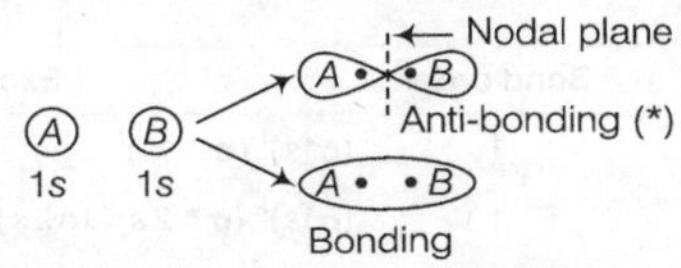

**14.** (d)

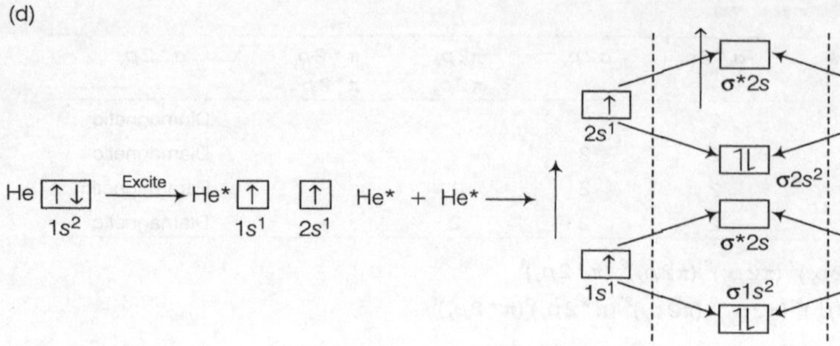


Number of electrons in bonding orbital = 4

and in anti-bonding orbital = 0

Bond order $= \frac{4}{2} = 2$

**15.** (a) $He_2$ (4 electrons) : $(\sigma 1s)^2\ (\sigma^* 1s)^2$

Bond order $= \frac{2-2}{2} = 0$

Electron is taken to next excited state that is $(\sigma 2s)$ then electronic configuration is

$$(\sigma 1s)^2\ (\sigma^* 1s)^1\ (\sigma 2s)^1$$

Number of electrons in bonding molecular orbital = 3

and in anti-bonding molecular orbital = 1

Thus, bond order = (3 – 1) /2 = 1

Thus, bond order increases by 1 unit.

**16.** (b) $(H_2)$, H—H

Bond order = 1

Bond energy of one (H—H) bond = 436 kJ $mol^{-1}$

$(H_2^+)$, $H \cdot H^+$ Bond order = 0.5

Then bond energy $= 436 \times \frac{1}{2} = 218$ kJ $mol^{-1}$

**17.** (d) $O_2^+$ $(15e^-)$ $O_2^-$ $(17e^-)$

One unpaired electron in $\pi^* 2p_y^1$ — One unpaired electron in $\pi^* 2p_z^1$

Thus, in each case, magnetic moment $= \sqrt{N(N+2)}$ BM $= \sqrt{3}$ BM

**18.** (b) $N_2(14e^-)$ : $(\sigma 1s)^2\ (\sigma^* 1s)^2 (\sigma\ 2s)^2 (\sigma^* 2s)^2 (\pi\ 2p_y)^2 (\pi 2p_z)^2\ (\sigma 2p_x)^2$ (bonding)

$$N_2 \longrightarrow N_2^+ + e^-$$

Electron from bonding molecular orbital of higher stability is lost (requires higher energy).

$O_2(16e^-)$ : $(\sigma 1s)^2\ (\sigma^* 1s)^2 (\sigma\ 2s)^2 (\sigma^* 2s)^2\ (\sigma 2p_x)^2\ (\pi\ 2p_y^2)\ (\pi^* 2p_y)^1\ (\pi 2p_z)\ (\pi^* 2p_z)^1$ (anti-bonding)

$$O_2 \longrightarrow O_2^+ + e^-$$

Electron from antibonding molecular orbital of lower stability is lost (requires lower energy).

$N_2$ has 14 electrons and $O_2$ has 16 electrons.

Thus, both Statements I and II are correct but Statement II is not the correct explanation of Statement I.

**19.** (a, b, d)

Isoelectronic species will generally have identical bond order :

| | | Bond order |
|---|---|---|
| (a) | NO (15e), $O_2^+$ (15e) | 2.5 |
| (b) | CO (14e), $NO^+$ (14e) | 3.0 |
| (c) | $N_2$ (14e), $O_2^{2-}$ (18e) | 3.0, 1.0 |
| (d) | $F_2$ (18e), $O_2^{2-}$ (18e) | 1.0 |

**20.** (a, b)

| | Bond order | Excited state | Bond order | |
|---|---|---|---|---|
| $H_2$ (2 electrons) $(\sigma 1s)^2\ (\sigma^* 1s)^0$ | 1 | $(\sigma 1s)^1\ (\sigma^* 1s)^1$ | 0 | Changes |
| $Li_2$ (6 electrons) $(\sigma 1s)^2\ (\sigma^* 1s)^2\ (\sigma 2s)^2\ (\sigma^* 2s)^0$ | 1 | $(\sigma 1s)^2\ (\sigma^* 2s)^2\ (\sigma 2s)^1\ (\sigma^* 2s)^1$ | 0 | Changes |
| $B_2$ (10 electrons) $(\sigma 1s)^2\ (\sigma^* 1s)^2$ $(\sigma 2s)^2\ (\sigma^* 2s)^2 (\pi 2p_y)^1\ (\pi 2p_z)^1$ | 1 | $(\sigma 1s)^2\ (\sigma^* 2s)^2 (\sigma 2s)^2\ (\sigma^* 2s)^2 (\pi 2p_y)^1\ (\pi 2p_z)^0$ $(\sigma 2p_x)^1$ | 1 | No change |
| $C_2$ (12 electrons) $(\sigma 1s)^2\ (\sigma^* 1s)^2$ $(\sigma 2s)^2\ (\sigma^* 2s)^2\ (\pi 2p_y)^2\ (\pi 2p_z)^2$ | 2 | $(\sigma 1s)^2\ (\sigma^* 1s)^2 (\sigma 2s)^2\ (\sigma^* 2s)^2$ $(\pi 2p_y)^2\ (\pi 2p_z)^1\ (\sigma 2p_x)^1$ | 2 | No change |

**21.** (a, d)

(a) CO (14$e$), no unpaired electron, magnetic moment = 0

$CO^+$ (13$e$), One electron is lost from bonding melecular orbital unpaired electron = 1, magnetic moment = $\sqrt{3}$ BM

(b) $N_2^-$ (15$e$) Unpaired electron = 1, magnetic moment = $\sqrt{3}$BM

$N_2^+$ (13$e$) Unpaired electron = 1, magnetic moment = $\sqrt{3}$BM

(c) Zn, $Zn^{2+}$, unpaired electron = 0

(d) $O_2$ (16$e$), unpaired electron = 2, magnetic moment = $\sqrt{8}$ BM

$O_2^-$ (15$e$), unpaired electron = 1, magnetic moment = $\sqrt{3}$BM

**Explanation for Q. Nos. (22-23)**

Total number of electrons = 16

[Including four of $(\sigma 1s)^2 (\sigma^* 1s)^2$ for inner orbitals]

Number of bonding electrons = 10,

Number of anti-bonding electrons = 6,

$$\text{Bond order} = \frac{10-6}{2} = 2$$

$$\underset{16e^-}{X_2} \longrightarrow \underset{14e^-}{X_2^{2+}} + 2e^-,$$

$$X_2 + 2e^- \longrightarrow X_2^{2-} \quad (18\,e^-)$$

| | Unpaired electrons | Magnetic moment | Bond order |
|---|---|---|---|
| $X_2$ | 2 | $\sqrt{8}$ BM | 2 |
| $X_2^{2+}$ | 0 | 0 | 3 |
| $X_2^{2-}$ | 0 | 0 | 1 |

Thus, $X_2^{2+}$ and $X_2^{2-}$ have same magnetic moment = 0

$X_2^{2+}$ and $X_2^{2-}$ have different bond order and bond energy.

**22.** (b) **23.** (a)

**24.** (d)

| Species | Electrons | Bond order | Unpaired electrons ($N$) | Magnetic moment $=\sqrt{N(N+2)}$BM |
|---|---|---|---|---|
| $N_2$ | 14 | 3 | 0 | 0 |
| $O_2$ | 16 | 2 | 2 | $\sqrt{8}$ BM |
| $N_2^+$ | 13 | 2.5 | 1 | $\sqrt{3}$ BM |
| $N_2^-$ | 15 | 2.5 | 1 | $\sqrt{3}$ BM |
| $O_2^-$ | 17 | 1.5 | 1 | $\sqrt{3}$ BM |
| $O_2^+$ | 15 | 2.5 | 1 | $\sqrt{3}$ BM |
| $O_2^{2-}$ | 18 | 1.0 | 0 | 0 |

Thus, (i) - (q,r), (ii) - (q,t),
(iii) - (q,r), (iv) - (p,s),
(v) - (q,s).

**25.** (7)

$O_2^-$ (17 electrons) has molecular orbital electronic configuration $(\sigma 1s)^2 (\sigma^* 1s)^2 (\sigma 2s)^2 (\sigma^* 2s)^2 (\sigma 2p_x)^2 (\pi 2p_y)^2 (\pi 2p_z)^2 (\pi^* 2p_y)^2 (\pi^* 2p_z)^1$

Underlined are anti-bonding molecular orbital

Thus, seven electrons are in anti-bonding molecular orbitals.

**26.** (5)

NO (15 electrons)

$(\sigma 1s)^2 (\sigma^* 1s)^2 (\sigma 2s)^2 (\sigma^* 2s)^2 (\sigma 2p_x)^2 (\pi 2p_y)^2 (\pi 2p_z)^2 (\pi^* 2p_y)^1$

MO without (*) are bonding MO.

## DPP-7 **Dipole Moment**

**1.** (i) **PLAN** Dipole moment of polyatomic species depends on the symmetry of the molecule or vector summation of individual moments.

(i) and (iv) are symmetrical thus $\mu = 0$.

In (b), bond-moments are towards the same direction.

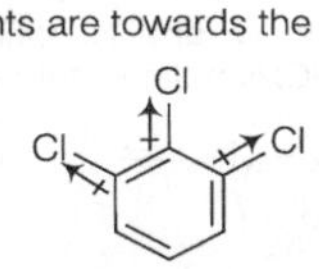


In (iii), dipole moment of (C—Cl) at $C_1$ is cancelled by that of at $C_4$. There is net dipole due to (C—Cl) at $C_2$.

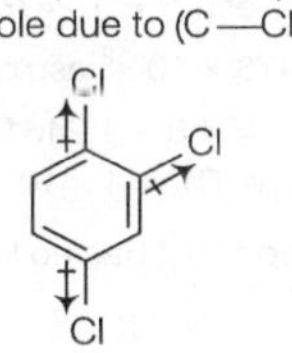


Thus, (i) = (iv) < (iii) < (ii)

**2.** **PLAN** Percentage ionic character

$$= \frac{\text{Observed value} \times 100}{\text{Calculated value}}$$

Observed dipole moment = 6.32 D

$0.156\text{ nm} = 0.156 \times 10^{-9}\text{ m} = 0.156 \times 10^{-7}\text{cm}$

Calculate value of dipole moment = $e \times l$

$= 4.8 \times 10^{-10}$ esu $\times 0.156 \times 10^{-7}$cm

$= 7.488 \times 10^{-18}$ esu cm

= 7.488 D

Thus, percentage ionic character

$$= \frac{6.32}{7.48} \times 100 = 84.4\%$$

**3.** 1 Debye = $3.336 \times 10^{-30}$Cm

also, 1 Debye = $1.0 \times 10^{-18}$esu cm

$$\therefore \quad 1\text{ esu cm} = \frac{3.336 \times 10^{-30}\text{Cm}}{1.0 \times 10^{-18}} = 3.336 \times 10^{-12}\text{Cm}$$

**4.** In $NH_3$, the orbital dipole due to lone pair is in the same direction as the resultant dipole moment of the N—H bonds, whereas, in $NF_3$, the orbital dipole is in, the direction opposite to the resultant dipole moment of the three N—F bonds. Thus, dipole moment of $NF_3$ is smaller than that of $NH_3$.

**5.** With increase in polarity of B—X bond, acidity also increases and should be thus in the order :

$$BF_3 > BCl_3 > BBr_3$$

But Lewis acidity is in reverse order :

$$BF_3 < BCl_3 < BBr_3$$

Three is lateral overlap of the vacant $2p$-orbital of B with one completely filled orbital of F leading to $p\pi$-$p\pi$ bonds between B and F. This B—F bond thus acquires double bond character. This also leads to compensate electron deficiency of boron and thus Lewis acid character of $BF_3$ is reduced. This $p\pi$-$p\pi$ bonding decreases going from $BF_3$ to $BBr_3$ and thus Lewis acidic nature increases in the order :

$$BF_3 < BCl_3 < BBr_3$$

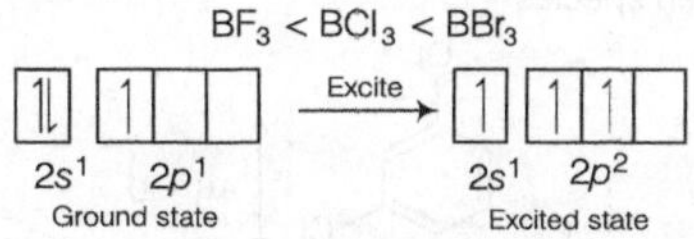


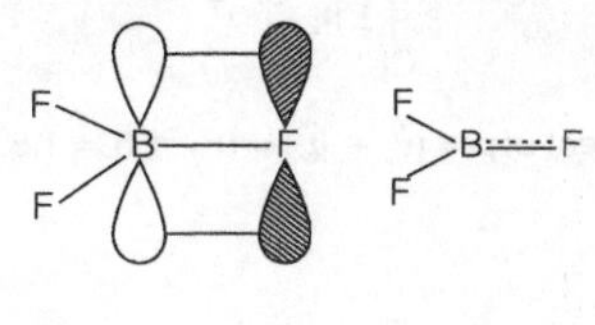

**6.** (d),

(a) $CH_2Cl_2$ has three different bond angles HCH, HCCl and ClCCl which are close but not equal to 109.5°. Since, electronegativity of H < C < Cl, bond moments do not cancel each other hence, $\mu \neq 0$.

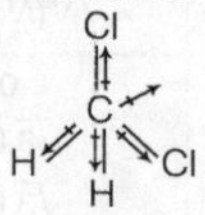

(b) 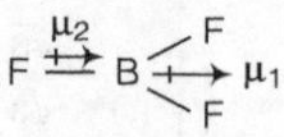

Resultant $\mu_1$ of (FBF) is cancelled by $\mu_2$ of (B—F).

$$\mu_{net} = \vec{\mu}_1 + \overleftarrow{\mu}_2 = \vec{\mu}_1 - \vec{\mu}_2 = 0$$

(c) 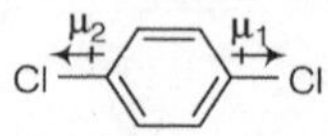 $\mu_{net} = 0$

(d) O=C=O $\quad \mu_{net} = 0$

**7.** (c) $\mu(CH_3Cl)$ 1.860 D (max)

$\mu(CH_3F)$ 1.847 D

$\mu(CH_3Br)$ 1.830 D

| C—F | C—Cl | C—Br | C—I | (Bond length pm) |
|---|---|---|---|---|
| 139 | 178 | 193 | 214 | |

$\longrightarrow$ Increases

I < Br < Cl < F (EN)

$\mu = e^- X$ bond length. 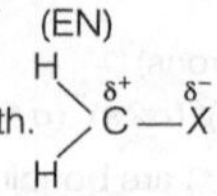

Bond length and (EN) oppose each other, with the larger halogens having longer bonds but weaker (EN). Thus $CH_3Cl > CH_3F > CH_3Br$.

**8.** (c)

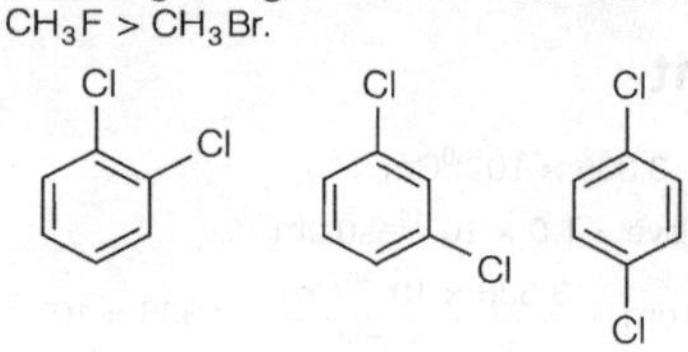

| *ortho* | *meta* | *para* |
|---|---|---|
| θ=60° | 120° | 180° |

$$\mu = 2\mu_{C-Cl}\cos\left(\frac{\theta}{2}\right) \Rightarrow \mu \propto \cos\left(\frac{\theta}{2}\right)$$

$\mu \propto \frac{1}{\theta}$ (since $\cos\theta \propto \frac{1}{\theta}$)

Thus, $\mu_p (= 0) < \mu_m < \mu_o$

Toluene with one $—CH_3$ is less polar then *m* - and *o*-dichlorobenzene. Thus, IV < I < III < II.

**9.** (b) If one considers following molecule, it is

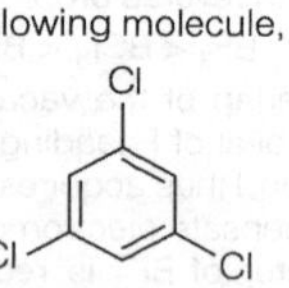

Symmetrical and thus $\mu = 0$

In the given species

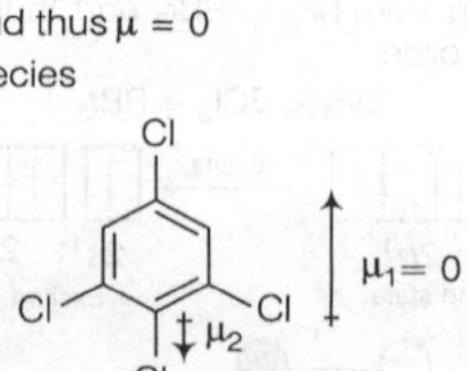

Dipole moment of $\mu = \vec{\mu}_1 + \vec{\mu}_2 = 0 + 1.5 = 1.5$ D

**10.** (c) $\mu = e \times l$

$\mu_{HCl} = e \times l(HCl)$

$1.03 \times 10^{-18}$ esu cm $= e \times 1.3 \times 10^{-8}$ cm

$$\text{Charge on HCl} = \frac{1.03 \times 10^{-18}}{1.30 \times 10^{-8}} \text{ esu}$$

$$\text{Charge on HI, } e = \frac{0.38 \times 10^{-18}}{1.6 \times 10^{-8}} \text{ esu}$$

$$\frac{e\,(HCl)}{e\,(HI)} = \frac{1.03 \times 1.6}{1.30 \times 0.38} = 3.336 : 1$$

**11.** (d)

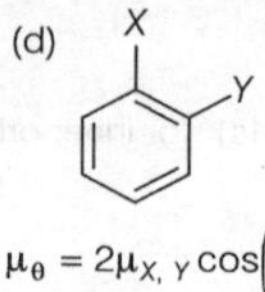

$$\mu_\theta = 2\mu_{X,Y}\cos\left(\frac{\theta}{2}\right)$$

where, $\theta = (X—C—Y)$ angle

$\mu_{X,Y} =$ vector moment of side *X* or *Y*

$$\mu_0 \propto \cos\left(\frac{\theta}{2}\right) \Rightarrow \mu_0 \propto \frac{1}{\theta}$$

Larger the value of θ, smaller the value of μ. Dipole-dipole repulsion (due to large groups (as Cl, Cl) increases θ, thus decreases μ.

Intramolecular H-bonding (as OH and $NO_2$) decreases θ, thus increases μ.

Repulsion

$(Cl—Cl) > (OH—OH) > (CH_3—CH_3)$

Thus, μ is minimum when $X = Y = Cl$

and μ is maximum when $X = OH$

$Y = NO_2$

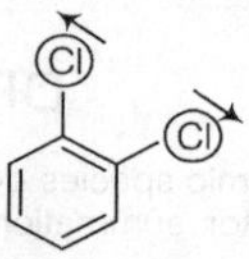

Repulsion increases θ

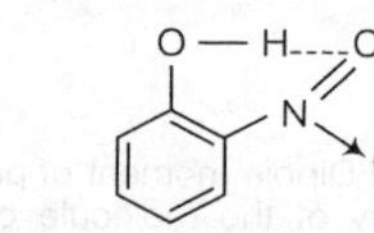

Intramolecular H-bonding decreases θ

**12.** (a) (EN) difference of Cl and H, makes Cl atom an electron withdrawing. Thus, HCl is polar in nature.

$$H—Cl \rightleftharpoons \overset{\delta+}{H}—\overset{\delta-}{Cl}$$

Dipole moment, $\mu = e \times l$

**13.** (a) $\mu = e \times l$

Given, $e = 4.8 \times 10^{-10}$ esu, $l = 1.2476 \times 10^{-8}$ cm

$\therefore \quad \mu = 4.8 \times 1.2476 \times 10^{-18}$ esu cm

$= 4.8 \times 1.246$ D $= 6.0$ D

1 Debye (D) $= 1 \times 10^{-18}$ esu cm

**14.** (a) $PCl_xF_{5-x}$ is non-polar if it has no lone pair on P.

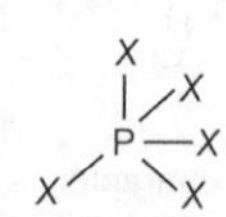

If $x = 0$, $PF_5$

$x = 3$, $PCl_2F_3$

$x = 5$, $PCl_5$

**15.** (c) van der Waals' force is the weakest force of attraction and covalent bond is the strongest. Correct order is

van der Waals' force — Weakest

Hydrogen bond

Coordinate bond

Covalent bond — Strongest

**16.** (d),

(a) 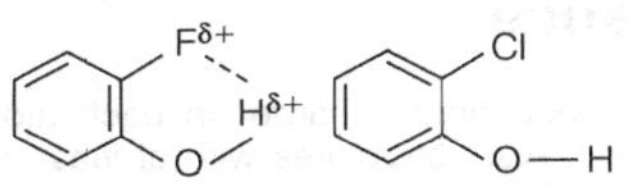

due to H-bonding angle decreases hence, $\mu$ increases. — no H-bonding

(b) as in (a) — no H-bonding.

(c) $\theta = 120°$ — $\theta = 180°$, $\cos 90 = 0$

Thus, $\mu = 0$

**17.** (c)

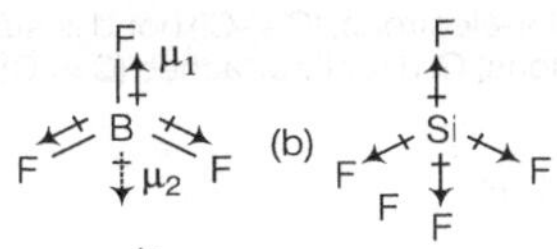

$\mu = \vec{\mu}_1 + \overleftarrow{\mu}_2 = 0$ — $\mu = 0$

$sp^2$ (triangular planar) — $sp^3$ (tetrahedral)

(c) $SF_4$ has one lone pair making it polar permanently.

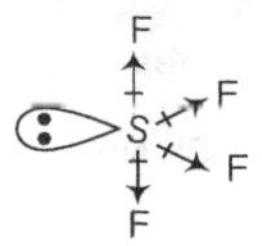

see-saw

$sp^3d$-hybridised S-atom.

(d) $XeF_4$ has two lone pairs in opposite side thus $\mu = 0$.

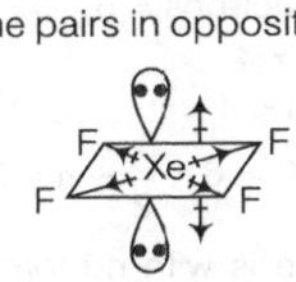

Xe is $sp^3d^2$ hybridised

(Square planar)

**18.** (a, d)

$SO_3$ 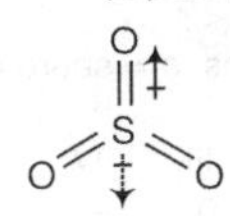 Symmetrical $\mu = 0$

$[Ni(CN)_4]^{2-}$ 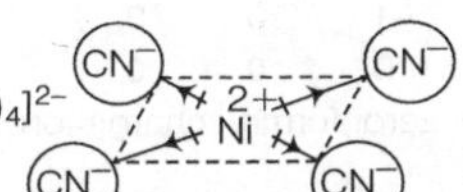 $\mu = 0$

$CCl_4$ 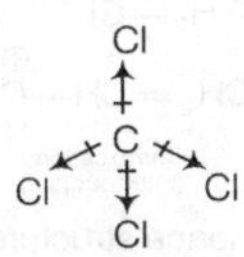 $\mu = 0$

$C_6H_6$ Symmetrical $\mu = 0$

$CO_2$ Symmetrical $\mu = 0$

CO, $SO_2$, $H_2O$, $NH_3$ and HCl have polarity thus, have dipole moments.

**19.** (b, d)

(a) 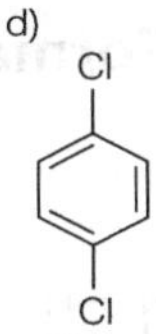 Symmetrical, $\mu = 0$

(b) 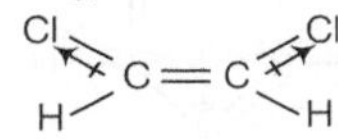 $\mu > 0$

(c) 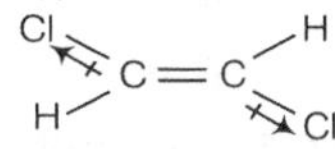 $\mu = 0$

(d) 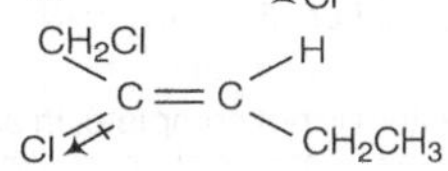 $\mu > 0$

**20.** (b, c)

(a) 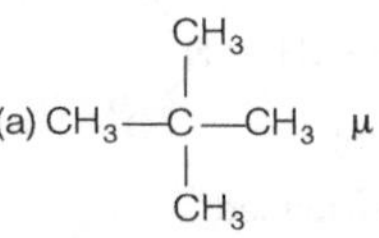 $\mu = 0$

(b) 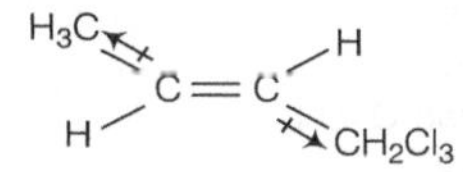 $\mu > 0$

(c) 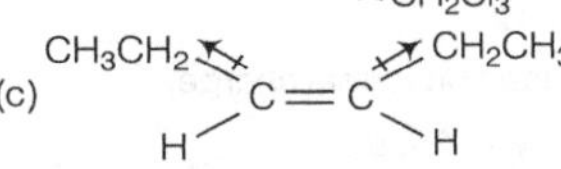 $\mu > 0$

(d) 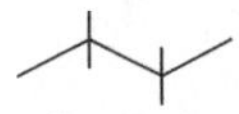 $\mu = 0$

Symmetry

**21.** (c) 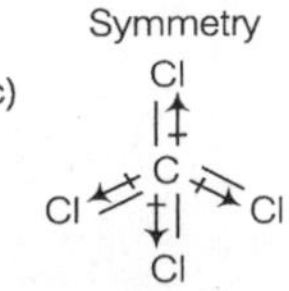

Resultant dipole moment of $CCl_4 = 0$.

**22.** (a) $CH_4 \longrightarrow \dot{C}H_3$ (free radical)

$\dot{C}H_3$ has one unpaired electron in unhybridised *p*-orbital. Thus, $sp^2$-hybridised paramagnetic.

**23.** (5) Due to electronegativity difference between the bonded atoms. ICl, NO, $SO_2$, $CH_2Cl_2$, OCS are polar compounds.

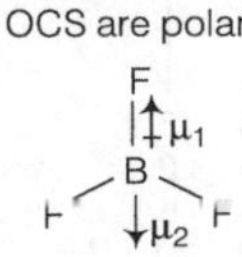

is non-polar due to $\vec{\mu}_1 + \overleftarrow{\mu}_2 = 0$.

$XeF_4$ $\mu = 0$

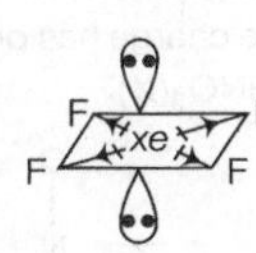

$CO_2$ 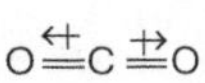 $\mu = 0$

**24.** (7) $CH_2Cl_2$, $H_2O$, $CHCl_3$, *o*-cresol, $SCl_2$, IBr and HCHO (in all seven) have **dipole moments**.

# DPP-8 Formal Charge and Resonance

**1.** (c)

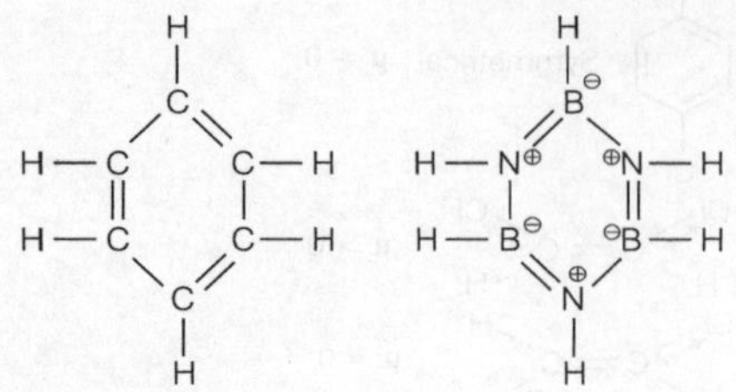

(C—H) bond lengths are different from (B—H) and (N—H) bond lengths. Both are planar.

**2.** (a) In structure (a), octet of N-atom is not complete. In all the other structures, octets of N and all the O-atoms are complete.

**3.** (c)

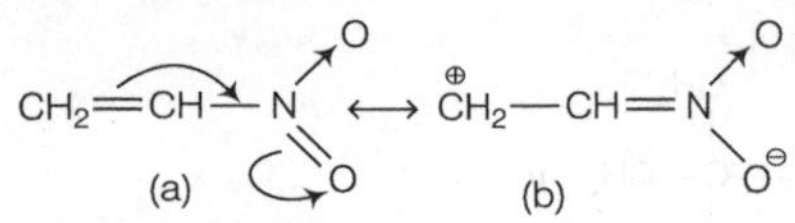

Due to resonance, both structures are formed.

**4.** (c), (a) Octet of C-incomplete.

(b) Octet of left C-incomplete.

(d) Valence electrons on N and O = 11.

but in the given structure = 12

**5.** (b) Structure is decided based on formal charge.

$$F = v - \frac{s}{2} - u$$

where, $v$ = valence electrons

$s$ = shared electrons

$u$ = unshared electrons

(a)

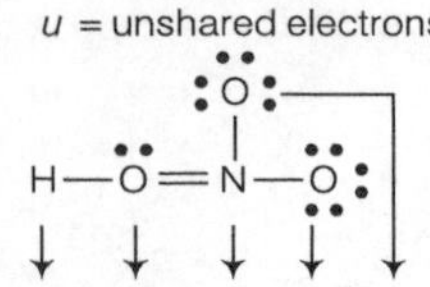

| | H | O | N | O | O |
|---|---|---|---|---|---|
| $v$ | 1 | 6 | 5 | 6 | 6 |
| $s$ | 2 | 6 | 8 | 2 | 2 |
| $u$ | 0 | 2 | 0 | 6 | 6 |
| F | 0 | 1 | 1 | −1 | −1 |

(b)

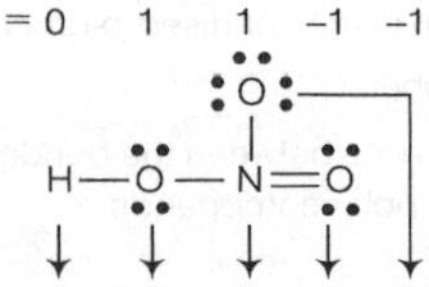

| | H | O | N | O | O |
|---|---|---|---|---|---|
| $v$ | 1 | 6 | 5 | 6 | 6 |
| $s$ | 2 | 4 | 8 | 4 | 2 |
| $u$ | 0 | 4 | 0 | 4 | 6 |
| F | 0 | 0 | 1 | 0 | −1 |

Structure with at least one neutral atom is favoured.

**6.** (d) Greater the resonating structures, greater the stability.

I : $NO_3^-$ ($Y$) with negative charge has greater delocalisation of $\pi$-electrons than $HNO_3$($X$).

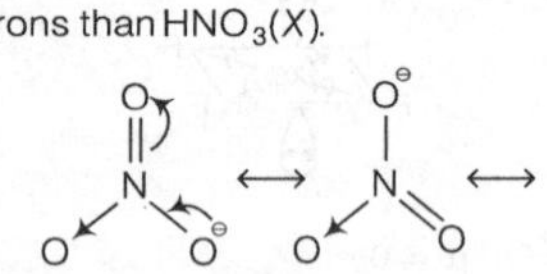

II :

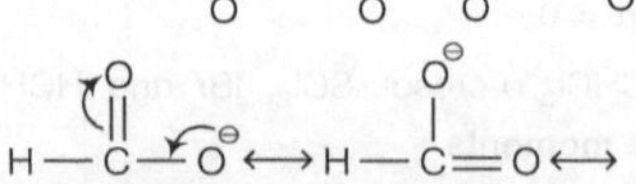

III :

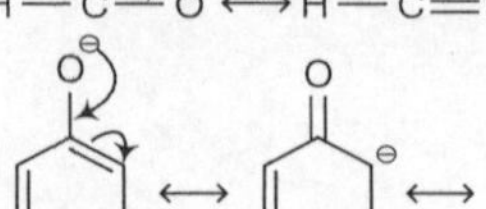

**7.** (a) Structure with least formal charge on each atom, makes greater contributions. Also, structures with at least one neutral atom is also favoured II and III identical.

| Structure | Formal charge on N left | middle | right |
|---|---|---|---|
| I | −1 | + 1 | −1 |
| II | −2 | + 1 | 0 |
| III | −2 | + 1 | 0 |

**8.** (c) $CH_2{=}CH{-}\ddot{\underset{\cdot\cdot}{Cl}}: \leftrightarrow \overset{\ominus}{C}H_2 - CH = \overset{\oplus}{Cl}$

Due to delocalisation of $\pi$-electrons, (C—Cl) bond is stable and it does not show $S_N$ reactions; Cl directly attached (C = C) bond, i.e. vinyl group.

(a) 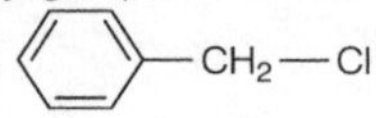 No

(b) 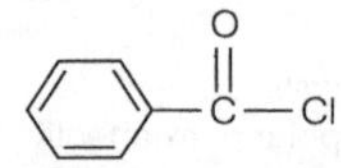 No

(c) 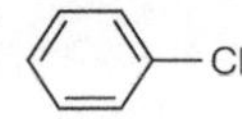 Yes

(d) $CH_2{=}CH - CH_2{-}Cl$ No

**9.** (c) Formal charge ($F$)

$$F = v - \frac{s}{2} - u$$

where, $v$ = valence electrons = 6

$s$ = shared electrons = 4

$u$ = unshared electron = 4

$$\therefore \quad F = 6 - \frac{4}{2} - 4 = 0$$

**10.** (d) The best structure is with no formal charge or least formal charge.

Formal charge ($F$) = $v - \frac{s}{2} - u$

where, $v$ = valence electrons, $s$ = share electrons, $u$ = unshared electrons.

| Structure | I | II | III | IV |
|---|---|---|---|---|
| O | 0 | −1 | 0 | −1 |
| Cl | +2 | +2 | 0 | +1 |
| N | −2 | −1 | 0 | 0 |
| Net | 0 | 0 | 0 | 0 |

Structure III with zero formal charge on each atom is most favoured.

**11.** (c) $S_N1$ means unimolecular (1) nucleophilic (N) substitution (S). Rate is independent of molar concentration of $H_2O$ but is dependend on molar concentration of $CH_2{=}CHCH_2{-}Cl$.

$$\underset{\text{I}}{CH_2{=}CH{-}\overset{14}{C}H_2Cl} \rightleftharpoons \underset{\text{Carbocation (Intermediate)}}{CH_2{=}CH{-}\overset{\oplus}{\underset{14}{C}}H_2}$$

Carbocation exists as resonance structures.

$$CH_2{=}CH{-}\overset{\oplus}{\underset{14}{C}}H_2 \rightleftharpoons \overset{\oplus}{C}H_2{-}CH{=}\underset{14}{C}H_2$$

Thus $H_2O$ (nucleophile) can attack at $C^{\oplus}$.

$$CH_2{=}CH{-}\overset{\oplus}{\underset{14}{C}}H_2 + H_2O \longrightarrow \overset{14}{C}H_2{=}CH{-}\overset{14}{C}H_2OH$$

($^{14}$C) at original position)

$$\overset{\oplus}{C}H_2{-}CH{=}\underset{14}{C}H_2 + H_2O \longrightarrow \overset{14}{C}H_2{=}CH{-}CH_2OH$$

($^{14}$C at changed position due to resonance).

**12.** (b)

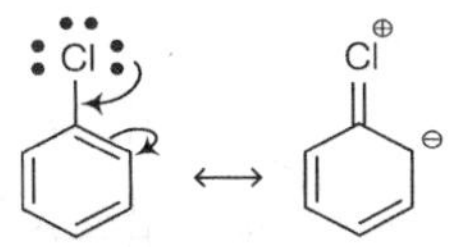

(C—Cl) bond has (C =Cl) double bond character, thus it is stable.

(a) 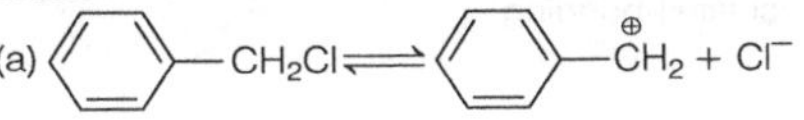

Benzyl carbocation is stable due to resonance, hence (C—Cl) bond is cleaved easily.

(b) $CH_2=CH-\ddot{C}l: \rightleftharpoons \overset{\ominus}{C}H_2-CH=\overset{\oplus}{C}l$
As in chlorobenzene

(c) $CH_2=CH-CH_2Cl \longrightarrow CH_2=CH-\overset{\oplus}{C}H_2 + Cl^{\ominus}$
Stable due to resonance, hence (C—Cl) bond is cleaved easily

**13.** (a) $CH_2=CH-\ddot{C}l: \longleftrightarrow \overset{\ominus}{C}H_2-CH=\overset{\oplus}{C}l$ Yes

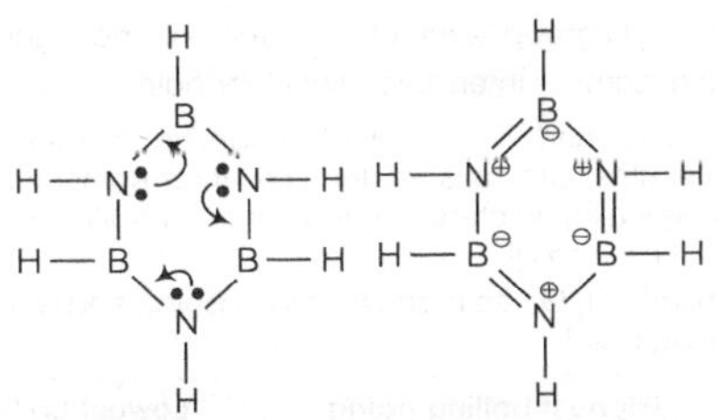

Yes

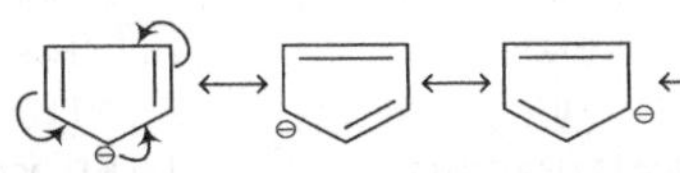 $\longleftrightarrow$ Yes

**14.** (a) Formal change ($F$) is given by

$$F = v - \frac{s}{2} - u$$

where, $v$ = valence electrons on the atom
$s$ = shared electrons
and $u$ = unshared electrons

| Atom | $v$ | $s$ | $u$ | $F$ |
|---|---|---|---|---|
| **Sulphur** | 6 | 6 | 2 | +1 |
| **Double bonded oxygen** | 6 | 4 | 4 | 0 |
| **Single bonded oxygen** | 6 | 2 | 6 | −1 |

This, its sturcture is $^{\ominus}O-\overset{\oplus}{S}=O$.

**15.** (a) **Plan** Oxygen is more electronegative than oxygen, hence the structure with a negative formal charge on oxygen atom is probably lower in energy than the structure that has a negative formal charge on N-atom is more accurate.
Thus (I) is more accurate than (II).

**16.** (a, b, c)
(a) All (O—O) bonds in $O_3$ are equivalent, thus it is resonance hybrid of I and II- True.

| | | | | | |
|---|---|---|---|---|---|
| $v$ = 6 | 6 | 6 | 6 | 6 | 6 |
| $s$ = 4 | 6 | 2 | 2 | 6 | 4 |
| $u$ = 4 | 2 | 6 | −1 | 1 | 0 |

Thus, both structures are acceptable.

(b) Bond order $= \frac{2+1}{2} = 1.5$ -true

(c) (O=O) in $O_2$ = 120.74 pm
(O—O) in $H_2O_2$ = 147.5 pm
(O—O) in $O_3$ = 127.8 pm
Thus, (c) is true.

(d) No unpaired electron in $O_3$ –diamagnetic.
Thus, (d) is false.

**17.** (b, c, d) Due to complete-delocalisation of $\pi$-electrons, benzene exists as resonance structures.

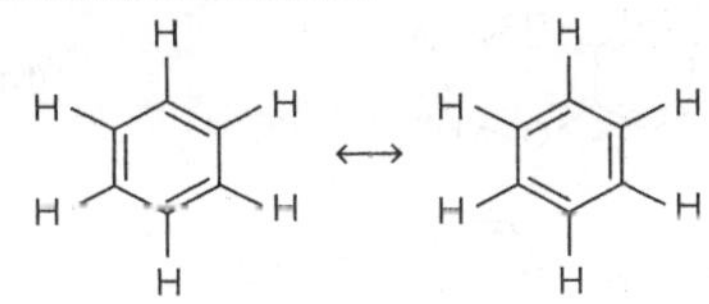

Thus, (C =C) bonds and (C—C) bonds are of equal length and stability.
All (C—H) bonds are equivalent.
C-atom is also $sp^2$-hybridised.

**Explanations for Q. Nos. (18-20)**

Bond length
$N-O > N-\overset{*}{O}, \overset{*}{O}-F = 142$ pm
double bond character — single bond character

$\angle ONO = 125°$ indicating $sp^2$- hybridised N -atom

($\overset{*}{O}$—F) bond length of 142 pm indicates single bond thus O-atom is $sp^3$-hybridised.

$\angle F-\overset{*}{O}-N = 105°$, thus $F\overset{*}{O}N$ plane is perpendicular to $O_2N\overset{*}{O}$ plane.

**18.** (c) **19.** (a) **20.** (d)

**21.** (6)Lone pair on N-atom is not delocalised in pyridine.

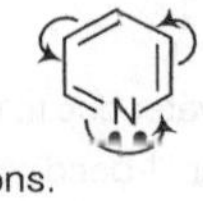

Thus total = 6$\pi$-electrons.

**22.** (5) Addition in conjugated diene can be 1, 2- or 1, 4-type

$$\underset{A}{CH_3-CH}=CH-CH=\underset{B}{CH-CH_3}$$

At position $A, B$ it is called 1, 4-addition.

# DPP-9 **Hydrogen Bonding**

**1. Plan** There are two types of H–bonding between

I. —H and electronegative atoms (*X*, O, N) are present in the same molecule of the substance–intramolecular H-bonding.

Extent of H-bonding is small, hence such compounds are volatile.

II. —H and electronegative atoms (*X*, O, N) are present in different molecules of same substance or different substances – intermolecular H-bonding.

Extent of H-bonding is large, hence such compounds have higher boiling point than that of type I and are less volatile.

(i) 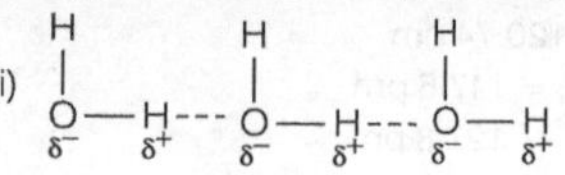 (type II)

(ii) 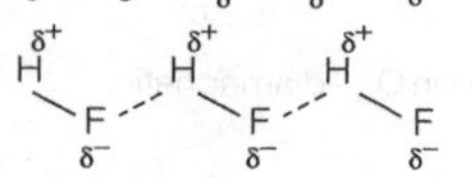 (type II)

(iii) 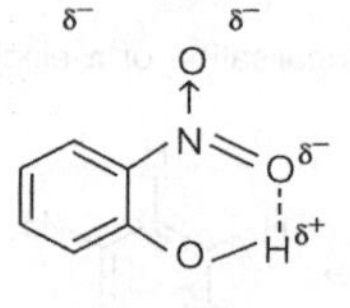 (type I)

(iv) 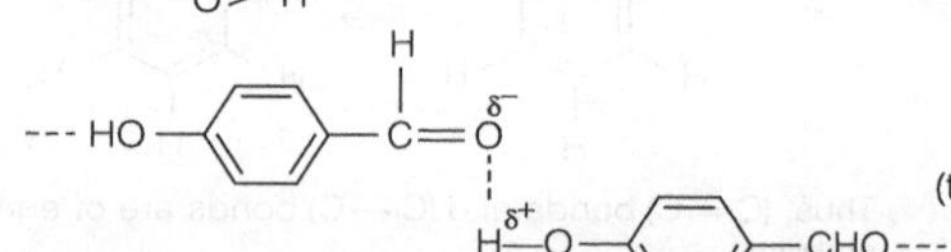 (type II)

**2.** $NH_3$ molecular are associated by intermolecular H-bonding. In case of $PH_3$, electronegativity of P is very small. Hence, no H-bonding and $PH_3$ exists in gaseous state. Boiling point of $PH_3 < AsH_3 < NH_3$.

**3.** (a) Polar H of carboxylic acid is attached to —OH group.

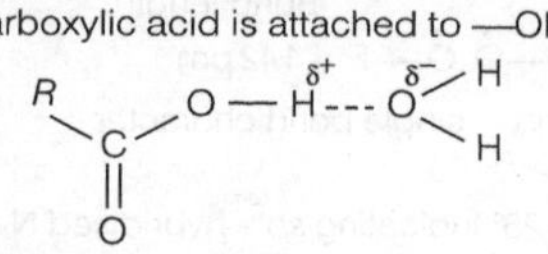

**4.** (d) 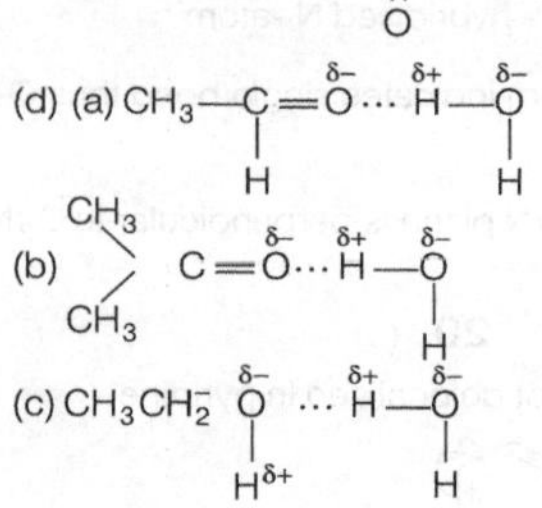

All these are soluble in water due to intermolecular H-bonding.

**5.** (a) Due to intermolecular H-bonding at two terminals boiling point of glycol is maximum.

**6.** (c)

(a) Intermolecular H-bonding, boiling point increases with molecular weight for compounds within the same family.

(b) H-bonding is absent is $CH_3CH_2CH_3$.

(c) Methanol is the larger molecule but $H_2O$ has more H-bonding because of two O— H bonds.

Thus, boiling point of $H_2O$ (II) > $CH_3OH$(I).

(d) Molecular weight of butanol is much higher and overcomes attractive forces of butanol.

**7.** (b) H-bonding is between polar $\overset{\delta+}{H}$ and electronegative O, N and F C-atom or H attached to C is not polar. Thus, (a), (c), (d) are incorrect.

**8.** (a)

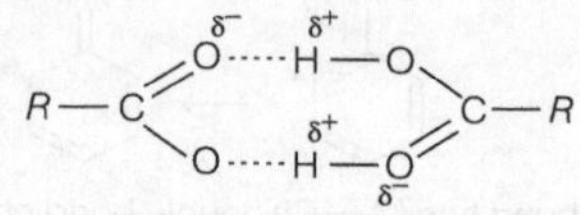

Thus, oxygen atom (1) is involved in dimer formation by intermolecular H-bonding.

**9.** (d)

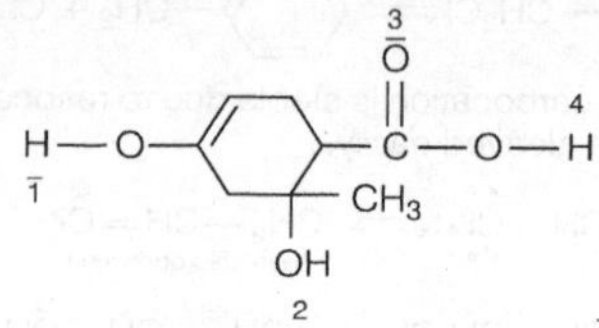

1. Enolic H, acidic and polar takes part in H-bonding.
2. 3° alcohol least acidic maximum basic.
3. Carbonyl oxygen is not involved.
4. Acidic and polar H, takes part in H-bonding.

**10.** (b) O—H group is attached to (ene) is most acidic and thus, has maximum intermolecular H–bonding.

**11.** (a) F, O and N are electronegative elements. Thus, HF molecules are associated by intermolecules H-bonding giving it a liquid state and thus highest boiling point out of HF, HCl, HBr and HI.

Similarly, $H_2O$ has highest boiling point, and $NH_3$ has highest boiling point.

| | Highest boiling poing | Lowest boiling point |
|---|---|---|
| I | HF | HCl |
| II | $H_2O$ | $H_2S$ |
| III | $NH_3$ | $PH_3$ |
| | due to H-bonding | Lack H-bonding |

**12.** (c) (EN) of S< O< F

Thus, H is attracted to larger extent by F atom making (H······F) bond shortest.

Order is $\overset{\delta+}{H}\cdots\cdots\overset{\delta-}{S} < \overset{\delta+}{H}\cdots\cdots\overset{\delta-}{O} < \overset{\delta+}{H}\cdots\cdots\overset{\delta-}{F}$

**13.** (a) $CH_3CH_2OH$ is in liquid state due to association as a result of intermolecular H-bonding.

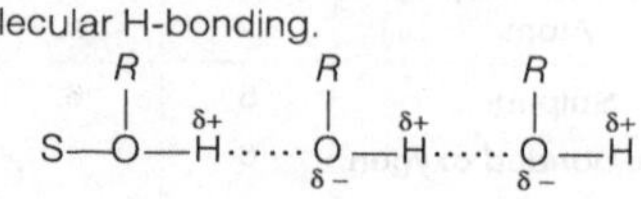

Vapours are formed due to overcoming of these H–bonding.

**14.** (d)

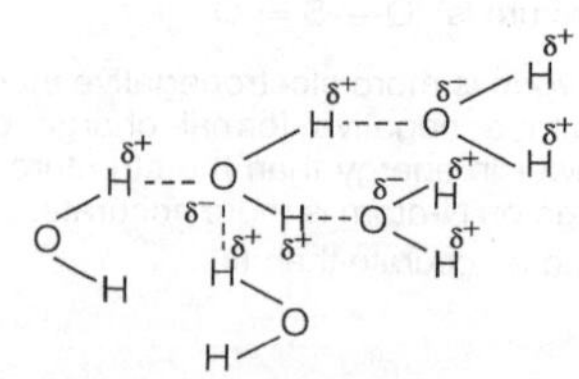

One water molecule is directly attached to four water molecules. Thus, coordination number of water is 4.

**15.** (b) HCl : EN of Cl is smaller, hence less polar than HF.

$CH_4$ : non-polar

$CH_3COO^-$ : H attached to C-atom is non-polar hence HCl, $CH_4$, $CH_3COO^-$ — lack H-bonding.

**16.** (a, b)

(a)

I. Intramolecular H-bonding.

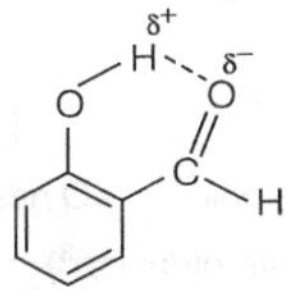

II. Intermolecular H-bonding makes boiling point higher than that of I. Thus, I is more volatile

(b)

I. Intramolecular H-bonding

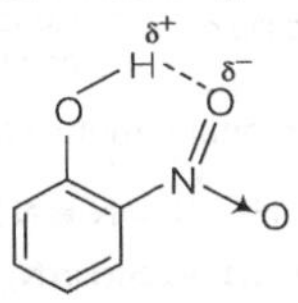

II. Intermolecular H-bonding makes boiling point higher than of *I* thus, I is more volatile.

(c) BP of $H_2O >> H_2S$

(d) BP of $CH_3CH_2OH$ (due to H-bonding) $>> CH_3—O—CH_3$

**17.** (a, b, d)

(a) When liquid water changes to ice, there is an air gap inside the cage formed due to intermolecular H–bonding. Thus, volume of ice > volume of water.

Since, $d = \frac{\text{mass}}{\text{volume}}$

Hence, density (ice) < density (water).

Thus, ice floats on water – hydrogen bonding plays its role.

(b) In gaseous phase, basicity of 1° < 2° < 3° amines due to increase of electron density at N-atom.

$RNH_2 < R_2NH < R_3N$

But in aqueous ($H_2O$) solvent, $RNH_2$ is stabilised by solvation. Thus 1° amine is more

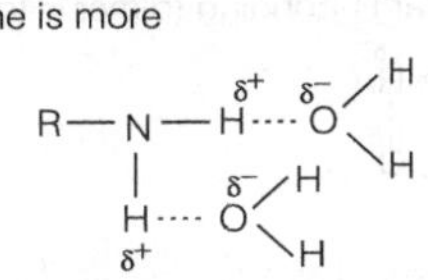

basic than 3° amine (3° < 1° <2°) amines.

(c) $I_+$ (positive inductive effect) explains this

(d) 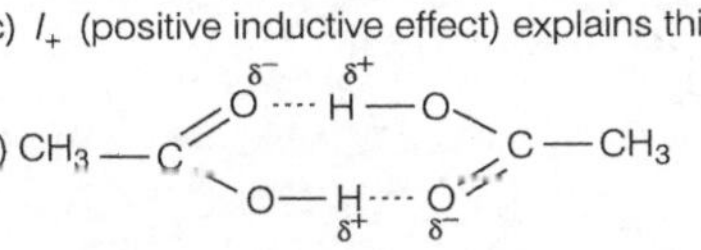

Dimer due to intermolecular H– bonding.

**18.** (a, c) (a) intramolecular H-bonding.

(a) 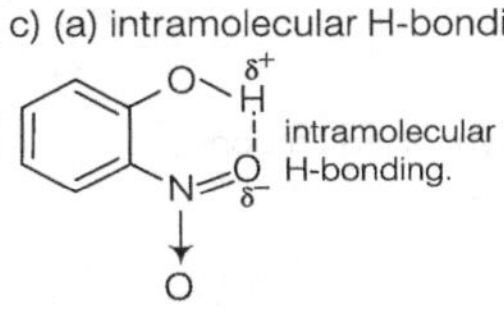

(b) 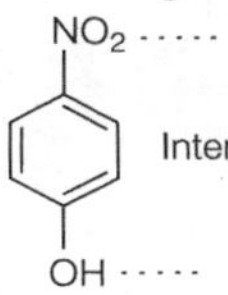 Intermolecular H-bonding

(c) 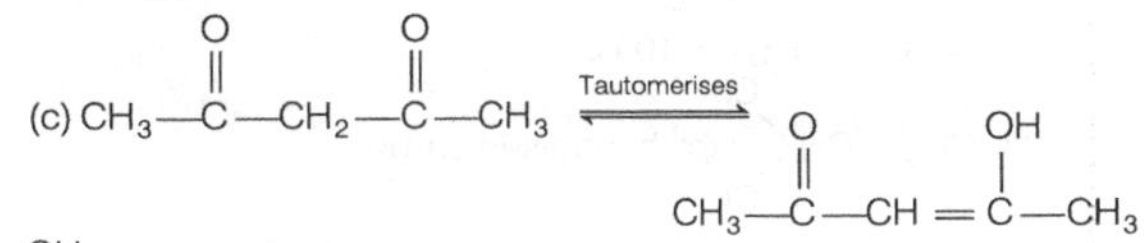

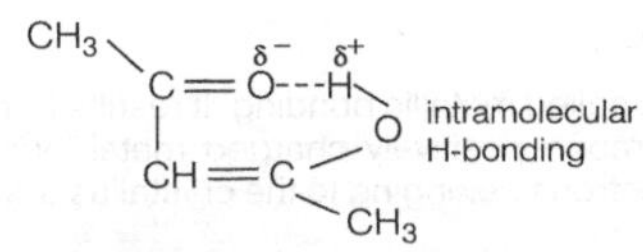

(d) Benzoic acid forms dimer by intermolecular H-bonding.

**19.** (c) I. $CCl_3CH(OH)_2$ with two —OH groups on same carbon atom is stable due to intramolecular H-bonding.

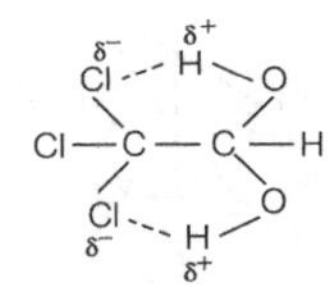

II. $CH_3CH(OH)_2 \longrightarrow CH_3\overset{O}{\overset{\|}{C}}H + H_2O$

III. $\underset{\text{stable}}{\begin{array}{c}CH_2OH \\ | \\ CH_2OH\end{array}}$

**20.** (b) *n*-butane lacks H-bonding. Hence, its boiling point is least.

Boiling point of $\underset{(78°C)}{CH_3CH_2OH} > \underset{(65°C)}{CH_3OH}$ due to difference in molecular weight.

Boiling point of $\underset{\text{Two polar }H^+}{H_2O} > \underset{\text{one polar }H^+}{CH_3CH_2OH}$

due to difference in polar $H^+$ for H-bonding.

Boiling point of $CH_3CH_2CH_2CH_2OH > H_2O$ due to much larger size of alcohol.

Thus, v < i < ii < iii < iv,

(i) → q, (ii) → r, (iii) → s, (iv) → t, (v) → p

## Revisal Problems (JEE Main)

**1.** (a) $KCN \rightleftharpoons K^+ + CN^-$ — $K^+$ and $CN^-$ – ionic; C and N – covalent

$NH_4Cl \rightleftharpoons NH_4^+ + Cl^-$ — $NH_4^+$ and $Cl^-$ – ionic

$NH_4^+$ has three covalent and one coordinate bond.

$H_2O$ : covalent bonding

: joined together by H-bonding

$H_2SO_4$ : ionic between $H^+$ and $SO_4^{2-}$

$SO_4^{2-}$ has covalent/coordinate bond.

**2.** (b) Species with odd number of electrons will certainly have unpaired electron in bonding, thus paramagnetic.

| | |
|---|---|
| $OH^-$ | 8 + 1 + 1 = 10 |
| OH | 8 + 1 = 9 |
| $NO_3$ | 7 + 24 = 31 |
| $SO_3$ | 8 + 24 = 32 |
| $SO_3^{2-}$ | 8 + 24 + 2 = 34 |
| $HO_2$ | 1 + 16 = 17 |

Thus, OH, $NO_3$ and $HO_2$ are paramagnetic.

**3.** (b)

I. 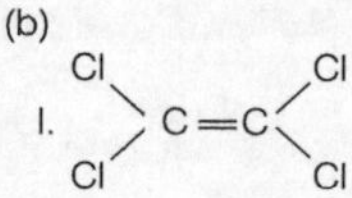 $sp^2$- hybridised carbon

$\rightarrow$ bond angle = 120°

II.

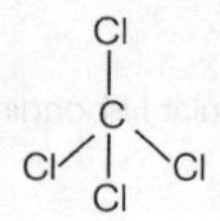

$sp^3$-hybridised carbon

$\Rightarrow$ bond angle = 109.5°

III. 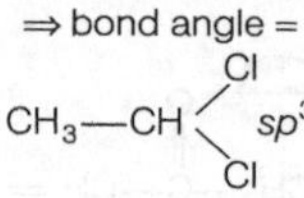 $sp^3$-hybridised carbon

$\Rightarrow$ bond angle = 109.5°

**4.** (c) Bonding in metals is called metallic bonding. It results from the electrical attractions among positively charged metal ions and mobile, delocalised electrons belonging to the crystal as a whole. Two theories :

(a) Band metal

(b) Electron-sea model

**5.** (a) I-atom is $sp^3d^3$ -hybridised

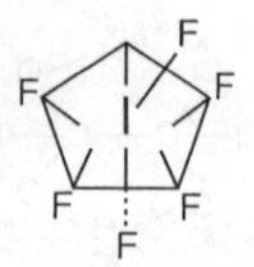

Bond angle = 72° (for five) and = 90° (for two)
No lone pair

**6.** (c) As $[Ar]3d^{10}\ 4s^2\ 4p^3\ 4d^0$

$As^{5+}$ $[Ar]\ 3d^{10}\ 4s^0\ 4p^0\ 4d^0$

$AsF_5$ – Arsenic is $sp^3d$ -hybridised.

Thus, five $sp^3d$-orbitals are formed due to intermixing of $4s$, $4p_x$, $4p_y$, $4p_z$, $d_{xy}$ orbitals giving five ($sp^3d$) orbitals of equivalent energy.

**7.** (d) $P(15) = 1s^2\ 2s^2\ 2p^6\ 3s^2\ 3p^3$

Valence electrons = 5

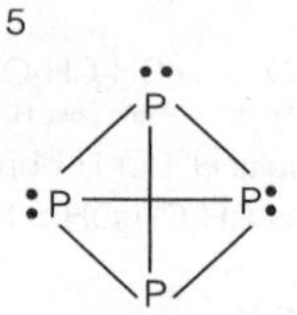

Each P-atom has three (P—P) bonds and one lone pair and is $sp^3$-hybridised.

Thus, $p$-character is $\frac{3}{4} \times 100 = 75\%$.

**8.** (b) $:\ddot{O}=\ddot{O}: \longrightarrow \ddot{\underset{..}{O}}:$

one lone pair    one σ bond
one π bond (not hybridised)
one (→) bond

Thus, it is $sp^2$-hybridised.

**9.** (a) $C_3H_4$ is propyne $CH_3—C\equiv CH$

$C_3^{4-}$ has following structure

$^{\ominus}\overset{\ominus}{\underset{\ominus}{C}}—C\equiv C^{\ominus}$

Thus, two σ and two π-bonds.

**10.** (c)

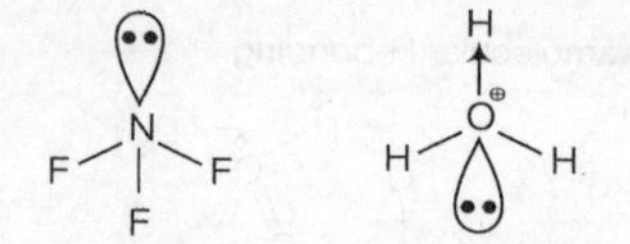

N has lone pair    O has lone pair

Both have pyramidal structure ($sp^3$)

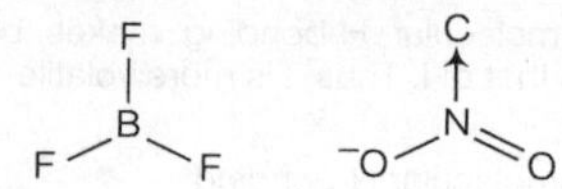

No lone pair on B    No lone pair on N

Both have triangular planar structure ($sp^2$)

$H—\ddot{N}=N\equiv N:$

Each of N-atom is $sp$-hybridised in $N_3H$.

**11.** (b) There are five bonding pairs or electron pairs. Thus, $sp^3d$-hybridisation. Trigonal bipyramidal structure.

Three bonding pairs are at 120° with one each other.

Two bonding pairs are at 90° with the plane in which first three bonding pairs exist.

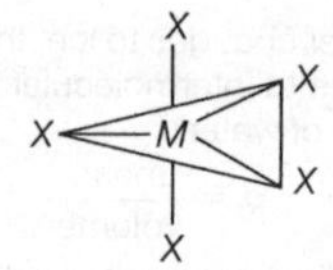

**12.** (b)

I. 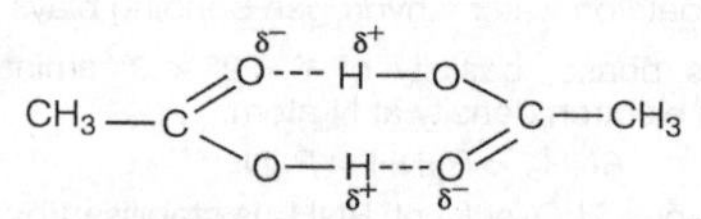

Interamolecular H-bonding (dimer is formed).

II. 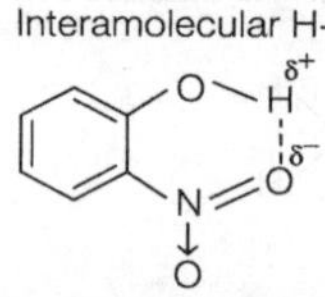

Intramolecular H-bonding (makes it more volatile than $m$-and $p$-isomers)

III. 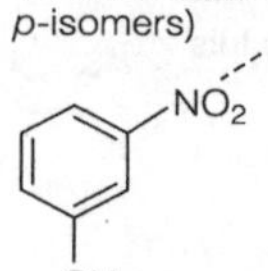

Intermolecular H-bonding.

IV. $H_3BO_3$ : Hexagonal structure is formed due to intermolecular H-bonding.

**13.** (b)

(a) $H_2N—NH_2$ (N—N) single bond.

(b) $N\equiv N$ (N—N) triple bond.

(c) 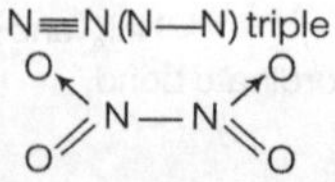 (N — N) single bond but longest due to dipole-dipole repulsion of N-atoms.

(d) $\overset{-}{N}=\overset{+}{N}=O$ (N—N) double bond

$$\text{Bond length} \propto \frac{1}{\text{Number of bonds}}$$

Thus, (N≡N) shortest.

**14.** (c) (a) Chelation is ring formation.

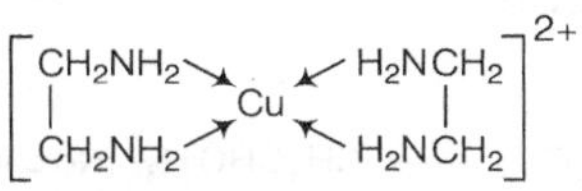


$CH_2NH_2$
|   is bidentate ligand.
$CH_2NH_2$

(b) 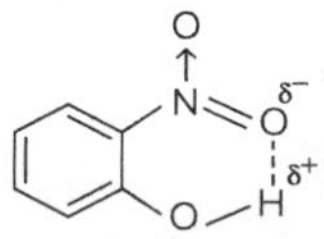


Intramolecular H–bonding.

**15.** (c) H—F $\quad \mu = e \times l = 1.91$ D

$H_2O$ θ = 104.5°, $\quad \mu = 1.84$ D

$H_2S$ θ = 92°, $\quad \mu = 1.88$ D

Thus, $HF > H_2S > H_2O$

**16.** (a) Going down the group, size of the cation increases. Since, the lattice energy decreases much more than the hydration energy with increasing size solubility of hydroxide increases (when size of cation and anion are of comparable size).

(b) and (c) If the anion is large (as $CO_3^{2-}$, $SO_4^{2-}$) compared to cation, the lattice energy will remain almost constant. Since, hydration energy decreases down a group, solubility will decrease hence (b) and (c) are incorrect.

**17.** (c) $KO_2$ has superoxide ion $O_2^-$ (17 $e^-$). Thus, one electron is uppaired.

$NO_2$ (23 $e$) – one unpaired electron.

NO (15 $e$) – one unpaired electron.

$\overset{+}{N}O_2$ (22 $e$) – unpaired electron

in $NO_2$ is lost when $\overset{+}{N}O_2$ is formed.

$O_2^{2-}$ (in $K_2O_2$ has 18 $e$) – no unpaired electron.

$O^{2-}$ (10 $e$) – no unpaired electron

**18.** (d) Compound is said to be ionic if (EN) difference between bonding atoms exceeds 1.7.

| | | |
|---|---|---|
| KCl | 3.0 − 0.8 = 2.2 | Yes |
| $PH_3$ | 2.1 − 2.1 = 0.0 | No |
| GeCl | 3.0 − 2.1 = 0.9 | No |
| $H_2O$ | 2.5 − 2.1 = 0.4 | No |
| CsF | 4.0 − 0.7 = 3.3 | Yes |

(**Note** *CsF is most ionic.*)

**19.** (b) $\quad Na(g) \longrightarrow Na^+(g) + e^-$,

$\Delta H$ = (IE) = 5.1 eV

$Cl(g) + e^- \longrightarrow Cl^-(g)$,

$\Delta H$ = (EA) = − 3.6 eV

Thus, energy required = 5.1 − 3.6 = 1.5 eV

**20.** (d) $H_2O$ and $F^-$ are weak ligands. Hence, unpaired electrons in $Fe^{3+}$ is (I) and (IV) are not affected. Thus outer $d$-orbitals are involved in hybridisation.

Fe = [Ar]

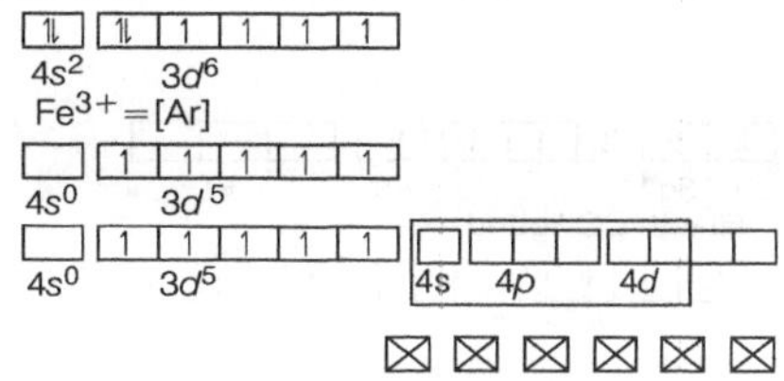


Six $sp^3d^2$-orbitals accept lone pair from weak ligands – octahedral geometry.

**21.** (a) 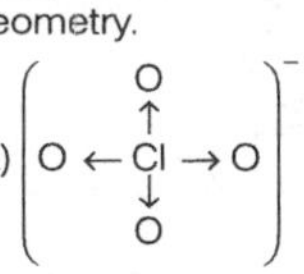


There are (Cl ⟶ O) coordinate bonds. Thus, bond length is shorter than the normal (Cl—O) bond length (1.72 A°). Thus, there is considerable double bond character. Thus, both statements I and II are correct and statement II is the correct explanation statement I.

**22.** (b) $N_2O_4$ is formed by dimerisation of $NO_2$.

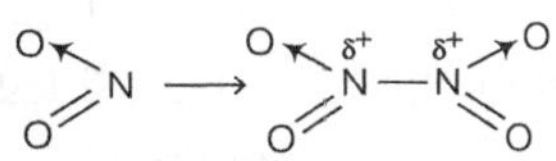


Unpaired electron is shared and (N—N) bond is formed. This results is formation of (δ +) on each N-atom. Thus, dipole-dipole repulsion increases (N—N) bond length (larger than its normal value (1.45 Å). Thus, both statements I and II are correct but statement II is not the correct explanation of statement I.

**23.** (b) $\quad I_2 + I^- \longrightarrow I_3^-$

Lewis acid $\quad$ Lewis base

There are two bond pairs and three lone pairs in the outer shell of central atom. To minimise the repulsive forces, the three lone pairs occupy the equatorial position. The ion is therefore linear in shape with a bond angle of exactly 180°.

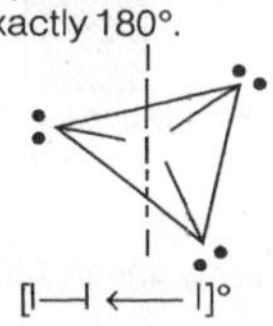

$[I—I \longleftarrow I]^{\circ}$

Thus, both statements I and II are correct but statement II is not the correct explanation of statement I.

**24.** (c) $I_3^-$ and $F_3^-$ are of same group. I can expand its octet but F cannot expand its octet, thus in $I_3^-$ octet rule is not violated but in $F_3^-$, octet rule is violated.

$F^-$ (electron rich) is a Lewis base.

Thus, statement I is correct and statement II is incorrect.

# Revisal Problems (JEE Advanced)

**1.** (b) $CN^-$ and CO are strong ligands. Thus, unpaired electrons in Ni or $Ni^{2+}$ are paired. $H_2O$ is a week ligand, thus unpaired electrons are not affected.

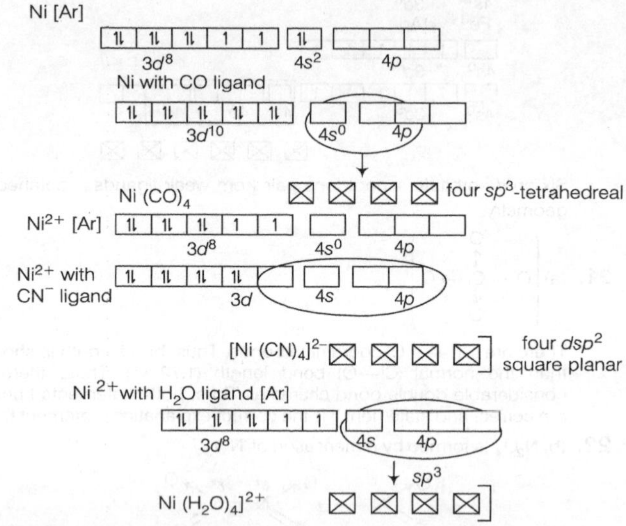


**Note** *There is no unpaired electron in* $[Ni(CO)_4]$ *— diamagnetic*

*Also,* $[Ni(CN)_4]^{2-}$ *– diamagnetic but* $[Ni(H_2O)_4]$ *has unpaired electrons. Thus, it is paramagnetic.*

**2.** (d) Based on Band-theory of metallic bonding, as energy increases, conducting nature decreases and element changes from conductor to insulator.

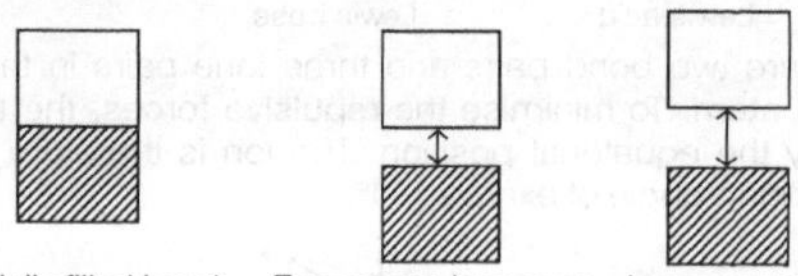

Partially filled band energy gap=0 (Metal) | Energy gap increases (Semi-conductor) | Larger energy gap (Insulator)

**3.** (c)

(i) $O{=}\overset{+}{N}{=}O$ *sp*-hybridised. linear, 180° (p)

(ii) 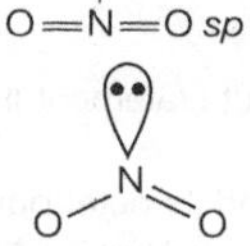 $sp^2$-hybridised one lone pair, 115°(s)

(iii) 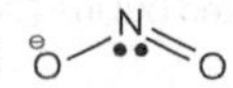 $sp^2$-hybridised 109°(t)

(iv) 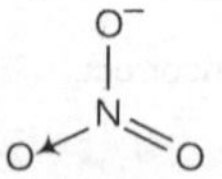 $sp^2$-hybridised 120°(r)

Thus, (i) → (p), (ii) → (s), (iii)→ (t), (iv) → (r)

**4.** (a) $\underset{sp^2}{AlH_3} \longrightarrow \underset{sp^3}{AlH_4^-}$

(b) $\underset{sp^3}{NH_3} \longrightarrow \underset{sp^3}{NH_4^+}$ (No change)

(c) 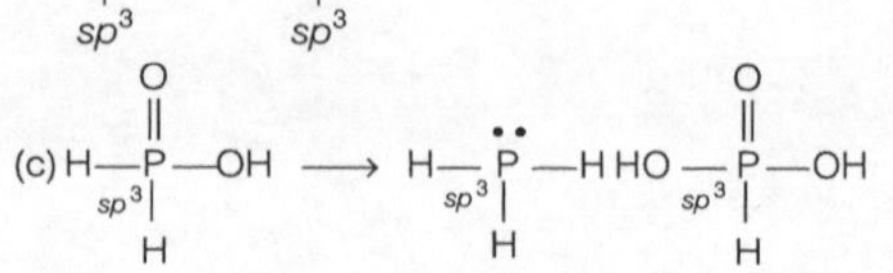

(No change)

(d) $\underset{sp^3}{CH_3CH_2CH_3} \longrightarrow \underset{sp^3}{CH_3\overset{Cl}{C}HCH_3}$ (No change)

**5.** (c) (*lp* – *lp*) and (*lp* – *bp*) repulsions are minimised by taking lone-pairs on equatorial position.

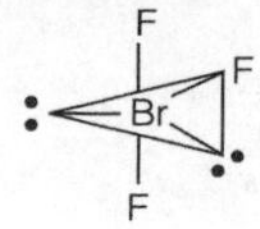

**6.** (c)

| Hybridisation | Structure | Bond angle |
|---|---|---|
| 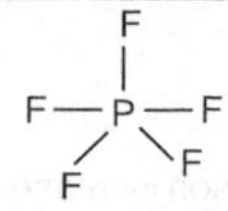 of P—$sp^3d$ | Bipyramidal | 120°, 90° |
|  of S—$sp^3d^2$ | Octahedral | 90° |

In $PF_5$, there are two types of angles thus has two types of bond lengths.

In $SF_6$, there is only one type of bond angle thus, only one type of bond.

There is no lone pair on P in $PF_5$.

Thus, statement I is correct and statement II is incorrect.

**7.** (c) 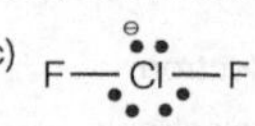

Lone pairs on Cl are at the equatorial positions to minimise repulsion.

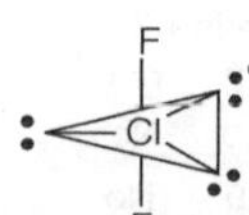

$sp^3d$ hybridisation

Two F-atoms are at the vertices thus linear structure.

Two lone pairs on Cl—$sp^3$ hybridisation thus bent.

Thus, statement I is correct but statement II is incorrect.

**8.** (a) C,Si—group 14. Thus, Statement I is correct larger size of Si results in poor sideway overlap of *s*-orbitals to form $\pi$-bond.

Thus, both statements I and II are correct and statement II is the correct explanation of statement I.

**9.** (a, b, c, d)

(a) 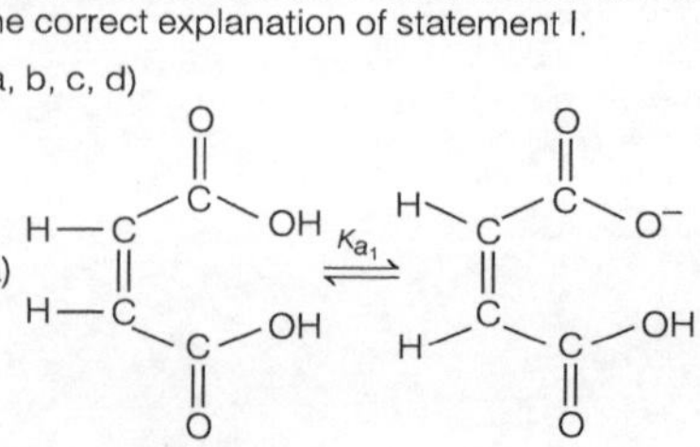

Polar

$\mu > 0$ (maleic acid)

I.

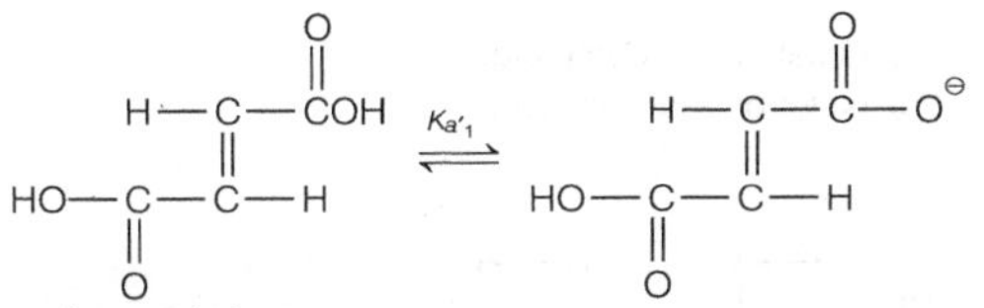

(less polar)
$\mu = 0$ (fumaric acid) II.

Maleic acid (*cis*-isomer) is more polar than its *trans*-isomer. Hence, maleic acid is more ionised than fumaric acid.

(b) Maleate ion (I) is stable due to intramolecular H–bonding. Hence, $K_{a_2}$ of maleic acid is smaller than that of fumaric acid in which intramolecular H–bonding is absent (II).

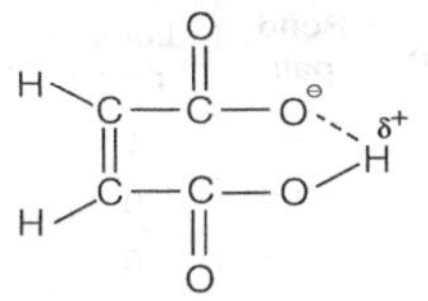

$K_{a_1}$ = First ionisation constant
$K_{a_2}$ = Second ionisation constant

(c) 2 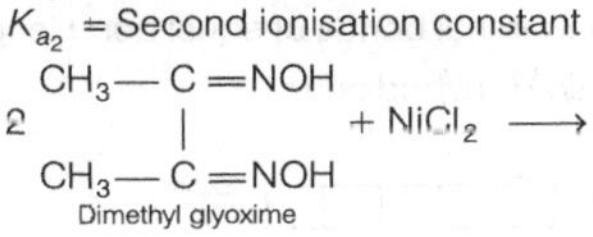

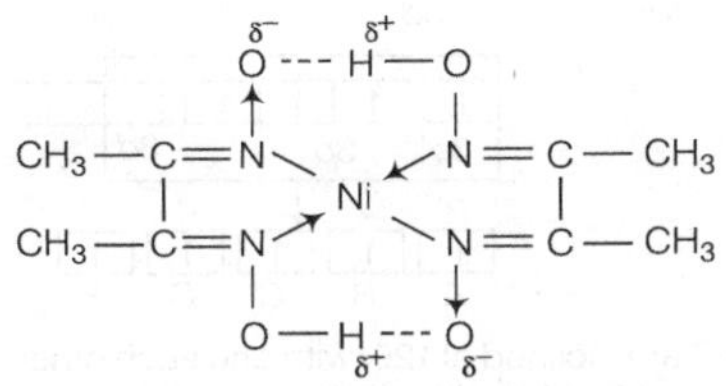

Nickel dimethyl glyoximate complex is stable due to intramolecular H–bonding.

(d) 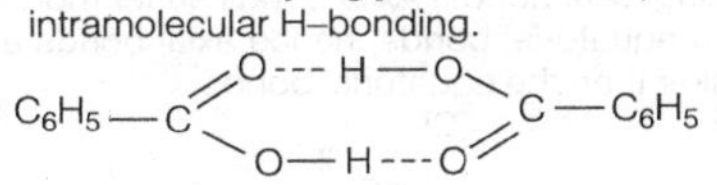 (Dimer)

**10.** (a, b, c, d) By the coulombic force of attraction ($F$) between cation and anion is

$$F = \frac{z_1 z_2 e^2}{Dr^2}$$

$z_1 = z_2 = (+1 \text{ or } -1)$ charges on cation or anion.
$r$ = internuclear distance
$D$ = Dielectric constant (same for $H_2O$ solvent).
If $F$ is small, ions goes into solution and thus makes salt soluble in water.

$$F \propto \frac{1}{r^2}$$

Larger ($r$) means smaller ($F$).

(a) $r$ increases hence solubility increases true.
(b) as in (a) true.
(c) If one of the ions is very big (as Cs), the solubility decreases with increasing size of the other ion.

F < Cl < Br < I,
size

Thus, CsI < CsBr <, CsCl < CsF
Thus, true, solubility

(d) $NO_3^-$ ion is larger.

size of $Li^+ < Na^+ < K^+ < Cs^+$

Hence, solubility decreases with increase in size of $M^+$
$CsNO_3 < KNO_3 < NaNO_3 < LiNO_3$ – true.

**11.** (a, b)
Based on MO theory
(a) $C_2$ (12 electrons) – all paired electrons
(b) $N_2$ (14 electrons) – all paired electrons
(c) $O_2$ (16 electrons) – two electrons unpaired
(d) $S_2$ (32 electrons) – two electrons unpaired
Thus, $C_2$ or $N_2$ are diamagnetic.

**12.** (a, b, c, d) Bond length $\propto \frac{1}{\text{Number of bonds (bond order)}}$

Bond energy $\propto$ Number of bonds
Magnetic moment = $\sqrt{N(N+2)}$ BM,
$N$ = Number of unpaired electrons

| Species | Electrons | MO Electronic configuration | Electrons in $N_B$ | Electrons in $N_A$ | Bond order $= \left(\frac{N_B - N_A}{2}\right)$ |
|---|---|---|---|---|---|
| $O_2^+$ | 15 | $\sigma 1s^2 \overset{*}{\sigma} 1s^2 \sigma 2s^2 \overset{*}{\sigma} 2s^2 \sigma 2p_x^2 \pi 2p_y^2 \pi 2p_z^2 \overset{*}{\pi} 2p_y^1$ | 10 | 5 | 2.5 |
| $O_2$ | 16 | $\sigma 1s^2 \overset{*}{\sigma} s^2 \sigma 2s^2 \overset{*}{\sigma} 2s^2 \sigma 2p_x^2 \pi 2p_y^2 \pi 2p_z^2 \pi^* 2p_y^1 \pi^* 2p_z^1$ | 10 | 6 | 2.0 |
| $O_2^-$ | 17 | $\sigma 1s^2 \sigma^* 1s^2 \ldots\ldots \pi^* 2p_y^2 \pi^* 2p_z^1$ | 10 | 7 | 1.5 |
| $O_2^{2-}$ | 18 | $\ldots\ldots \pi^* 2p_y^2 \pi^* 2p_z^2$ | 10 | 8 | 1.0 |

Bond order = $O_2^{2-} < O_2^-, O_2, O_2^+$
Bond length = $O_2^{2-} > O_2^- > O_2 > O_2^+$
Bond energy as bond order magnetic moment $O_2^{2-} < \bar{O}_2 = O_2^+ < O_2$

**13.** (b, c, d)

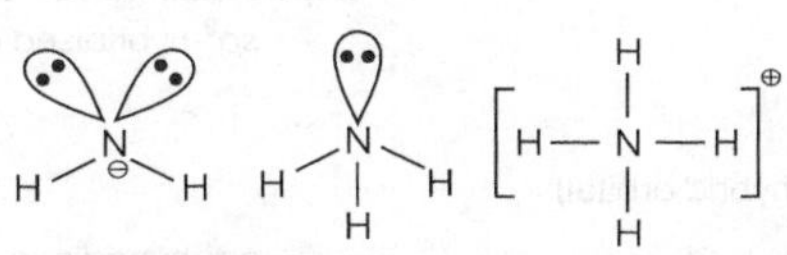

($lp$-$lp$), ($lp$-$bp$) repulsion ←

Bond angle → Maximum — Thus, (a) correct

| | | | | |
|---|---|---|---|---|
| Lone-pair | 2 | 1 | 0 | Thus, (b) incorrect |
| (N—H) Bond length | same | – | – | Thus, (c) incorrect |
| Hybridisation | $sp^3$ | $sp^3$ | $sp^3$ | Thus, (d) incorrect |

**14.** (a, c, d)

(a) $I_3^-$ $sp^3d$

(b) $PCl_3$ $sp^3$

(c) $XeF_2$ $sp^3d$

(d) $SF_4$ $sp^3d$

**15.** (b, c)

(a) Smaller the energy gap, larger the conduction.

(b) Unpaired electrons in graphite is responsible for conduction used in bonding in hexagonal layer.

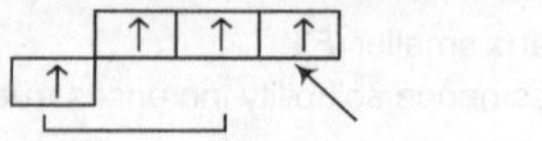


(c) $Li_2$ (6 electrons) $(\sigma 1s)^2\ (\overset{*}{\sigma}\ 1s)^2\ (\sigma\ 2s)^2$

Bonding electrons $= 4$

Antibonding electrons $= 2$

$$\text{Bond order} = \frac{4-2}{2} = 1$$

Thus, $Li_2$ is stable.

(d) Diamond is non-conducting graphite is conducting.

**16.** (b, c)

(a) 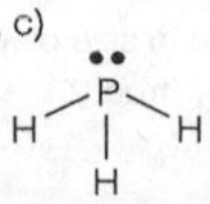 $sp^3$ due to one lone pair

(b) 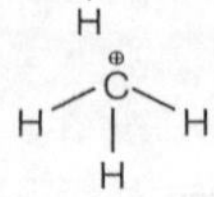 $sp^2$

(c) 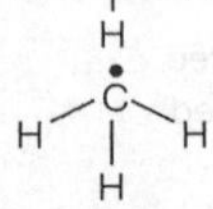

It has three (C—H) σ-bonds and singly occupied *p*-orbital. Thus, $sp^2$.

(d) $O \leftarrow \overset{\oplus}{N}{=}O$ $sp$

**17.** (c, d)

I : O=C=O
sp (C)

II : 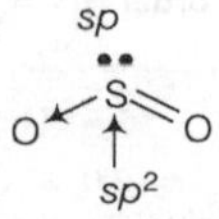

$sp^2$ (one lone pair is a part of hybrid orbital).

III : 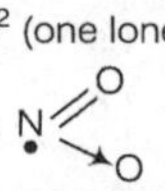

(one unpaired electron is a part of hybrid orbtial).

IV. H—C≡C—H (both C: sp)

V : 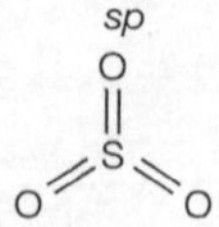 $sp^2$

VI. H—$\ddot{N}$(—H)—H $sp^3$ (one lone pair is a part of hybrid orbital)

**18.** (b, c) (a)

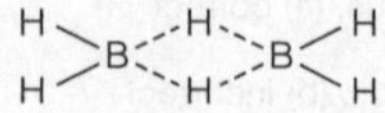

(B—H) bond has only one electron thus B is electron deficient.

(b) Octet of B is complete.

(c) Octet of B is complete.

(d) 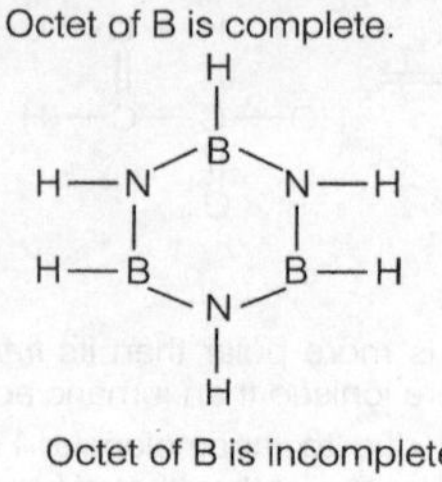

Octet of B is incomplete.

**19.** (b)

| Species | Hybridisation | Bond pair | Lone pair | Bonds | |
|---|---|---|---|---|---|
| | | | | Equatorial | Axial |
| $SF_4$ | $sp^3d$ | 4 | 1 | 4 | 0 |
| $IF_7$ | $sp^3d^3$ | 7 | 0 | 5 | 2 |
| $PF_5$ | $sp^3d$ | 5 | 0 | 3 | 2 |
| $ClO_4^-$ | $sp^3$ | 4 | 0 | 3 | 1 |

In $IF_7$, there are five equatorial bonds of equal bond length.

**20.** (a, b, c) P-atoms are $sp^3d$ -hybridised.

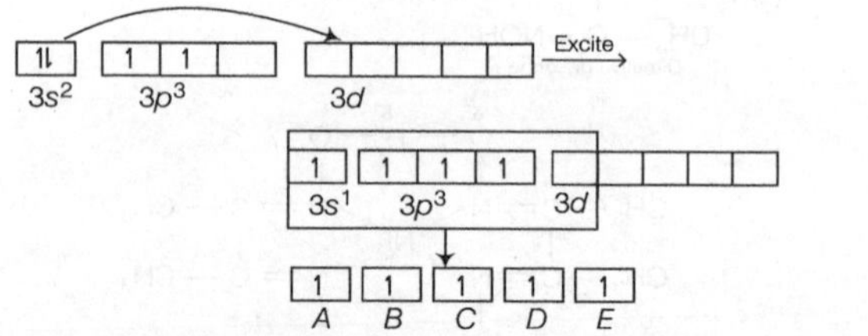


*A*, *B* and *C* are inclined at 120° with one each other form equatorial bonds. *D* and *E* at 90° with (*ABC*) plane. *D* is above and *E* is below this plane forming axial bonds. Axial bonds suffer more repulsive interaction from equatorial bonds, hence axial bonds are longer and sligtly weaker than the equatorial bonds.

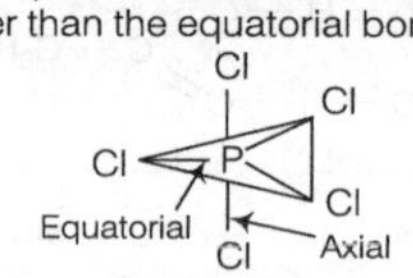


**21.** (b, d) (a) $Cl_2\overset{sp^2}{C}{=}CCl_2$ Planar

(b) N≡C—C≡N Linear (C: sp)

(c) $F_3C—C≡CN$ ($sp^3$, $sp$, $sp$)

(d) $[S—C≡N]^-$ Linear ($sp$, $sp$)

**Explanations for Q. Nos. (22-23)**

$sp^2$-hybridised carbon bond angle = 120°

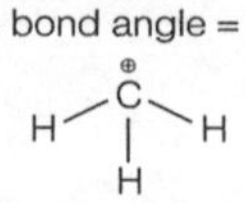

$sp^3$-hybridised carbon bond angle < 109°28′

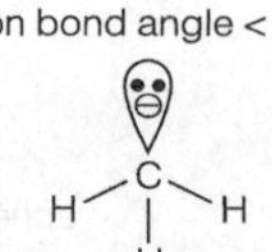

Unpaired electron is in unhybridised $p_z$-orbital.

Thus $sp^2$-hybridised bond angle = 120°.

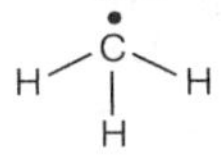

Thus (b), (c)

**22.** (b) **23.** (c)

## Explanation for Q. Nos. (24-25)

**Explanation** Each N-atom has one lone pair and three σ-bonds — $sp^3$ hybridised.

X : 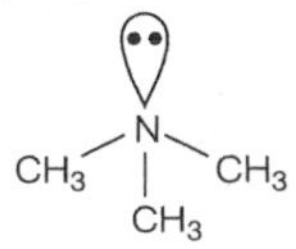

($lp - bp$) repulsion makes it. Trigonal pyramidal.

$Y$ : $N[Si(CH_3)_3]_3$ Lone pair on N-atom is transferred to the empty $d$-orbitals of silicon ($p\pi - d\pi$ overlapping) and its Lewis base character is lost.

Thus, $Y$ has trigonal planar structure.

$Z$ : 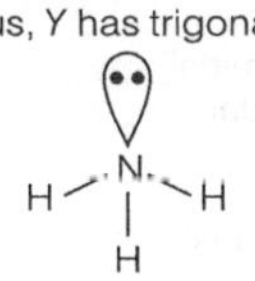

as in ($X$) trigonal pyramidal.

| Special | Structure |
|---|---|
| $X$ | II |
| $Y$ | I |
| $Z$ | II |

**24.** (b) **25.** (c)

**26.** (b)

(i) $2O_2^- \longrightarrow O_2 + O_2^{2-}$

$O_2^- \longrightarrow O_2 + O_2^{2-}$ (Oxidation, Reduction)

$ON\left(-\frac{1}{2}\right)$ 0 –1 per atom

$O_2^-$ undergoes disproportionation.

**(i) → (p, s)**

(ii) $2CrO_4^{2-} + 2H^+ \longrightarrow Cr_2O_7^{2-} + H_2O$

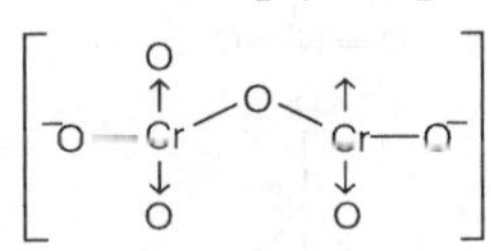

Cr-metal ion is bridged and has tetrahedral structure. **(ii) → (r)**

(iii) $\underset{+7}{Mn}O_4^- + \underset{+3}{N}O_2^- + H^+ \longrightarrow \underset{+5}{N}O_3^- + \underset{+2}{Mn^{2+}}$

$MnO_4^-$ is reduced.

$NO_3^-$ is oxidised thus a redox reaction.

$NO_3^-$ has triangular planar structure.

Thus, **(iii) → (p,q)**

(iv) $NO_3^- + H_2SO_4 + Fe^{2+} \longrightarrow Fe^{3+} + NO_2^- + H_2O + SO_4^{2-}$

$Fe^{2+}$ is oxidised and

$NO_3^-$ is reduced thus redox reaction.

Thus, **(iv) → (p)**

**27.** (c)

(i) $B_2$ (10 electrons)

$\sigma 1s^2\ \overset{*}{\sigma}1s^2\ \sigma 2s^2\ \overset{*}{\sigma}2s^2 \pi 2p_y^1\ \pi 2p_z^1$

Two unpaired electrons-paramagnetic.

Bonding electrons = 6

Anit-bonding electrons = 4

Bond order = $\frac{6-4}{2} = 1$

(B—B) bond is formed by mixing of $s$ and $p$-orbitals of valence shell.

$B_2 \xrightarrow{\text{Oxidation}} B_2O_3$

$B_2 \xrightarrow{\text{Reduction}} BH_3$

Thus, **(i) → (p, q, r, t)**

(ii) $N_2$ (14 electrons)

$N_2(\sigma 1s^2\ \overset{*}{\sigma}1s^2\ \sigma 2s^2\ \overset{*}{\sigma}2s^2\ \pi 2p_y^2 \pi 2p_z^2 \sigma p_x^2)$

No unpaired electron – diamagnetic

Bonding electrons = 10

Anti-bonding electrons = 4

Bond-order = $\frac{10-2}{2} = 3$

$H_3N \xleftarrow{\text{Reduction}} N_2 \xrightarrow{\text{Oxidation}} N_2O$

Thus, **(ii) → (q, r, s, t)**

(iii) $O_2^-$ (17 electrons)

$(\sigma 1s^2\ \sigma^*\ 1s^2\ \sigma 2s^2\ \sigma^*\ 2s^2\ \sigma 2p_x^2\ \pi 2p_y^2$
$\pi 2p_z^2\ \pi^*\ 2p_y^2\ \pi^*\ 2p_z^1)$

One unpaired electrons thus paramagnetic

Bonding electrons = 10

Anti-bonding electrons = 7

Bond order = $\frac{10-7}{2} = 1.5$

$2O_2^- \longrightarrow O_2 + O_2^{2-}$ (Reduction, Oxidation)

$O_2^-$ is oxidised and reduced.

Thus, **(iii) → (p, q, r, t)**

(iv) $O_2$ (16 electrons)

There are two electrons in anti-bonding MO ($\overset{*}{\pi}\ 2p_y^1$ and $\overset{*}{\pi} 2p_z^1$).

Thus, it is paramagnetic.

Bond order = 2

$O_2 \xrightarrow{\text{Reduction}} O_2^{2-}$

$O_2 + 2F_2 \xrightarrow{\text{Oxidation}} 2OF_2$ (ON of O = + 2 in $OF_2$)

Thus, **(iv) → (p, q, r, s, t)**

**28.** (4) $p = \frac{n}{V}RT = \frac{w}{mV}RT$

Since $\frac{w}{V} = gL^{-1} = \text{density}$

Hence $p = \frac{d}{m}RT$

$m$ (HF–associated) $= \frac{dRT}{p} = \frac{3.17 \times 0.0821 \times 300}{1.00} = 78.08\ gL^{-1}$

$m$ (HF) of single unit = 20 $gL^{-1}$

Thus, number of HF molecules = 78/20 ≈ 4

**29.** (8)

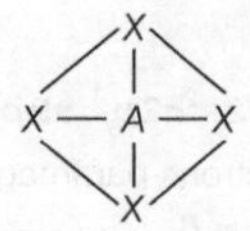


Distance between $A$ and $X$ = 4 units

Thus, distance between two $X$-atoms = $4 \times 2$ = 8 units.

**30.** (7)

$Na(11): 1s^2 2s^2 2p^6 3s^1$

Electron in (3 s) orbital is a conduction band. Each orbital has one electron, thus one energy level.

$$26.8 \text{ mg Na} = 26.8 \times 10^{-3} \text{ g Na-atoms}$$
$$= \frac{26.8 \times 10^{-3}}{23} \text{ g mol Na}$$
$$= \frac{26.8 \times 10^{-3} \times 6.02 \times 10^{23}}{23} \text{ Na-atoms}$$
$$= 7.0 \times 10^{20} \text{ atoms}$$
$$= 7.0 \times 10^{20} \text{ conduction bands.}$$

Thus, $n = 7$

## JEE Main and AIEEE Archive

**1.** (b) $CsI_3$ contains $Cs^+$ and $I_3^-$ ions.

**2.** (d) In quinol 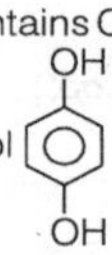 and thioquinol, 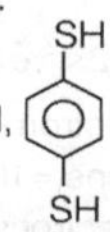 —OH and —SH groups do not cancel their dipole moments and hence exists as different conformations.

**3.** (a, b) $C_2 (6 + 6 = 12) = \sigma 1s^2, \overset{*}{\sigma} 1s^2, \sigma 2s^2, \overset{*}{\sigma} 2s^2, \pi 2p_y^2 \approx \pi 2p_z^2$

Since, all the electrons are paired, it is a diamagnetic specie.

$N_2 (7 + 7 = 14) = \sigma 1s^2, \overset{*}{\sigma} 1s^2, \sigma 2s^2, \overset{*}{\sigma} 2s^2, \pi 2p_y^2 \approx \pi 2p_z^2, \sigma 2p_x^2$

It is also a diamagnetic specie because of the absence of unpaired electrons.

$O_2 (8 + 8 = 16)$

or $S_2 = \sigma 1s^2, \overset{*}{\sigma} 1s^2, \sigma 2s^2, \overset{*}{\sigma} 2s^2,$

$\sigma 2p_x^2, \pi 2p_y^2 \approx \pi 2p_z^2, \overset{*}{\pi} 2p_y^1 \approx \overset{*}{\pi} 2p_z^1$

Since, $O_2$ or $S_2$ contain 2 unpaired electrons, therefore they are paramagnetic.

**4.** (c) Species having zero or negative bond order do not exist.

$$H_2^{2+} (1 + 1 - 2 = 0) = \sigma 1s^0$$

Bond order = 0

$He_2 (2 + 2 = 4) = \sigma 1s^2, \overset{*}{\sigma} 1s^2$

$$\text{Bond order} = \frac{N_b - N_a}{2} = \frac{2 - 2}{2} = 0$$

So, both $H_2^{2+}$ and $He_2$ do not exist.

**5.** (b) $Li_2 (3 + 3 = 6) = \sigma 1s^2, \overset{*}{\sigma} 1s^2, \sigma 2s^2$

$$\text{Bond order} = \frac{N_b - N_a}{2} = \frac{4 - 2}{2} = 1$$

$Li_2^+ (3 + 3 - 1 = 5) = \sigma 1s^2, \overset{*}{\sigma} 1s^2, \sigma 2s^1$

$$\text{Bond order} = \frac{3 - 2}{2} = \frac{1}{2} = 0.5$$

$Li_2^- (3 + 3 + 1 = 7) = \sigma 1s^2, \overset{*}{\sigma} 1s^2, \sigma 2s^2, \overset{*}{\sigma} 2s^1$

$$\text{Bond order} = \frac{4 - 3}{2} = \frac{1}{2} = 0.5$$

Stability order is $Li_2 > Li_2^+ > Li_2^-$ (because $Li_2^-$ have more number of electrons in antibonding orbitals which destabilises the species).

**6.** (c) N, P, As, Sb — Group = 15; Valence electrons = 5

$MCl_3$ has $sp^3$ hybridised $M$-element with one lone pair.

Lone-pair and bond pair repulsion decreases bond angle. However, the bond pairs of electrons are much farther away from the central atom than they are in $NCl_3$. Thus, lone pair causes even greater distortion in $PCl_3$, $AsCl_3$ and $SbCl_3$.

Thus, bond angle decreases from $NCl_3$ (maximum) to $SbCl_3$ (minimum).

**7.** (c)

| | | |
|---|---|---|
| (a) | $CO_3^{2-}, NO_3^-$ | Triangular planar |
| (b) | $PCl_4^+, SiCl_4$ | Tetrahedral |
| (c) | $PF_5$ | Trigonal bipyramidal |
| | $BrF_5$ | Square pyramidal |
| (d) | $AlF_6^{3-}, SF_6$ | Octahedral |

**8.** (c) Covalent character is favoured by Fajans' rule.

(i) Larger the charge on the ions,
(ii) Smaller the size of cations,
(iii) Larger the size of anions,
(iv) Cation with 18-electrons structure (e.g. $Cu^+$),

then larger the polarising power and the covalent character is favoured.

In all the given compounds, anion is same ($Cl^-$), hence polarising power is decided by size and charge of cation.

$Al^{3+}$ with maximum charge and smallest size has maximum polarising power hence, $AlCl_3$ is maximum covalent.

**9.** (b) Count $\sigma$-bond, lone pairs and unpaired electrons or count number of atoms directly attached, lone pairs and unpaired electrons.

This is = 2, then $sp$ = 5, then $sp^3d$
= 3, then $sp^2$ = 6, then $sp^3d^2$
= 4, then $sp^3$

$NO_3^-$ (I)

$NO_2^+$ (II)

$NH_4^+$ (III)

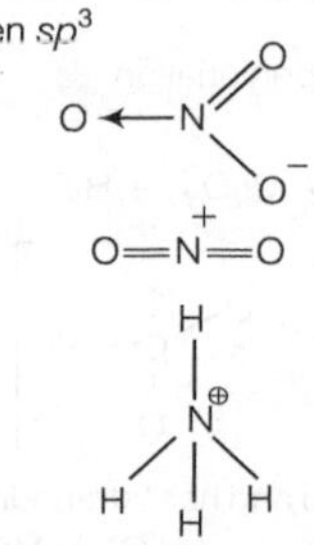

| | $\sigma$-bond | Lone pair | Unpaired electron | Total |
|---|---|---|---|---|
| I. | 3 | × | × | 3 ($sp^2$) |
| II. | 2 | × | × | 2 ($sp$) |
| II. | 2 | × | × | 4 ($sp^3$) |

**10.** (d) $IF_7$ Outer orbital diagram of I

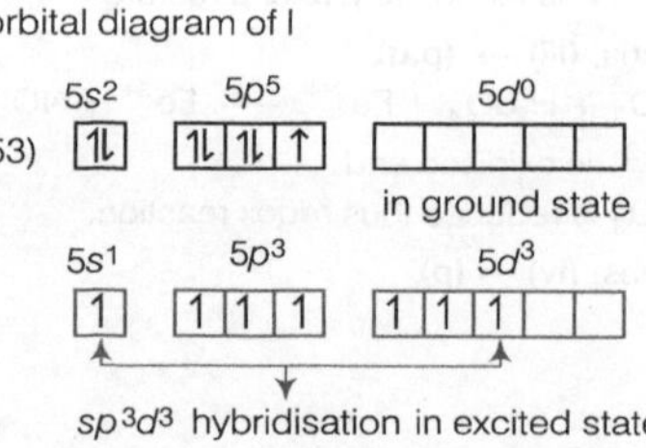

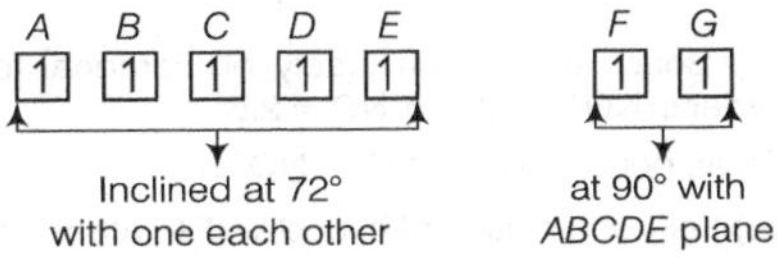


Seven $sp^3d^3$ hybrid orbitalsforming *s*-bonds with F-atoms,

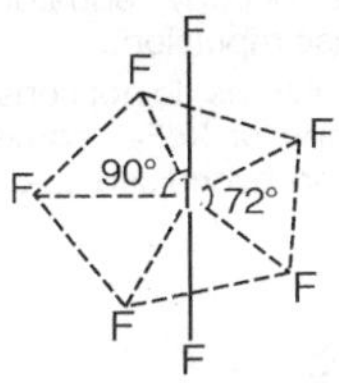


Pentagonal-bipyramidal structure

**11.** (d)

| | Species | Hybridisation of Xe | Lone pair on Xe | Bonding pairs |
|---|---|---|---|---|
| (a) | $XeO_3$ | $sp^3$ | 1 | 3 |
| (b) | $XeF_4$ | $sp^3d^2$ | 2 | 4 |
| (c) | $XeF_6$ | $sp^3d^3$ | 1 | 6 |
| (d) | $XeF_2$ | $sp^3d$ | 3 (max) | 2 |

**12.** (a) $CaC_2 \rightleftharpoons Ca^{2+} + \underset{\text{Carbide ion}}{C_2^{2-}}$

In carbide ion, two carbon atoms are joined by triple bond (It is isoelectronic with N2).

$$—C \equiv C—$$

with two $\pi$ and one $\sigma$-bonds.

**13.** (a) Higher the bond order, shorter is the bond length $O_2^{2+}$ has the shortest bond length ( BO = 3).
Bond order of remaining species are
$O_2^+$ (2.5), $O_2^-$ (1.5) and $O_2^{2-}$(1)

**14.** (b) Isoelectronic species have same number of electrons, $NO^+$, $C_2^{2-}$, $CN^-$ and $N_2$ all have 14 electrons.

**15.** (a) $CN^-$ and $NO^+$ both have same number of electrons and same bond order (3).

**16.** (a) The correct option is $O_2^{2-}$. This species has 18 electrons, which are filled in such a way that all molecular orbitals are fully filled, so diamagnetic.

$$\sigma 1s^2, \overset{*}{\sigma} 1s^2, \sigma 2s^2, \overset{*}{\sigma} 2s^2, \sigma 2p_x^2, \pi 2p_y^2 \approx \pi 2p_z^2, \overset{*}{\pi} 2p_y^2 \approx \overset{*}{\pi} 2p_z^2$$

**17.** (b) $NO \longrightarrow NO^+ + e^-$

($NO^+$) Total electrons = 14

$$\sigma 1s^2, \sigma^* 1s^2, \sigma 2s^2, \sigma^* 2s^2, \pi 2p_y^2 \approx \pi 2p_z^2, \sigma 2p_x^2$$

Diamagnetic, Bond order $= \dfrac{10-4}{2} = 3$

(NO) Total electrons = 15

$$\sigma 1s^2, \overset{*}{\sigma} 1s^2, \sigma 2s^2, \overset{*}{\sigma} 2s^2, \sigma 2p_x^2, \pi 2p_y^2 \approx \pi 2p_z^2, \overset{*}{\pi} 2p_y^1$$

Paramagnetic, Border $= \dfrac{10-5}{2} = 2.5$

Electron is taken away from non-bonding molecular orbital, thats why bond order increases.

**18.** (b) Since, H-bond is the electrostatic force of attraction between H-atom and electronegative element. Thus, fluorine being the most electronegative element forms stronger H-bonds. On the other hand, H—F possesses larger electronegative difference than O—H.
Therefore, H-F molecule is more polarised than O—H, due to which H-atom of H—F bears larger amount of partial charge than that of O—H. This suggests H-atom of H—F attracts more strongly F-atom than that of O—H.
Hence, H—F forms more stronger H-bonds with fluorine than that of O—H.

**19.** (a) $SF_4$ has trigonal bipyramidal geometry, lone pair of electrons repels the axial bond pair and decreases the bond angle to 173°.

$SiF_4$

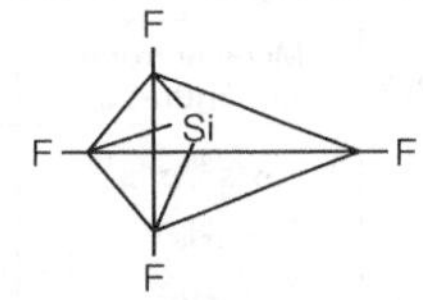


Regular tetrahedral

$XeF_4$

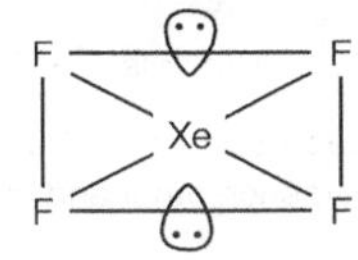


Square planar

$BF_4^-$

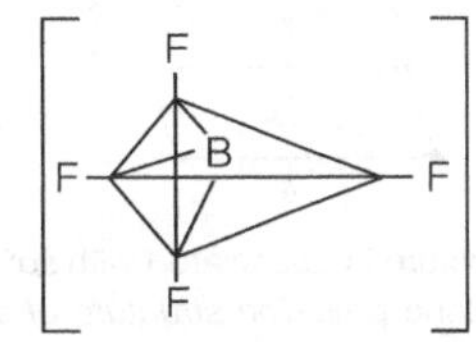


Tetrahedral

**20.** (d) $NH_3 > PH_3 > AsH_3 > SbH_3$

As the electronegativity of central atom increases, bonded electron polarise towards central atom more, so repulsion increases and bond angle increases.

**21.** (a) Molecular orbital configuration of given molecules are

$O_2^{2-}$ (Total number of electrons = 18)

$$\sigma 1s^2, \overset{*}{\sigma} 1s^2, \sigma 2s^2, \overset{*}{\sigma} 2s^2, \sigma 2p_x^2, \pi 2p_y^2 \approx \pi 2p_z^2, \overset{*}{\pi} 2p_y^2 \approx \overset{*}{\pi} 2p_z^2$$

$B_2$( Total number of electrons = 10)

$$\sigma 1s^2, \sigma^* 1s^2, \sigma 2s^2, \sigma^* 2s^2, \pi 2p_y^1 \approx \pi 2p_z^1$$

$N_2^+$ ( Total number of electrons = 13)

$$\sigma 1s^2, \sigma^* 1s^2, \sigma 2s^2, \sigma^* 2s^2, \pi 2p_y^2 \approx \pi 2p_z^2, \sigma 2p_x^1$$

$O_2$( Total number of electrons = 16)

$$\sigma 1s^2, \sigma^* 1s^2, \sigma 2s^2, \sigma^* 2s^2, \sigma 2p_x^2, \pi 2p_y^2 \approx \pi 2p_z^1, \pi^* 2p_y^1 \approx 2\pi^* 2p_z^1$$

Thus, $O_2^{2-}$ does not contain any unpaired electron.

**22.** (a) Greater the charge and smaller the radius, larger the polarising power and thus, greater the covalent nature. This leads to increase in lattice energy.

**23.** (a)

| Molecule | Structure | Hybridisation of central atom | Lone pair |
|---|---|---|---|
| $SF_4$ | 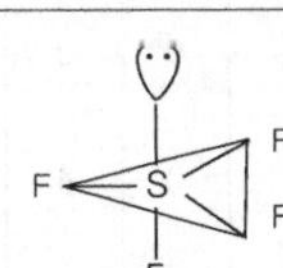 | $sp^3d$ | one |
| $CF_4$ | 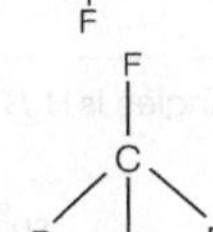 | $sp^3$ | zero |
| $XeF_4$ | 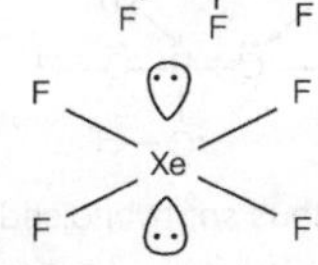 | $sp^3d^2$ | two |

**24.** (c)
A species is said to be diamagnetic, if it has all paired electrons.

| Species | Electrons | MO electronic configuration | Magnetic behaviour |
|---|---|---|---|
| $H_2^-$ | 3 | $\sigma 1s^2, \overset{*}{\sigma} 1s^1$ | Paramagnetic |
| $H_2^+$ | 1 | $\sigma 1s^1$ | Paramagnetic |
| $H_2$ | 2 | $\sigma 1s^2$ | Diamagnetic |
| $He_2^+$ | 3 | $\sigma 1s^2, \overset{*}{\sigma} 1s^1$ | Paramagnetic |

**25.** (d) $sp^3d^2$ hybridisation has octahedral structure such that four hybrid orbitals are at 90° *wrt* each other and others two at 90° with first four.

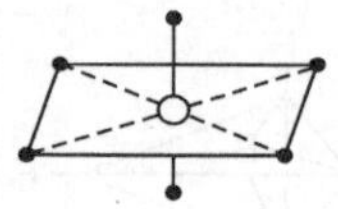

**26.** (c) Tetrahedral structure is associated with $sp^3$ hybridised central atom without any lone pair. *The structure of all the compounds given are as follows*

$[Ni(CN)_4]^{2-}$ 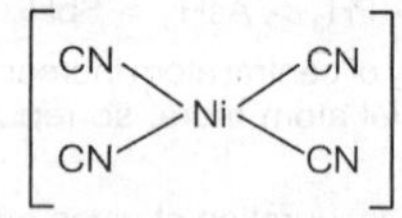 Square planar

$XeF_4$ 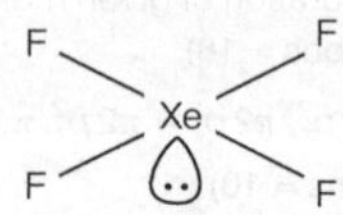 Distorted octahedral

$SF_4$ 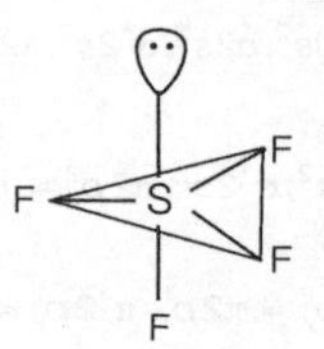 Distorted trigonal bipyramidal

$BF_4^-$ 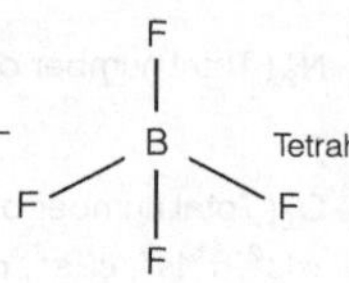 Tetrahedral

**27.** (c)

| Species | Structure | *lp* | *bp* | VSEPR | Bond angle |
|---|---|---|---|---|---|
| $H_2S$ | 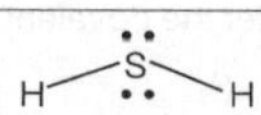 | 2 | 2 | *lp-lp* *lp-bp* *bp-bp* | 90° |
| $NH_3$ | 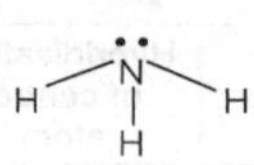 | 1 | 3 | *lp-bp* *bp* →*3bp* | 107° |
| $BF_3$ | 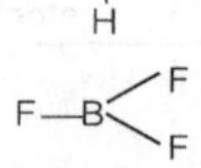 | 0 | 3 | *bp-bp* | 120° |
| $SiH_4$ | 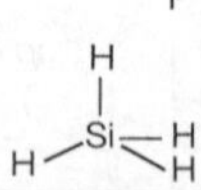 | 0 | 4 | *bp-bp* | 109° 28′ |

Thus, the order of bond angles is $H_2S < NH_3 < SiH_4 < BF_3$.

**28.** (b) $H_3BO_3$ has structure

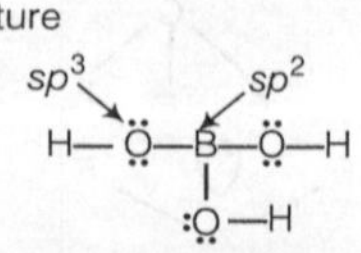

Boron has three bonds, thus $sp^2$ hybridised. Each oxygen has two bonds and two lone pairs, hence $sp^3$ hybridised.

**29.** (b) Bond length is inversely proportional to bond order. Bond order in $NO^+ = 3$ and $NO = 2.5$.
Thus, bond length in $NO > NO^+$.

**30.** (b) $XeF_2$ contains $sp^3d$ hybridised Xe atom with three lone pairs of electrons. Fluorine atoms occupy axial positions while three lone pairs of elelctrons occupy equatorial positions in trigonal bipyramid to minimise repulsion.
Since, lone pair of electrons do not considered for the shape of the molecule, thus, shape of $XeF_2$ comes out to be linear. While C- atom of $CO_2$ is *sp* hybridised and thus, possesses a linear shape.

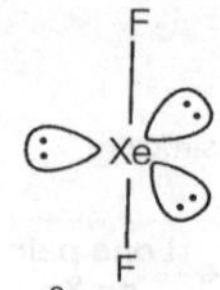

$sp^3d$ hybridised Xe-atom

O═C═O

*sp* hybridised C-atom

**31.** (c)

| Molecule | Hybridisation | Repulsion | Bond angle |
|---|---|---|---|
| $SO_2$ | $sp^2$ | $\frac{lp-bp,\ bp-bp}{lp-lp,\ bp-lp}$ | 119.5° |
| $OH_2$ | $sp^3$ | $bp-bp$ | 104.5° |
| $SH_2$ | $sp^3$ | —do— | 90° |
| $NH_3$ | $sp^3$ | $lp-bp,\ bp-bp$ | 107° |

**32.** (c) Both $NO_2$ and $O_3$ have angular geometry, thus possess permanent dipole moment. While $SiF_4$ possesses regular tetrahedral structure and $CO_2$ possesses a linear structure. In $SiF_4$ and $CO_2$, the individual dipoles on the Si—F and O bonds cancel and the molecules have no diple moment.

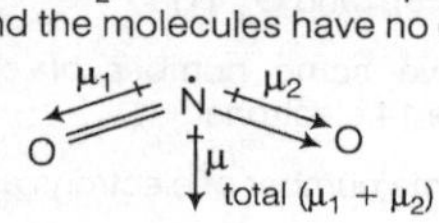

Thus, $\mu_{total} \neq 0$

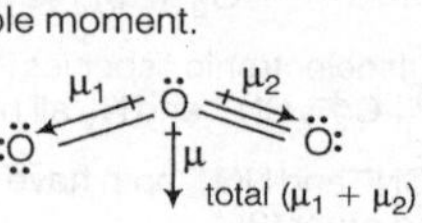

Thus, $\mu_{total} \neq 0$

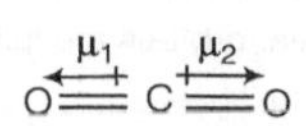

$\mu_{total} = 0\ (\mu_1 = -\mu_2)$

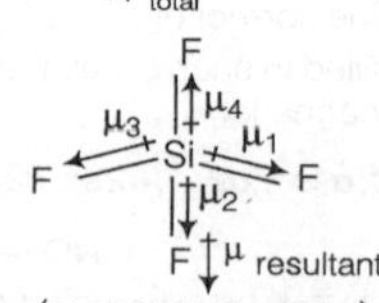

$(\mu_1 + \mu_2 + \mu_3 = -\mu_4)$
$\therefore \mu_{total} = 0$

**33.** (c)

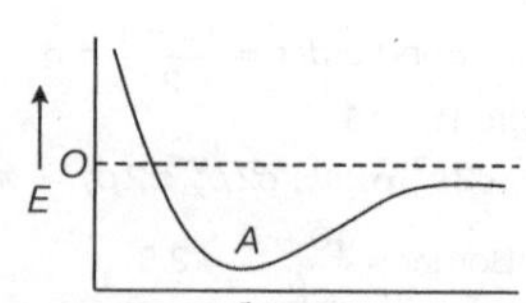

As repulsive forces decrease, bond formation starts. At point *A*, minimum energy is attained and molecule is stable.

**34.** (a) Lone pairs $= \dfrac{\text{Valence electrons} - \text{Bonding pairs}}{2}$

| Species | Valence electrons on Xe | Bonding Pairs | Lone Pairs |
|---|---|---|---|
| $XeF_2$ | 8 | 2 | 3 |
| $XeF_4$ | 8 | 4 | 2 |
| $XeF_6$ | 8 | 6 | 1 |

**35.** (d) 109° 28′, it means pure regular tetrahedral geometry.

(a) $NH_3$, $sp^3$, pyramidal (107°) (b) $H_2O$, $sp^3$, bent (104.5°).
(c) $CH_5^+$, does not exist (d) $NH_4^+$, $sp^3$, tetrahedral (109.5°).

**36.** (a)

(a) $AlH_3$ $sp^2$ 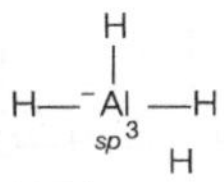 Change in hybridisation as well as in shape

(b) $H_2O$ $sp^3$ (2 *bp* +2 *lp*) — $H_3O^+$ $sp^3$ (3*bp* + 1 *lp*) — Change in shape only, bent to pyramidal

(c) $NH_3$ $sp^3$ (3 *bp* + 1 *lp*) — $NH_4^+$ $sp^3$ (4 *bp*) — Change in shape only, pyramidal to tetrahedral

**37.** (d) Higher the bond order, greater is the bond strength.

| Species | Bond order |
|---|---|
| $O_2^{2-}$ | 1 |
| $O_2^-$ | 1.5 |
| $O_2$ | 2.0 |
| $O_2^+$ | 2.5 |

## JEE Advanced and IIT JEE Archive

**1.** (c) If (2*s*-2*p*) mixing is not operative, then

| | σ1s | σ* 1s | σ2s | σ* 2s | σ2$p_x$ | π2$p_y$ π2$p_z$ | π* 2$p_y$ π* 2$p_z$ | σ* 2$p_x$ |
|---|---|---|---|---|---|---|---|---|
| $Be_2$(8) | 2 | 2 | 2 | 2 | | | | Diamagnetic |
| $B_2$ (10) | 2 | 2 | 2 | 2 | 2 | | | Diamagnetic |
| $C_2$(12) | 2 | 2 | 2 | 2 | 2 | 1 | – | Paramagnetic |
| $N_2$(14) | 2 | 2 | 2 | 2 | 2 | 2 | – | Diamagnetic |

**2.** (a, b, d)

(a) When liquid water changes to ice, there is an air gap inside the cage formed due to intermolecular H–bonding. Thus, volume of ice > volume of water.

Since, $$d = \frac{\text{mass}}{\text{volume}}$$

Hence, density (ice) < density (water).

Thus ice floats on water – hydrogen bonding plays its role.

(b) In gaseous phase, basicity of 1° < 2° < 3° amines due to increase of electron density at N-atom.

$$RNH_2 < R_2NH < R_3N$$

But in aqueous ($H_2O$) solvent, $RNH_2$ is stabilised by solvation. Thus, 1° amine is more

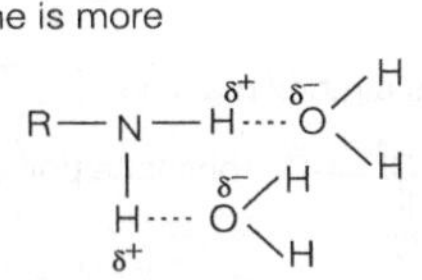

basic than 3° amine (3° < 1° < 2°) amines.

(c) $I_+$ (positive inductive effect) explains this

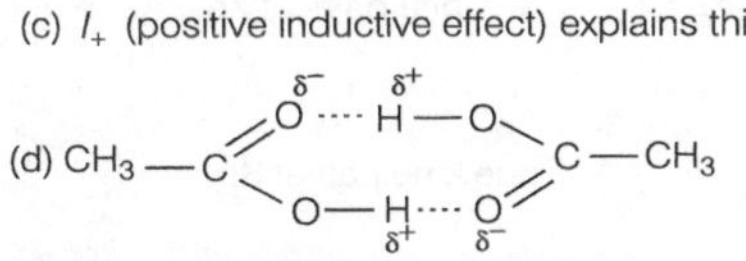

Dimer due to intermolecular H- bonding.

**3.** (c) **Plan** This problem includes basic concept of bonding. It can be solved by using the concept of molecular orbital theory.

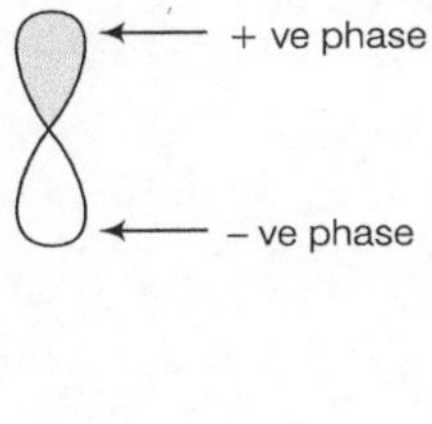


Any orbital has two phase +ve and –ve. In the following diagram +ve phase is shown by darkening the lobes and –ve by without darkening the lobes.

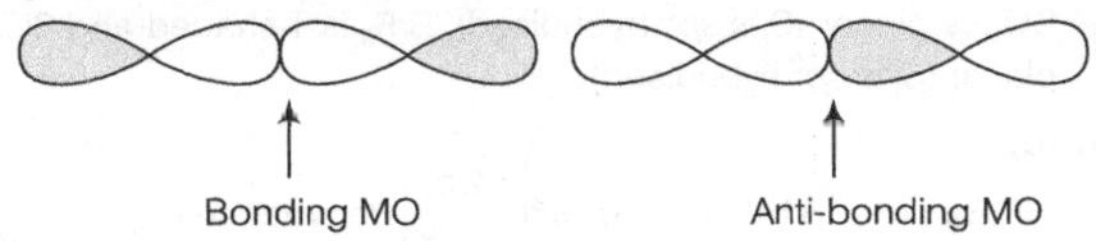


When two same phase overlap with each other it forms bonding molecular orbital otherwise anti-bonding.

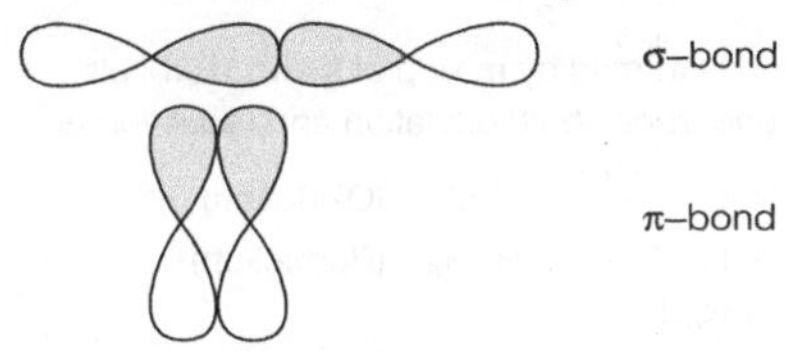


On the basis of above two concepts correct matching can be done as shown below

(i) 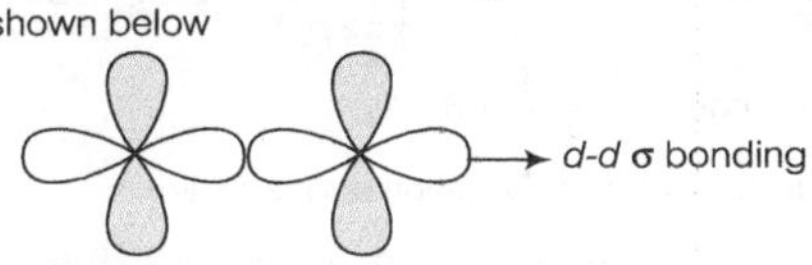


∴ (i) → (q)

(ii) 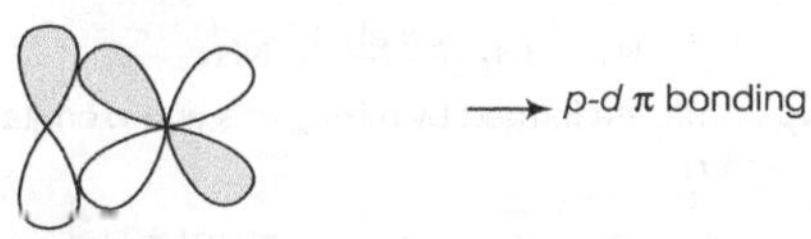


∴ (ii) → (r)

(iii) 

∴ (iii) → (p)

(iv) 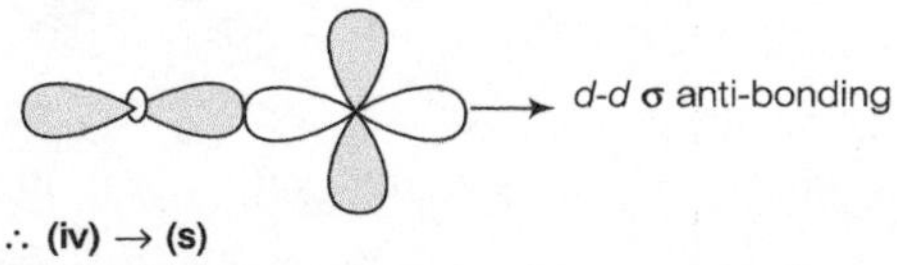


∴ (iv) → (s)

Hence, (c) is the correct option.

**4.** (a) For molecules lighter than $O_2$, the increasing order order of energies of molecular orbitals is :

$$\sigma 1s\, \overset{*}{\sigma}\, 1s\, \sigma\, 2s\, \overset{*}{\sigma} 2s \begin{bmatrix} \pi 2p_y \\ \pi 2p_z \end{bmatrix} \sigma 2p_x\, \overset{*}{\sigma} 2p_x \begin{bmatrix} \overset{*}{\pi} 2p_y \\ \overset{*}{\pi} 2p_z \end{bmatrix} .....$$

Where, $\pi 2p_y$ and $\pi 2p_z$ are degenerate molecular orbitals, first singly occupied and then pairing starts if Hund's rule is obeyed. If Hund's rule is violated in $B_2$, electronic arrangement would be :

$$\sigma 1s^2\, \overset{*}{\sigma}\, 1s^2\, \sigma 2s^2\, \overset{*}{\sigma}\, 2s^2 \begin{bmatrix} \pi 2p_y^2 \\ \pi 2p_z \end{bmatrix} ...$$

No unpaired electron-diamagnetic.

$$\text{Bond order} = \frac{\text{bonding electrons} - \text{anti-bonding electrons}}{2}$$

$$= \frac{6-4}{2} = 1$$

**5.** (d)

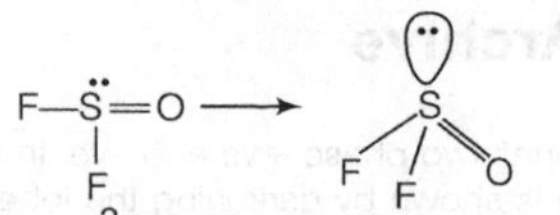

S is $sp^3$ hybridised  Pyramidal

$SO_3$ is planar (S is $sp^2$ hybridised), $BrF_3$ is T-shaped and $SiO_3^{2-}$ is planar (Si is $sp^2$ hybridised).

**6.** (d)

(i) $B_2 : \sigma 1s^2 \overset{*}{\sigma} 1s^2\ \sigma 2s^2\ \overset{*}{\sigma} 2s^2 \left| \begin{matrix} \pi 2p_y^1 \\ \pi 2p_z^1 \end{matrix} \right|$ Paramagnetic.

Bond order $= \dfrac{6-4}{2} = 1$

Bond is formed by mixing of *s* and *p* orbitals.

$B_2$ undergoes both oxidation and reduction as

$B_2 + O_2 \xrightarrow{\text{Heat}} B_2O_3$ (Oxidation)

$B_2 + H_2 \longrightarrow B_2H_6$ (Reduction)

**(i) → (q,r)**

(ii) $N_2 : \sigma 1s^2 \overset{*}{\sigma} 1s^2\ \sigma 2s^2 \overset{*}{\sigma} 2s^2 \left| \begin{matrix} \pi 2p_y^2 \\ \pi 2p_z^2 \end{matrix} \right| \sigma 2p_x^2$ Diamagnetic.

Bond order $= \dfrac{10-4}{2} = 3 > 2$

$N_2$ undergoes both oxidation and reduction as

$$N_2 + O_2 \xrightarrow{\Delta} NO$$

$$N_2 + 3H_2 \xrightarrow{\text{Catalyst}} NH_3$$

In $N_2$, bonds are formed by mixing of *s* and *p* orbitals.

**(ii) → (q,r)**

(iii) $O_2^- : \sigma 1s^2 \overset{*}{\sigma} 1s^2\ \sigma 2s^2\ \overset{*}{\sigma} 2s^2 \sigma 2p_x^2 \left| \begin{matrix} \pi 2p_y^2 \\ \pi 2p_z^2 \end{matrix} \right| \begin{matrix} \overset{*}{\pi} 2p_y^2 \\ \overset{*}{\pi} 2p_z^1 \end{matrix} \Bigg| \overset{*}{\sigma} 2p_x^0$

Paramagnetic with bond-order = 15. $O_2^-$ undergoes both oxidation and reduction and bond involves mixing of *s* and *p*-orbitals.

**(iii) → (p,q,r,t)**

(iv) $O_2 : \sigma 1s^2 \overset{*}{\sigma} 1s^2\ \sigma 2s^2\ \overset{*}{\sigma} 2s^2 \sigma 2p_x^2 \left| \begin{matrix} \pi 2p_y^2 \\ \pi 2p_z^2 \end{matrix} \right| \begin{matrix} \overset{*}{\pi} 2p_y^1 \\ \overset{*}{\pi} 2p_z^1 \end{matrix} \Bigg| \overset{*}{\sigma} 2p_x^0$

Paramagnetic with bond order = 2.

$O_2$ undergoes reduction and the bond involves mixing of *s* and *p*-orbitals.

**(iv) → (p,r,s,t)**

**7.** (b)

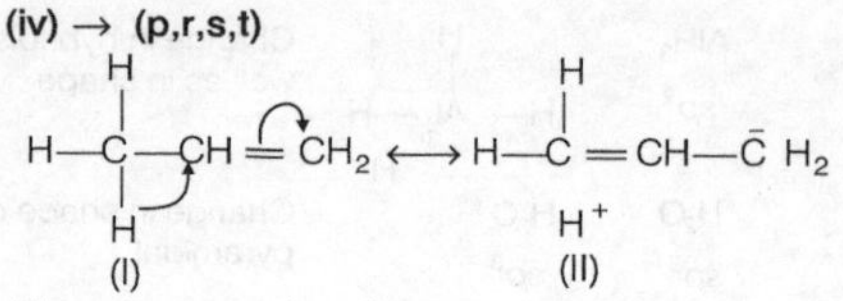

I and II are hyperconjugation structures of propene and involves σ-electrons of C—H bond and *p*-orbitals of pi-bond in delocalisation.

**8.** (a) The bond order of CO = 3. $NO^+$, $CN^-$ and $N_2$ are isoelectronic with CO, have the same bond orders as CO. $NO^-$ ($16e^-$) has bond order of 2.

**9.** (d) $O_2^-$ in $KO_2$ has 17 electrons, species with odd electrons are always paramagnetic.

**10.** (c) (i) In the reaction : $O_2^- \longrightarrow O_2 + O_2^{2-}$

Oxygen on reactant side is in $-\frac{1}{2}$ oxidation state. In product side, one of the oxygen is in zero oxidation state, i.e. oxidised while the other oxygen is in −1 oxidation state, i.e. reduced. Hence, in the above reaction, oxygen ($O^{-1/2}$) is simultaneously oxidised and reduced (disproportionated).

**(i) → (s)**

(ii) In acid medium, $CrO_4^{2-}$ is converted into $Cr_2O_7^{2-}$ which is a dimeric, bridged tetrahedral :

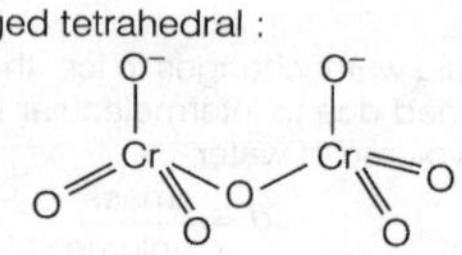

**(ii) → (r)**

(iii) $MnO_4^- + NO_2^- + H^+ \longrightarrow Mn^{2+} + NO_3^-$

The above is a redox reaction and a product $NO_3^-$ has trigonal planar structure.

**(iii) → (q)**

(iv) $NO_3^- + H_2SO_4 + Fe^{2+} \longrightarrow Fe^+ + NO$

**(iv) → (p)**

The above is a redox reaction.

**11.** (d) $ClO_3^-$ : 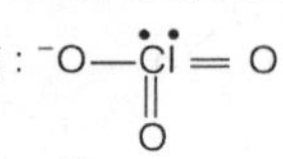 one lone pair at Cl.

$XeF_4$ : 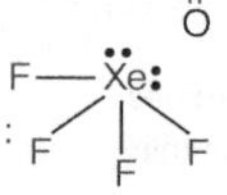 two lone pairs at Xe.

$SF_4$ : 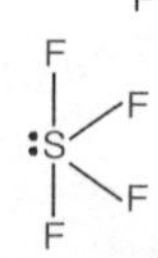 one lone pair at S.

$I_3^-$ : 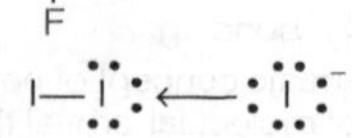 three lone pairs at central iodine.

# 4. Stoichiometry

## DPP-1 Mole Concept

**1.** (i) **Plan** 1 g (Alloy) $\xrightarrow{\text{X per cent}}$ g of Mg $\xrightarrow[\text{atomic weight}]{\text{Divide by}}$ moles of Mg

Multiply by $N_0$ = atoms of Mg

$$1 \text{ g alloy} = 1 \times \frac{24}{100} \text{ g Mg} = 0.24 \text{ g}$$

$$\text{Moles of Mg} = \frac{0.24}{24} = 0.01 \text{ mol}$$

$$\text{Atoms of Mg} = 0.01 \times N_0 = 0.01 \times 6.02 \times 10^{23} = 6.02 \times 10^{21} \text{ Mg atoms}$$

(ii) **Plan** At NTP 1 mole of every ideal gas = 22.4 L

0.8 L is the volume of = 1.0 g

$\therefore$ 22.4 L is the volume $= \frac{1}{0.8} \times 22.4 = 28.0$ g

Thus, molar mass = 28.0 g

Moles of diatomic gas $= \frac{1}{28}$ in 1 g (0.8 L)

Molecules in 1 g $= \frac{N_0}{28}$

Atoms in 1 g $= \frac{2N_0}{28} = \frac{2 \times 6.02 \times 10^{23}}{28} = 4.3 \times 10^{22}$

**Alternately** 22.4L gas at NTP = 1 mole

0.8L gas at NTP $= \frac{0.8}{22.4} = \frac{1}{28}$ mole

1 mole = $N_0$ molecules = 2 $N_0$ atoms of diatomic gas

$\frac{1}{28}$ mole $= \frac{2 \times 6.02 \times 10^{23}}{28} = 4.3 \times 10^{22}$ atoms

(iii) **Plan** Volume of 1 mole $SO_3$ at STP = 22.4 L

or Volume of 80 g $SO_3$ at STP = 22.4 L

Volume of 20 g $SO_3$ at STP $= \frac{22.4}{80} \times 20 = 5.6$ L

5.6 L of $SO_x$ has mass = 16 g

$\therefore$ 22.4 L of $SO_x$ has mass $= \frac{16}{5.6} \times 22.4 = 64.0 \text{ g mol}^{-1}$

$$S + x(O) = 64.0$$

$$32 + 16x = 64.0$$

$\therefore$ $x = 2.0$

Thus, $SO_x = SO_2$

$16 \text{ g } SO_2 = \frac{16}{64} \text{ mol} = 0.25 \text{ mol}$

$= 0.25 N_0$ molecules of $SO_2$

$= 0.25 \times 2N_0$ atoms of oxygen

$= 0.5 \times 6.02 \times 10^{23} = 3.01 \times 10^{23}$ atoms

**2.** (i) **Plan**

(a) Let, atomic mass = $A$

Then, 1 mole of atom has = $N_0$ atoms

Mass of one atom of $_6C = \frac{A}{N_0} = \frac{A}{6.02 \times 10^{23}}$

$$\frac{A}{6.02 \times 10^{23}} = 2.32481 \times 10^{-23}$$

$$A = 6.02 \times 10^{23} \times 2.32481 \times 10^{-23} = 14.0 \text{ g mol}^{-1} = \frac{14.0}{N_0} = 14 \text{ amu or } \mu$$

(b) Thus, atom is $^{14}_{6}C$; Proton = 6

Neutron + Proton = Atomic mass

$\therefore$ Neutron = 14 − 6 = 8 ; Proton = Electron = 6

(ii) **Plan** $4.0 \times 10^{-23}$ g is the mass of = 1 atom

$\therefore$ 1 g is the mass of = $2.50 \times 10^{22}$ atoms

**3.** (i) **Plan** Mole $= \frac{\text{Mass (in grams)}}{\text{Molar mass}}$

Molar mass ($C_{12}H_{22}O_{11}$) = 342 g mol$^{-1}$

$$6.84 \text{ g } C_{12}H_{22}O_{11} = \frac{6.84 \text{ g}}{342 \text{ g mol}^{-1}} = 0.02 \text{ mol}$$

(ii) **Plan** Mole $= \frac{\text{Mass (in gram)}}{\text{Molar mass}}$

Molar mass (urea) = 60 g mol$^{-1}$

$\therefore$ Mass (in grams) = mole × molar mass = 0.01 × 60 = 0.60 g

(iii) **Plan** Mole $= \frac{\text{Mass (g)}}{\text{Molar mass}} = x$

$\therefore$ 1 mole solute contains = $N_0$ molecule

Thus, $x$ mole contains = $N_0x$ molecules

Electrons in 1 molecule = $y$

Thus, electrons in $N_0x$ molecules = $N_0xy$ = number of coins

Electrons in molecule of

$$CH_4 = 6 + 4 = 10$$

$$1.6 \mu g = 1.6 \times 10^{-6} \text{ g } CH_4$$

$\therefore$ Moles $= \frac{1.6 \times 10^{-6}}{16} = 1 \times 10^{-7}$ mole

Molecules = $6.02 \times 10^{23} \times 1 \times 10^{-7} = 6.02 \times 10^{16}$

Electron or coins = $6.02 \times 10^{16} \times 10 = 6.02 \times 10^{17}$

**4.** (i) **Plan** Convert % into moles → molar ratio → empirical formula → molecular formula

| | % | Moles | Ratio |
|---|---|---|---|
| N | 30.43 | 2.17 | 1 |
| O | 69.57 | 4.35 | 2 |

Formula = $NO_2$

Empirical formula weight = 14 + 32 = 46

Molecular mass = 92 g mol$^{-1}$

Thus, $NO_2$ dimerises forming $N_2O_4$(92) oxide is $N_2O_4$.

(ii) **Plan** Percentage $= \frac{\text{Mass of element}}{\text{Molar mass}} \times 100$

| $N_2COH_4$ (60) | $C_2H_4O_2$ (60) |
|---|---|
| $N = \frac{28 \times 100}{60} = 46.67$ | $C = \frac{24 \times 100}{60} = 40.00$ |
| $C = \frac{12 \times 100}{60} = 20.00$ | $H = \frac{4 \times 100}{60} = 6.66$ |
| $H = \frac{4 \times 100}{60} = 6.66$ | $O = \frac{32 \times 100}{60} = 53.33$ |
| $O = \frac{16 \times 100}{60} = 26.66$ | |

**5.** (d) 1 mole $CH_2O$ = 1 + 4 + 1 = 6 moles of atoms = 6$N_0$ atoms

1 mole $SO_2$ = 3 moles of atoms = 3$N_0$ atoms

6 $N_0$ atoms are in = 1 mole

3 $N_0$ atoms are in = 0.5 mole

**6.** (b)

| Species | Mass | Moles | Molecules |
|---|---|---|---|
| $O_2$ | 1 g | 1/32 | $N_0$/32 |
| $N_2$ | 4 g | 4/28 | 4$N_0$/28 |

Ratio of molecules $\frac{N(O_2)}{N(N_2)} = \frac{1/32}{1/7} = \frac{7}{32}$

**7.** (d) 1 mole $CO_2$ = $N_0$ molecules

Electrons in one molecule $CO_2$ = 6 + 8 × 2 = 22

Thus, electrons in $N_0$ molecules = 22$N_0$

**8.** (a)

| Species | Amount | Moles | Molecules | Atoms |
|---|---|---|---|---|
| $CH_4$ | 1.6 g | 0.1 | 0.1 $N_0$ | 0.5 $N_0$ (Maximum) |
| $NH_3$ | 1.7 g | 0.1 | 0.1 $N_0$ | 0.4 $N_0$ |
| $H_2O$ | 1.8 g | 0.1 | 0.1 $N_0$ | 0.3 $N_0$ |

**9.** (a) 1 kg electrons = 1000 g electrons

$$\text{Moles} = \frac{\text{mass}}{\text{mass of 1 mol of electrons}}$$

Mass of 1 mole of electrons

$$= 6.02 \times 10^{23} \text{ mol}^{-1} \times 9.1 \times 10^{-31} \text{ kg}$$
$$= 5.4782 \times 10^{-7} \text{ kg mol}^{-1}$$
$$= 5.4782 \times 10^{-4} \text{ g mol}^{-1}$$

$$\therefore \text{Number of moles} = \frac{1000}{5.4782 \times 10^{-4}} = 1.8254 \times 10^6$$

**10.** (b) Average atomic mass $= \frac{\Sigma m_i X_i}{\Sigma X}$

where, $m$ = atomic weight,
$X$ = their percentages

For, $\text{Fe} = \frac{A_1X_1 + A_2X_2 + A_3X_3}{X_1 + X_2 + X_3}$

$$= \frac{54 \times 5 + 56 \times 90 + 57 \times 5}{5 + 90 + 5}$$
$$= 55.95$$

**11.** (b) 1 mole of any substance = $N_0$ atoms = $6.02 \times 10^{23}$ atoms

$6.02 \times 10^{23}$ atoms = 1 mole

$$\therefore \quad 1 \text{ atom} = \frac{1}{6.02 \times 10^{23}} = 1.66 \times 10^{-24} \text{ mol}$$

**12.** (d) For an ideal gas $pV = nRT$

$$\therefore \quad n = \frac{pV}{RT}$$
$$= \frac{10 \text{ atm} \times 50 \text{ L}}{0.0821 \text{ L atm mol}^{-1}\text{K}^{-1} \times 300 \text{ K}}$$
$$= 20.3 \text{ moles}$$

**13.** (a) $\frac{\text{mass}}{V \text{ (volume)}} = \text{Density } (d)$

$$\therefore \quad \text{Mass} = V \times d = 0.105 \times 11.34 = 1.1907 \text{ g}$$

$$\text{Moles of lead} = \frac{1.1907}{206} = 5.78 \times 10^{-3}$$

1 mole contains $N_0$ atoms

$\therefore$ $5.78 \times 10^{-3}$ mole contains $= 5.78 \times 10^{-3} \times N_0$

$$= 5.78 \times 10^{-3} \times 6.02 \times 10^{23}$$
$$= 3.48 \times 10^{21} \text{ atoms}$$

**14.** (b) 1 mole contains = $6.02 \times 10^{23}$ atoms

$$\therefore 6.50 \times 10^{23} \text{ Cd atoms are in} = \frac{6.5 \times 10^{23}}{6.02 \times 10^{23}} \text{ mole Cd}$$
$$= 1.08 \text{ mole Cd} = 1.08 \times 112 \text{ g Cd}$$
$$= 120.96 \text{ g Cd} = \frac{120.96 \times 100}{8} \text{ g}$$

(alloy containing 8% Cd) = 1512$g$ = 1.512 kg

**15.** (d) Molar mass of $S_8 = 32 \times 8 = 256 \text{ g mol}^{-1}$

Volume = $0.568 \text{ mm}^3 = 0.568 \times 10^{-3} \text{ cm}^3$

Mass = volume $\times d = 0.568 \times 10^{-3} \text{ cm}^3 \times 2.07 \text{ g/cm}^3$

$$= 1.17576 \times 10^{-3} \text{ g}$$

$$\text{Moles} = \frac{1.17576 \times 10^{-3}}{256} = 4.5928 \times 10^{-6}$$

$S_8$ molecules $= 4.5928 \times 10^{-6} \times 6.02 \times 10^{23} = 2.765 \times 10^{18}$

**16.** (c) $M^{2+}$ has electrons = as in $SO_2$

$$= 16 + 16 = 32$$

$M^{2+}$ has protons $= 32 + 2$

$$= 34 \ (P)$$

Thus, atomic number of $M^{2+} = 34 = Z$

Neutrons in $M^{2+} = (Z + 2)$

$$= 34 + 2 = 36 \ (N)$$

Ionic mass = $N + P$

$$= 36 + 34 = 70$$

**17.** (d) Ionic mass of $X^{3-} = 31$

Thus, $(N + P) = 31$

$\therefore$ $P = 31 - N = 31 - 16 = 15$

Thus, electrons in neutral atom = 15

Electrons in $X^{3-}$ is = 15 + 3 = 18

**18.** (b) Lead pencil uses carbon (graphite)

Thus, 0.0001 g carbon $= \frac{0.0001}{12}$ mole carbon

1 mole carbon = $6.02 \times 10^{23}$ atoms

$$\therefore \quad \frac{0.0001}{12} \text{ mole carbon} = 6.02 \times 10^{23} \times \frac{0.0001}{12}$$
$$= 5.02 \times 10^{18} \text{ atoms}$$

**19.** (a) 0.05 mL $H_2O$ = 0.05 g $H_2O$ $= \frac{0.05}{18}$ mole $H_2O$

1 mole contains = $N_0$ water molecules

$$\therefore \quad \frac{0.05}{18} \text{ mole contains} = \frac{0.05}{18} N_0 \text{ molecules}$$
$$= \frac{0.05 \times 6.02 \times 10^{23}}{18} = 1.67 \times 10^{21}$$

**20.** (c) 1 g $H_2O$ = 1 mL $H_2O$

18 g $H_2O$ = 18 mL $H_2O$ = $N_0$ molecules

Thus, volume of one molecule $= \frac{18}{N_0} = \frac{18}{6.02 \times 10^{23}}$ mL

But, $V = \frac{4}{3}\pi r^3$

$$\therefore \quad \frac{4}{3}\pi r^3 = \frac{18}{6.02 \times 10^{23}}$$

$$\therefore \quad r^3 = \frac{3 \times 18}{4\pi \times 6.02 \times 10^{23}} \text{ mL}$$

$$r = \sqrt[3]{\frac{3 \times 18}{4\pi \times 6.02 \times 10^{23}}}$$
$$= 1.925 \times 10^{-8} \text{ cm} = 1.925 \text{ Å}$$

**21.** (d) Empirical weight of $CF_2$ = 50

Molecular formula = $(CF_2)_n$

where, $n = \frac{\text{molecular weight}}{\text{empirical weight}}$

1.70 g $F_2$ occupies = 1 L volume

38 g $F_2$ occupies $= \frac{38}{1.70} = 22.4$ L (under given condition)

Thus, mass of $(CF_2)_n$ that occupies 22.4 L

$$= 22.4 \times 8.93$$
$$= 200 \text{ g mol}^{-1}$$

Thus, molecular weight of $(CF_2)_n = 200 \text{ g mol}^{-1}$

$$\therefore \quad n = \frac{200}{50} = 4$$

$\therefore$ Molecular formula = $(CF_2)_4 = C_4F_8$

**22.** (b) $2N_2O + 3O_2 \longrightarrow 4NO_2$

| $2N_2O$ | $3O_2$ | $4NO_2$ |
|---|---|---|
| 2 mol | 3 mol | 4 mol |
| 2 volume | 3 volume | 4 volume |
| 2 L | 3 L | 4 L |

**23.** (a)

| Compound X | % | Moles | Ratio |
|---|---|---|---|
| N (14) | 33.28 | 2.38 | 1 |
| Si (25) | 66.72 | 2.38 | 1 |

X is Si N

| Compound Y | % | Moles | Ratio | Whole number ratio |
|---|---|---|---|---|
| N (14) | 39.94 | 2.88 | 0.75 | 3 |
| Si (28) | 60.06 | 2.14 | 1.00 | 4 |

Y is $Si_3N_4$

Thus, law of multiple proportion is followed.

**24.** (b) (i) $n_{CH_4} = \frac{1}{16} = \frac{5}{80}$ (s) (ii) $n_{O_2} = \frac{1}{32} = \frac{2.5}{80}$ (r)

(iii) $n_{SO_2} = \frac{1}{64} = \frac{1.25}{80}$ (q) (iv) $n_{SO_3} = \frac{1}{80} = \frac{1}{80}$ (p)

Thus, (i) → (s), (ii) → (r), (iii) → (q), (iv) → (p)

**Explanations for Q. Nos. (25-26)**

| | Moles | Ratio |
|---|---|---|
| $Na_2SO_3 = 50\%$ | 50 / 126 | 1 |
| $H_2O = 50\%$ | 50 /18 | 7 |

Thus, salt is $Na_2SO_3 \cdot 7H_2O$

Water molecules = 7

(a) Number of atoms in anhydrous salt

$= (Na_2SO_3) = 2 + 1 + 3 = 6$

Water molecules = 7

Thus, (a) is incorrect.

(b) $Na_2SO_3 \cdot 7H_2O = 27$ atoms

Thus, (b) is incorrect.

(c) $Na_2SO_3 \cdot 7H_2O = 10$ oxygen atoms

Thus, (c) is correct.

**25.** (c) **26.** (c)

**Explanations for Q. Nos. (27-28)**

One mole of $X$ has $= N_0$ atoms $= 6.02 \times 10^{23}$

Mass of one atom of $X = 6.642 \times 10^{-23}$ g

Thus, mass of $6.02 \times 10^{23}$ atoms $= 6.642 \times 10^{-23} \times 6.02 \times 10^{23}$

$= 40.0 \text{ g mol}^{-1}$

Thus, atomic weight of $X = 40.0 \text{ g mol}^{-1}$

$$\text{moles} = \frac{\text{mass (g)}}{\text{atomic weight (g mol}^{-1})} = \frac{40 \times 1000}{40} = 1000 \text{ moles}$$

**27.** (c) **28.** (b)

**29.** (8)

$x$ moles of $CH_4 = x\,N_0$ molecules of $CH_4$

$= x\,N_0 \times 5$ atoms of carbon and hydrogen

$5x\,N_0 = 2.408 \times 10^{25}$

$$\therefore \quad x = \frac{2.408 \times 10^{25}}{5N_0} = \frac{2.408 \times 10^{25}}{5 \times 6.02 \times 10^{23}} = 8$$

**30.** (3) Molar mass of dye $= 20000 \text{ g mol}^{-1}$

$$Fe\ (0.84\%) = 20000 \times \frac{0.84}{100} = 168 \text{ g}$$

$$\text{Thus, atoms of Fe} = \frac{168}{56} = 3$$

**31.** (6) Volume of 1 g alkene = 0.8 L at STP

Alkene in $C_nH_{2n}$ of molar mass $= 12x + 2x = 14x$

0.8 L of gas is the volume = 1 g $C_nH_{2n}$

$$\therefore \ 22.4 \text{ L of gas is the volume of} = \frac{22.4}{0.8} = 28.0\text{g}$$

$\therefore \quad 14n = 28$

$\therefore \quad n = 2$

$\therefore$ Atoms of carbon and hydrogen in

$C_nH_{2n} = C_2H_4 = 2 + 4 = 6$

## DPP-2 Chemical Calculations

**1. Plan** Impurity present in the ore is not converted into zinc and thus, actual yield is less than the theoretical yield

$$\underset{97\text{ g}}{ZnS} \longrightarrow \underset{65\text{ g}}{Zn}$$

(i) $5.32 \times 10^3\text{g ZnS} = \frac{65}{97} \times 5.32 \times 10^3\text{g Zn} = 3565 \text{ g} = 3.565 \text{ kg}$

(ii) Actual yield = 3.30 kg

$$\text{Thus, } \% \text{ yield} = \frac{\text{actual yield}}{\text{theoretical yield}} \times 100$$

$$= \frac{3.30}{3.565} \times 100 = 92.57\%$$

**2. Plan** $2SO_2 + O_2 + 2H_2O \longrightarrow 2H_2SO_4$

Ratio 2 1 2 2

Limiting reactant is present in smallest quantity

Quantity of the product is determined based on the limiting reactant.

**Required** $SO_2 = 400 \text{ g} = \frac{400}{64} \text{ mol} = 6.25 \text{ mol}$

$O_2 = 175 \text{ g} = \frac{175}{32} \text{ mol} = 5.47 \text{ mol}$

$H_2O = 125 \text{ g} = \frac{125}{18} \text{ mol} = 6.94 \text{ mol}$

| $SO_2$ | $O_2$ | $H_2O$ |
|---|---|---|
| 2 | 1 | 2 |
| **6.25** | 3.125 | 6.25 |
| 10.94 | **5.47** | 10.94 |
| 6.94 | 3.47 | **6.94** |

(i) Thus, $SO_2$ is the limiting reactant.

(ii) $H_2SO_4$ formed = 6.25 mol = 612.5 g

(iii) $O_2$ reacted $= \frac{6.25}{2}$ mol = 3.125 mol = (3.125 × 32) g

$O_2$ unreacted = 175 g − 3.125 × 32 g = 75 g

$H_2O$ reacted = 6.25 mol = 6.25 × 18 g

$H_2O$ unreacted = (125 − 6.25 × 18) g = 12.5 g

**3. Plan** $2Al + 6HCl \longrightarrow 2AlCl_3 + 3H_2$

2 mol 3 mol

2 × 27 g = 54 g ... 3 × 2 g = 6 g

g Al in alloy ...... g $H_2$

Mass of alloy $= V \times d = 0.691 \text{ cm}^3 \times 2.85 \text{ g cm}^{-3} = 1.96935$ g

Al in alloy 93.7% $= 1.96935 \times \frac{93.7}{100} = 1.84528$ g

1.84528 g Al in alloy gives $H_2 = \frac{6}{54} \times 1.84528 = 0.2050$ g

**4. Plan** $2Al + 6HCl \longrightarrow 2AlCl_3 + 3H_2$

2 mol 6 mol

2 × 27 g 6 × 36.5 g

1.35 g Al reacts with $= \frac{6 \times 36.5}{54} \times 1.35$ g HCl = 5.475 g HCl

100 g HCl solution has = 28.0 g HCl

100 × 1.14 mL HCl solution has = 28.0 g HCl

28.0 g HCl is in = 114 mL HCl solution

5.475 g HCl is in $= \frac{114}{28} \times 5.475$ mL HCl

= 22.3 mL HCl solution

**5. Plan** 1 mole $Ca(OH)_2 \equiv$ 2 moles NaOH

$\equiv$ 2 moles $Cl_2$

$\equiv$ 1 mole $Ca(OCl)_2$

Convert grams of $Ca(OH)_2$ into moles

$$740\text{ g} = \frac{740\text{ g}}{74\text{ g mol}^{-1}} = 10\text{ moles}$$

$\therefore$ NaOH needed = 20 moles

$Cl_2$ needed = 20 moles

$Ca(OCl)_2$ formed = 10 moles

**6.** (d) $N_2 + 3H_2 \longrightarrow 2NH_3$

As per reaction : 1 mol 3 mol 2 mol

0.5 mol $\longrightarrow$ 1 mol $NH_3$

$N_2$ gives 3 mol $H_2$ gives $\longrightarrow$ 2 mol $NH_3$

1.5 mol $H_2$ gives $\longrightarrow$ 1 mol $NH_3$

**Note** *$N_2$ and $H_2$ are in actual stoichiometric ratio, hence completely react to form $NH_3$.*

There is no limiting reactant.

**7.** (d) $NH_3 + HCl \longrightarrow NH_4Cl$

1 mol 1 mol 1 mol

17 g 36.5 g 53.5 g

17 g 17 g ?

Taken

17 g of $NH_3$ gives = 53.5 g $NH_4Cl$

36.5 g HCl gives = 53.5 g $NH_4Cl$

$$17\text{ g HCl gives} = \frac{53.5}{36.5} \times 17\text{ g } NH_4Cl$$

$= 24.92$ g $NH_4Cl$

**Note** *Reactant which gives product in smaller quantity is the limiting reactant.*

$$\text{Some } NH_3 \text{ is left unreacted} = \left(17 - \frac{17 \times 17}{36.5}\right)\text{ g}$$

$= 9.082$ g

**8.** (c) $Fe_2O_3 + 3CO \longrightarrow 2Fe + 3CO_2$

160 g $Fe_2O_3$ give = 112 g Fe

$$480\text{ g } Fe_2O_3 \text{ give} = \frac{112}{160} \times 480\text{ g Fe} = 336\text{ g Fe (Theoretical)}$$

But actual yield is = 252.0 g

$$\text{Thus, \% yield} = \frac{\text{actual yield} \times 100}{\text{theoretical yield}} = \frac{252}{336} \times 100 = 75\%$$

**9.** (d)

| | $3BaCl_2$ | $+ 2Na_3PO_4 \longrightarrow$ | $Ba_3(PO_4)_2$ | $+ 6NaCl$ |
|---|---|---|---|---|
| Stoichiometry | 3 moles | 2 moles | 1 mole | |
| From | 0.5 mole | — | 0.5/3 = 0.17 mole | |
| From | — | 0.2 mole | $0.2 \times \frac{1}{2} = 0.10$ mole | |

Limiting reactant is $Na_3PO_4$ and $BaCl_2$ is in excess.

Thus, 0.2 mole $Na_3PO_4$ gives = 0.10 mole of $Ba_3(PO_4)_2$.

**Note** *Yield is taken from the reactant which gives smaller quantity.*

**10.** (b) $BaCl_2 + K_2SO_4 \longrightarrow BaSO_4 + 2KCl$ ($BaSO_4$: White ppt.)

| $BaCl_2$ | $K_2SO_4$ | $BaSO_4$ |
|---|---|---|
| 1 mol | 1 mol | 1 mol |
| 20.8 g | 174 g | 233 g |
| 100 mL | 50 mL | |
| of 20.8% | of 17.4% | |
| 20.8 g | 8.7 g | |
| 0.1 mol | 0.05 mol | |

1 mol $BaCl_2$ gives = 1 mol $BaSO_4$

0.1 mol $BaCl_2$ gives = 0.1 mol $BaSO_4$

0.05 mol $K_2SO_4$ gives = 0.05 mol $BaSO_4$

Thus, $K_2SO_4$ is the limiting reagent.

$BaSO_4$ formed = 0.05 mol = 11.65 g

**11.** (c) $3Si + 2N_2 \xrightarrow{\Delta} Si_3N_4$

(3× 28) g 140 g

140 g of $Si_3N_4$ is obtained from = 84 g Si if yield is 100%

$$= 84 \times \frac{100}{75}\text{ g Si if yield is 75\%} = 112.0\text{ g Si}$$

**12.** (b) Dibromide is $MBr_2$

$MBr_2 + 2AgNO_3 \longrightarrow 2AgBr\downarrow + M(NO_3)_2$ (AgBr: Yellow pricipitate)

1 mol 2 mol

$(M + 160)$ g $MBr_2$ gives = 2(188) g AgBr

$$1.84\text{ g } MBr_2 \text{ gives} = \frac{376 \times 1.84}{(M + 160)}\text{ g AgBr} = 3.76\text{ (actual)}$$

$$\therefore \frac{376 \times 1.84}{M + 160} = 3.76 \Rightarrow M = 24$$

Thus, $MBr_2 = 24 + 160 = 184$ g $mol^{-1}$.

**13.** (d) $Ag_2CO_3 \xrightarrow{\Delta} 2Ag\downarrow + CO_2\uparrow + \frac{1}{2}O_2\uparrow$ (2Ag: Residue; $CO_2$, $O_2$: Gases escape)

1 mole 2 moles

$Ag_2CO_3 \longrightarrow 2Ag$

276 g 216 g

2.76 g 2.16 g Ag (Residue)

**14.** (d) $4NH_3(g) + 5O_2(g) \longrightarrow 4NO(g) + 6H_2O(l)$

4 moles $NH_3$ gives $\longrightarrow$ 4 moles NO

1 moles $NH_3$ gives $\longrightarrow$ 1 moles NO

5 moles $O_2$ gives $\longrightarrow$ 4 moles NO

$$1\text{ moles } O_2 \text{ gives} = \frac{4}{5} = 0.8\text{ moles NO}$$

Thus, $O_2$ is the limiting reactant, amount of NO formed is decided by $O_2$.

NO = 0.8 mol = 0.8 × 30 g = 24 g

**15.** (a) $CaH_2 + 2H_2O \longrightarrow Ca(OH)_2 + 2H_2$

1 mole 2 mol 2 mol

Given 1 mole 1 mol

1 mol $CaH_2$ gives = 2 mol $H_2$

2 mol $H_2O$ gives = 1 mol $H_2$

1 mol $H_2O$ gives = 1 mol $H_2$

Thus, $H_2O$ is the limiting reagent

**Note** *$CaH_2$ is left unreacted = 0.5 mole.*

## Explanations for Q. Nos. (16-17)

$NaNO_3 + 3NaNH_2 \longrightarrow NaN_3 + 3NaOH + NH_3$

1 mol 3 mol

15 g 15 g

$\frac{15}{85} = 0.1765$ mol, $\frac{15}{39} = 0.3846$ mol

1 mole $NaNO_3$ reacts with = 3 moles $NaNH_2$

0.1765 mole $NaNO_3$ reacts with = 3 × 0.1765 mole $NaNH_2$

= 0.5295 mol $NaNH_2$

But $NaNH_2$ taken = 0.3846 mole

Hence, $NaNH_2$ is the limiting reactant

Amount of $NaN_3$ is decided by $NaNH_2$

3 moles $NaNH_2$ give = 1 mole $NaN_3$

$$0.3846\text{ mole gives} = \frac{1}{3} \times 0.3842\text{ mol } NaN_3 = 0.1281\text{ mole } NaN_3$$

= 0.1281 × 65 g $NaN_3$ = 8.32 g $NaN_3$

$2NaN_3 \longrightarrow 3N_2$

2 moles $NaNo_3$ gives = 3 moles $N_2$ gas

$$\frac{8.32}{65}\text{ mole } NaN_3 \text{ give} = \frac{3}{2} \times \frac{8.32}{65}\text{ mole } N_2 = 0.192\text{ mole } N_2$$

1 mole $N_2$ at NTP = 22.4 L

$\therefore$ 0.192 mole $N_2$ at NTP = 4.30 L

**16.** (d) **17.** (a)

**18.** (a) $CaSO_4 = 0.791\text{ g} = \frac{0.791\text{ g}}{136\text{ g mol}^{-1}} = 5.816 \times 10^{-3}$

$H_2O = 1 - 0.791 = 0.209$ g

$$= \frac{0.209\text{ g}}{18\text{ g mol}^{-1}} = 11.61 \times 10^{-3}$$

**19.** (c)

| | | Ratio |
|---|---|---|
| $CaSO_4 = 5.816 \times 10^{-3}$ | | 1 |
| $H_2O = 11.61 \times 10^{-3}$ | | 2 |

Thus, molecular formula = $CaSO_4 \cdot 2H_2O$.

**20.** (5) Ga = 1.25 g; $Ga_xO_y$ = 1.68 g

Oxygen combined with 1.25 g Ga = 1.68 − 1.25 = 0.43 g

$$x = \text{moles of Ga} = \frac{1.25}{70}$$

$$y = \text{moles of O} = \frac{0.43}{16}$$

$$\frac{x}{y} = \frac{1.25/70}{0.43/16} = \frac{0.0179}{0.0268} = \frac{1}{1.50} = \frac{2}{3}$$

Thus, $(x + y) = 5$

**21.** (7) $MgSO_4 \cdot xH_2O$ = 1.687 g

$MgSO_4$ = 0.824 g

$H_2O$ = 1.687 − 0.824 = 0.863 g

| | Moles | Ratio |
|---|---|---|
| $MgSO_4$ (120) | $\frac{0.824}{120} = 6.87 \times 10^{-3}$ | 1 |
| $H_2O$ (18) | $\frac{0.863}{18} = 0.0479$ | 7 |

Thus, epsom salt is $MgSO_4 \cdot 7H_2O$

**22.** (8)

$$X\,Br_2 \longrightarrow XCl_2$$

$(x + 160)$ g  $(X + 71)$ g

3.36 g  1.58 g

Moles $\frac{3.36}{(X + 160)} = \frac{1.58}{(X + 71)}$

$$\frac{3.36}{(X + 160)} = \frac{1.58}{(X + 71)}$$

$$1.58\,(X + 160) = 3.36\,(X + 71)$$

$$(3.36 - 1.58)X = (160 \times 1.58 - 71 \times 3.36)$$

$$1.78X = 14.24$$

$$X = 8$$

**23.** (c) (i) $C + O_2 \longrightarrow CO_2$

12 g  32 g  44 g

$$1\text{ g C gives} \longrightarrow \frac{44}{12}\text{ g } CO_2$$

$$1\text{ g } O_2 \text{ gives} \longrightarrow \frac{44}{32}\text{ g } CO_2 < \frac{44}{12}$$

Thus, $O_2$ is the limiting reagent

$CO_2$ = 1.375 g from $O_3$ **(i) → (r)**

(ii) $2C + 2H_2 \longrightarrow C_2H_4$

2 mol  2 mol  1 mol

24 g  4 g  28 g

1 g  1 g  ?

$$C_2H_4 \text{ from 1 g C} = \frac{28}{24} = 1.167\text{ g}$$

$$C_2H_4 \text{ from 1 g } H_2 = \frac{28}{4} = 7.0\text{ g}$$

Thus, carbon is the limiting reagent

$C_2H_4$ = 1.17 *g* **(ii) → (q)**

(iii) $H_2 + O_2 \longrightarrow H_2O_2$

1 mol  1 mol  1 mol

2 g  32 g  34 g

1 g  1 g  ?

$$1\text{ g } H_2 \text{ gives} = \frac{34}{2} = 17\text{ g } H_2O_2$$

$$1\text{ g } O_2 \text{ gives} = \frac{34}{32}\text{ g } H_2O_2 = 1.0625\text{ g}$$

Thus, $O_2$ is the limiting reagent

$H_2O_2$ = 1.0625 g **(iii) → (p)**

(iv) $2H_2 + O_2 \longrightarrow 2H_2O$

2 mol  1 mol  2 mol

4 g  32g  36 g

1 g  1  g ?

$$1\text{ g } H_2 \text{ gives} = \frac{36}{4} = 9.0\text{ g } H_2O$$

$$1\text{ g } O_2 \text{ gives} = \frac{36}{32} = 1.125\text{ g } H_2O$$ **(iv) → (p)**

**24.** (c)

(i) $CO + \frac{1}{2}O_2 \longrightarrow CO_2$

1 mol 0.5 mol  **(i) → (p)**

(ii) $N_2 + 2O_2 \longrightarrow N_2O_4$

1 mol 2 mol  **(ii) → (r)**

(iii) $P_4 + 5O_2 \longrightarrow 2P_2O_5$

1 mol 5 mol  **(iii) → (s)**

(iv) $Cl_2 + 1.5\,O_2 \longrightarrow Cl_2O_3$

1.5 mol  **(iv) → (q)**

# DPP-3 Concentration Factors

**1.** (i) **Plan** $\quad \text{Molarity} = \dfrac{\text{moles of solute}}{\text{volume of solution (L)}}$

$1\text{ m}^3$ of water has = 1000 kg water

$= 1000 \times 10^3$ g water

$$= \frac{1 \times 10^6}{18}\text{ moles water}$$

$1\text{ m}^3 = 1000\text{ dm}^3 = 1000\text{ L}$

$$\therefore \quad \text{Molarity} = \frac{\text{Moles}}{\text{Volume (L)}} = \frac{\frac{1 \times 10^6}{18}}{1000} = \frac{10^3}{18} = 55.55\text{ mol L}^{-1}$$

(ii) **Plan** $\quad \text{Molality of solution} = \dfrac{\text{moles of solute}}{\text{mass of solvent (kg)}}$

100 mL of $H_2SO_4$ solution = 100 mL × 1.84 g $\text{mL}^{-1}$

= 184 g solution

$$H_2SO_4 \text{ in 100 mL solution} = 93\text{ g} = \frac{93}{98}\text{ mole}$$

Water in 100 mL solution = 184 − 93 = 91 g = 0.091 kg

$$\therefore \quad \text{Molality} = \frac{93/98\text{ mole}}{0.091\text{ kg}} = 10.43\text{ kg mol}^{-1}$$

**2.** **Plan** $\quad \text{Molarity} = \dfrac{\text{moles of solute}}{\text{volume (L)}}$

Normality = molarity × basicity of acid

$$\text{Molality} = \frac{\text{moles of solute}}{\text{mass of solvent (kg)}}$$

$H_2SO_4$ is 98% (w/w).

It means 100 g $H_2SO_4$ solution has = 98 g $H_2SO_4$

Thus, water = 100 − 98 = 2 g = 0.002 kg

$$H_2SO_4 = 98\text{ g} = \frac{98}{98} = 1\text{ mol}$$

$$100\text{ g } H_2SO_4 \text{ solution} = \frac{100}{1.8}\text{ mL} = \frac{1}{18}\text{ L}$$

$$\therefore \quad \text{Molarity} = \frac{\text{mol}}{\text{L}} = \frac{1}{1/18} = 18\text{ mol L}^{-1}$$

Normality of dibasic $H_2SO_4$ = 18 × 2 = 36 N

$$\text{Molality} = \frac{\text{mole}}{\text{kg}} = \frac{1}{0.002}$$

$$= 500\text{ mol kg}^{-1}$$

**3. Plan** Mole fraction of component 1 = $\frac{n_1}{n_1 + n_2}$

(i) Ethanol = 46% = 46 g in 100 g solution = $\frac{46}{46}$ = 1 mole

$$\text{Water} = 100 - 46 = 54\text{ g} = \frac{54}{18} = 3\text{ moles}$$

Total moles = 1 + 3 = 4 moles

Thus, $X_{water} = \frac{3}{4} = 0.75$

$$X_{ethanol} = \frac{1}{4} = 0.25 = (1 - X_{water})$$

(ii) Molality when water is solute = $\frac{\text{moles of water}}{\text{mass of ethanol (kg)}} = \frac{3}{0.046}$

$= 65.22\text{ mol kg}^{-1}$

Molality when ethanol is solute = $\frac{\text{moles of ethanol}}{\text{mass of water (kg)}}$

$= \frac{1}{0.054} = 18.52\text{ mol kg}^{-1}$

**4.** (d) Ethanol = $9.2\text{g} = \frac{9.2\text{ g}}{46\text{ g mol}^{-1}} = 0.2$ mol

$$H_2O = 18.0\text{ g} = \frac{18.0\text{ g}}{18.0\text{ g mol}^{-1}} = 1.0\text{ mol}$$

Total moles = 0.2 + 1.0 = 1.2 mol

Mole fraction of ethanol = $\frac{\text{moles of ethanol}}{\text{total moles}} = \frac{0.2}{1.2} = 0.167$

**5.** (c) Urea = $120\text{ g} = \frac{120}{60} = 2$ mol, Water = 1000 g

Total mass of solution = 1120 g

Total volume of solution = $\frac{\text{mass}}{\text{density}} = \frac{1120}{1.15}\text{ mL} = \frac{112}{115}\text{ L}$

Molarity = $\frac{\text{mol}}{\text{L}} = \frac{2.0}{112/115} = \frac{230}{112} = 2.053$ M

**6.** (a) Resultant molarity on mixing two or more strong acids

$$M\text{ (resultant)} = \frac{\Sigma MV}{\Sigma V}$$

| Acid (HCl) | $M\,V$ |
|---|---|
| 0.5 M | 750 mL |
| 2.0 M | 250 mL |

$\therefore$ $M\text{ (resultant)} = \frac{M_1V_1 + M_2V_2}{V_1 + V_2}$

$= \frac{0.5 \times 750 + 2 \times 250}{750 + 250} = 0.875$ M

**7.** (d),

(a) Molality and mole fraction do not include volume expression. Hence, on changing temperature, these values remain unchanged.

(Volume changes on heating) : Correct

(b) If $d$ = 1 g/mL

1000 g = 1000 mL

Hence, molarity = molality.

Thus, true

(c) $H_3PO_2 \rightleftharpoons H^+ + H_2PO_2^-$

It is monobasic acid, hence

1 M = 1 N.

Thus, true

**8.** (d) One equivalent of magnesium = 0.5 mol

Thus, magnesium is divalent.

Thus, $\underset{48\text{ g}}{2Mg} + \underset{32\text{ g}}{O_2} \longrightarrow 2MgO$

21 g Mg = $\frac{32}{48} \times 21 = 14$ g $O_2$ gas

$1.4\text{ g }O_2 = 1.0$ L

$\therefore$ $14\text{ g }O_2 = 10.0$ L

**9.** (d) Molarity = $\frac{\text{moles of solute}}{\text{volume (in L)}}$

$Na_2CO_3 = 26.5\text{ g} = \frac{26.5}{106}\text{ mol} = 0.25$ mol

Volume = 250 mL = 0.250 L

$\therefore$ $M\,(Na_2CO_3) = \frac{0.25}{0.25} = 1$ M

| $Na_2CO_3$ | $\rightleftharpoons$ | $2Na^+$ | + | $CO_3^{2-}$ |
|---|---|---|---|---|
| 1 M | | 0 | | 0 |
| 0 M | | 2 M | | 1 M |

$\therefore$ $[Na^+] = 2$ M

$[CO_3^{2-}] = 1$ M

**10.** (b) Moles of $K_2SO_4$ in 135 mL solution = $\frac{0.188 \times 135}{1000}$

Thus, resultant molarity of 105 mL (= 0.105 L) solution = $\frac{0.188 \times 135}{1000 \times 0.105} = 0.242$ M

**11.** (d) 1 M solution means

1000 mL solution has 1 mole KCl = 74.5 g KCl

(a) Incorrect

(b) $M = \frac{74.5/74.5}{0.250} = 4$ M

Thus, incorrect.

(c) 7.45 mg KCl = $7.45 \times 10^{-3}$ g KCl in 1 mL (0.001 L)

$\therefore$ $M = \frac{\frac{7.45 \times 10^{-3}}{74.5}}{0.001\text{ L}} = 0.1$ M

Thus, incorrect.

(d) $\frac{372.5}{74.5} = 5$ mol in 5.00 L

$\therefore$ $M = \frac{5}{5} = 1$ M

Thus, correct .

**12.** (d) $\underset{\substack{2\text{ mol} \\ 1\text{ mol}}}{2Na(s)} + 2H_2O(l) \longrightarrow \underset{\substack{2\text{ mol} \\ 1\text{ mol}}}{2NaOH} + H_2(g)$

NaOH in 155 mL of 0.175 M solution = $\frac{155 \times 0.175}{1000}$ g

= 0.027125 mol

Thus, Na taken = 0.027125 mol = 0.027125 × 23 g = 0.624 g

**13.** (d) Concentration is 25% by mass of solvent

Thus, 100 g solvent ($H_2O$) has urea = 25 g

Solvent = 100 g

Solute = 25 g

Thus, Solution = 125 g

125 g solution has urea = 25 g

100 g solution has urea = $\frac{25}{125} \times 100 = 20\%$

**14.** (b) $X_{glucose} = 0.1 = \frac{n_1}{n_1 + n_2}$, $X_{H_2O} = 0.9 = \frac{n_2}{n_1 + n_2}$

$\therefore$ $\frac{X_{glucose}}{X_{H_2O}} = \frac{0.1}{0.9} = \frac{1}{9}$

If glucose is = 1 mole

Then, $H_2O$ is = 9 moles = 9 × 18 = 162 g = 0.162 kg

Molality = $\frac{\text{moles of solute}}{\text{mass of solvent (in kg)}} = \frac{1}{0.162} = 6.173$ molal

Alternately, Molality = $\frac{1000X_1}{(1 - X_1)M_2}$

$X_1$ = mole fraction, $M_2$ = molar mass of solvent

Molality = $\frac{1000 \times 0.1}{0.9 \times 18}$

= 6.173

**15.** (a) Urea = $x$ moles = $60x$ g

Water = $x$ moles = $18x$ g

Solution = $(60x + 18x) = 78x$ g

Percentage by mass of solution = $\frac{60x \times 100}{78x} = 76.92\%$

**16.** (c) $K_2SO_4 = 17.4$ ppm = 17.4 g in $10^6$ mL solution

Moles of $K_2SO_4 = \frac{17.4}{174} = 0.1$ mol

Volume = $10^6$ mL = $10^3$ L

$\therefore$ Molarity of $K_2SO_4 = \frac{0.1}{10^3} = 1 \times 10^{-4}$ M

$$\underset{1\text{ mol}}{K_2SO_4} \rightleftharpoons \underset{2\text{ mol}}{2K^+} + \underset{1\text{ mol}}{SO_4^{2-}}$$

$\therefore$ $[K^+] = 2 \times 1 \times 10^{-4} = 2 \times 10^{-4}$ M

**17.** (d) Solution is 40% by weight of solution

Thus, NaOH = 40 g = $\frac{40}{40} = 1$ mol

$H_2O = 100 - 40 = 60$ g = 0.060 kg

Molality = $\frac{\text{Moles of solute}}{\text{Mass of solvent (in kg)}} = \frac{1.0}{0.060} = 16.67$ mol kg$^{-1}$

**18.** (b) $CH_3COONa$ = 0.50 mol kg$^{-1}$ water

Solute = 0.50 mole = 0.50 × 82 g = 41 g

Solvent ($H_2O$) = 1 kg = 1000 g

$\therefore$ Solution = 1000 + 41 = 1041 g

0.50 mole $CH_3COONa$ is in 1041 g

$\therefore$ 0.15 mole $CH_3COONa$ is in = $\frac{1041}{0.50} \times 0.15 = 312.3$ g

**19.** (b) 1 L of solution has = 120 g residue (urea)

$= \frac{120}{60} = 2$ mol urea

1 L solution = 1000 mL = 1000 × 1.2 g

= 1200 g urea solution

Thus, water = 1200 − 120 g = 1080 g water

Molality = $\frac{\text{Moles of urea}}{\text{Mass of water (in kg)}}$

$= \frac{2\text{ mol}}{1.080\text{ kg}} = 1.852$ mol kg$^{-1}$

**20.** (b) Moles of ethanol = $n_1$, Moles of water = $n_2$

$X_{\text{ethanol}} = \frac{n_1}{n_1 + n_2} = 0.040$, $X_{H_2O} = \frac{n_2}{n_1 + n_2} = 0.96$

$\therefore$ $\frac{X_{\text{ethanol}}}{X_{H_2O}} = \frac{0.040}{0.96} = \frac{1}{24} = \frac{n_1}{n_2}$

If ethanol is = 1.0 mole

Then, $H_2O$ is = 24.0 moles

$H_2O$ = 24 × 18 g = 24 × 18 mL = 432 mL

Thus, volume of solution = 0.432 L

Molarity = $\frac{\text{mol}}{\text{L}} = \frac{1}{0.432} = 2.3148$ M

**21.** (c)
$$\underset{2\text{ mol} = 68\text{ g}}{2H_2O_2} \longrightarrow 2H_2O + \underset{\substack{1\text{ mol} = 22.4\text{ L }O_2 \\ \text{at NTP}}}{O_2}$$

If volume strength is ***x* volume**, it means

1 mL of $H_2O_2$ solution at NTP gives $x$ mL $O_2$ gas.

By above reaction, 22400 mL $O_2$ is from 68 g $H_2O_2$

Thus, $x$ mL $O_2$ is from $\frac{68x}{22400}$ g $H_2O_2$

This amount is in 1 mL $H_2O_2$ solution.

Thus, concentration

$= \frac{68x}{22400}$ g mL$^{-1}$ $= \frac{68x \times 1000}{22400}$ gL$^{-1}$ $= \frac{68x}{22.4}$ gL$^{-1}$

Molarity = $\frac{\text{concentration in gL}^{-1}}{\text{molar mass}} = \frac{\frac{68x}{22.4}}{34} = \frac{x}{11.2}$ mol L$^{-1}$

Normality = $2 \times M = \frac{x}{5.6}$ equivalent L$^{-1}$

Given, $1.5 = \frac{x}{5.6} \Rightarrow x = 1.5 \times 5.6 = 8.4$ volume

**22.** (b)
$$\underset{2\text{ mol}}{2Al} + \underset{6\text{ mol}}{6HCl} \longrightarrow 2AlCl_3 + 3H_2$$

HCl in 0.05 mL (1 drop) = $\frac{0.05 \times 12.0}{1000}$ mol = $6 \times 10^{-4}$ mol

Al reacted = $\frac{1}{3}$ of HCl = $2 \times 10^{-4}$ mol

$= 2 \times 10^{-4} \times 27$ g $= 5.4 \times 10^{-3}$ g

Density = $\frac{\text{mass}}{\text{volume}}$

$\therefore$ Mass = $V \times d$

If, area = $A$

Then, volume = $A \times$ thickness = $\left(A \times \frac{0.10}{10}\right)$ cm$^3$

$\therefore$ Mass = $A \times \frac{0.10}{10} \times 2.7$

$\therefore$ $A \times \frac{0.10}{10} \times 2.7 = 5.4 \times 10^{-3}$

$A = 0.2$ cm$^2$

**23.** (d) 1 mL of 0.01 M solution ($P/Q/R$) is diluted to 100 mL.

Thus, molar concentration of dilute solution

$$\underset{\text{Original}}{M_1V_1} = \underset{\text{Dilute}}{M_2V_2}$$

$0.01 \times 1 = M_2 \times 100$

$M_2 = 1 \times 10^{-4}$ M

$P$ : Iron (II) sulphate (VI) is $FeSO_4$.

$$FeSO_4 \rightleftharpoons Fe^{2+} + SO_4^{2-}$$

$\therefore$ $[Fe^{2+}]_P = 1 \times 10^{-4}$ M

$Q$ : Iron (III) sulphate (IV) is $Fe_2(SO_3)_3$.

$$Fe_2(SO_3)_3 \rightleftharpoons 2Fe^{3+} + 3SO_3^{2-}$$

+3 +4

Thus, $[Fe^{3+}]_Q = 2 \times 10^{-4}$ M

$R$ : Iron (III) phosphate (V) is

$Fe^{3+}PO_4^{3-}$

+3 +5

$$FePO_4 \rightleftharpoons Fe^{3+} + PO_4^{3-}$$

$\therefore$ $[Fe^{3+}]_R = 1 \times 10^{-4}$ M

**24.** (a) 2.37 mg N per mL = $2.37 \times 10^3$ mg N per L

$\frac{2.37 \times 10^3}{10^3}$ g N per L

2.37 g N per L

$\frac{2.37}{14}$ mol N per L

$\frac{2.37}{14 \times 2}$ mol ($NH_4NO_3$) per L

(Since, every $NH_4NO_3$ has 2 N-atoms).

$\therefore$ Molarity of diluted solution is 0.0846 M

Thus, $M_1V_1 = M_2V_2$

Original Dilute

$0.715 \times V = 0.0846 \times 1000$

$\therefore$ $V = \frac{0.0846 \times 1000}{0.715}$

= 118.4 mL

**25.** (b) $Na_2SO_4 \rightleftharpoons 2Na^+ + SO_4^{2-}$

1 mole  2 moles  1 mole

$Na_2SO_4 = 6.0$ micromoles

$= 6.0 \times 10^{-6}$ mole

$\therefore$ $Na^+$ ions $= 2 \times 6.0 \times 10^{-6}$ mole

$= 2 \times 6.0 \times 10^{-6} \times 23$ g

in 250 mL solution (250 g $H_2O$ being dilute solution)

250 g $H_2O$ has $Na^+$ ions $= 2 \times 6.0 \times 10^{-6} \times 23$ g

$10^6$g (per million) $H_2O$ has $Na^+$ ions $= \dfrac{2 \times 6.0 \times 10^{-6} \times 23 \times 10^6}{250}$

$= 1.104$ g

$= 1.104$ ppm

**26.** (a) Weight of sugar solution = 214.2 g

Weight of sugar = 34.2 g = $\dfrac{34.2}{342} = 0.1$ mol

Weight of water = 214.2 − 34.2

= 180.0 g = 0.180 kg

Moles of water = $\dfrac{180.0}{18} = 10.0$

Kilomoles = $\dfrac{10}{1000} = 0.01$ kilomole

**(i) → (r)**

Volume of solution = $\dfrac{\text{mass (solution)}}{\text{density}}$

$= \dfrac{214.2 \text{ g}}{1.2 \text{ g mL}^{-1}}$

= 178.5 mL = 0.1785 L

Total moles (sugar + water) = 10.1

Mole fraction of sugar = $\dfrac{0.1}{10.1} = 9.9 \times 10^{-3}$

**(ii) → (q)**

Molality of solution = $\dfrac{\text{moles of solute}}{\text{mass of water (in kg)}} = \dfrac{0.1}{0.180}$

$= 5.55 \times 10^{-4}$ mol kg$^{-1}$

**(iii) → (p)**

Molarity of solution = $\dfrac{\text{moles of solute}}{\text{volume (in L)}}$

$= \dfrac{0.1 \text{ mol}}{0.1785 \text{ L}} = 0.56$ M

**(iv) → (s)**

**27.** (c)

| Element | % | Moles | Ratio | Formula |
|---|---|---|---|---|
| O (16) | 48.46 | 3.03 | 4 | |
| P (31) | 23.45 | 0.76 | 1 | $O_4PN_2H_9$ |
| N (14) | 21.21 | 1.52 | 2 | |
| H (1) | 6.87 | 6.87 | 9 | |

Being a fertiliser, it contains $NH_3$ or $NH_4^+$ salt (being ionic).

thus, cation is $NH_4^+$ and anion is $HPO_4^{2-}$.

Hence, the ionic salt is $(NH_4)_2HPO_4$.

**28.** (a) $(NH_4)_2HPO_4 \rightleftharpoons HPO_4^{2-} + 2NH_4^+$

$HPO_4^{2-} \equiv 1H^+$

Thus, 1 M = 1 N

**29.** (b) $(NH_4)_2HPO_4 \equiv 2NH_3$

132 g ≡ 34 g

% of $NH_3 = \dfrac{34}{132} \times 100$

= 25.76%

**Explanations for Q. Nos. (30-31)**

Glauber's salt is $Na_2SO_4 \cdot 10H_2O = 8.0575 \times 10^{-2}$ kg

Molar mass = 322 g mol$^{-1}$

Moles of Glauber's salt = $\dfrac{8.075 \times 10^{-2} \times 1000 \text{ g}}{322 \text{ g mol}^{-1}}$

$= 0.25$ mol in 1 dm$^3$ (1 L) solution

Thus, molarity = $\dfrac{\text{mol}}{\text{L}} = \dfrac{0.25}{1.0} = 0.25$ M

Molality ($m$) is related to molarity ($M$), molar mass ($m_1$) and density ($d$) of solution by equation :

$$m = \frac{1000M}{(1000d - Mm_1)} \quad d \text{ is in g/mL}$$

$d = 1077.2$ kg m$^{-3}$ = $1077.2 \times 10^3$ g m$^{-3}$

$= \dfrac{1077.2 \times 10^3}{10^6}$ g mL$^{-1}$ = 1.0772 g mL$^{-1}$

$\therefore$ $m = \dfrac{1000 \times 0.25}{(1000 \times 1.0772 - 0.25 \times 322)} = 0.25$ molal

**Note** $d = 1 \text{ g cm}^{-3} = 1 \text{ kg dm}^{-3} = 1000 \text{ kg m}^{-3}$

**30.** (b) **31.** (b)

**32.** (8) w/w % concentration = 29.2

Thus, 100 g solution of HCl has HCl = 29.2 g

Density = 1.25 g mL$^{-1}$

Thus, 100 g HCl solution = $\dfrac{100 \text{ g}}{1.25 \text{ g/mL}} = \dfrac{100}{1.25}$ mL

$= \dfrac{100}{1.25} \times \dfrac{1}{1000}$ L $= \dfrac{1}{12.5}$ L

Molarity of stock solution = $\dfrac{29.2/36.5 \text{ mol}}{\dfrac{1}{12.5} \text{ L}} = 10$ M

For dilution, $M_1V_1 = M_2V_2$

$10 \times V_1 = 0.4 \times 200$

$V_1 = 8$ mL

**33.** (8) There is no change in volume on dissolution.

Thus, 1000 mL solvent = 1000 mL solution

= 1000 × 0.4 g solvent = 400 g solvent = 0.4 kg solvent

Moles of solute = 3.2 mol in 1000 mL

or 0.4 kg solvent

Molality = $\dfrac{3.2}{0.4} = 8$ mol kg$^{-1}$

**34.** (2) 1000 mL solution = 4.0 moles urea = 240 g urea

Since, $\dfrac{m}{V}$ = density ($d$)

$\therefore$ Mass = $V \times d$

$1000 \times d$ g solution = 240 g urea

Solvent ($H_2O$) = $(1000d - 240)$ g = $\left(\dfrac{1000d - 240}{1000}\right)$ kg

Molality = $\dfrac{4 \text{ mol} \times 1000}{(1000d - 240)}$ mol kg$^{-1}$

Given, molality = 2.273 mol kg$^{-1}$

$\therefore$ $\dfrac{4000}{(1000d - 240)} = 2.273$

$\dfrac{4000}{2.273} = (1000d - 240)$

$1760 = 1000d - 240$

$1000d = 2000$

$\therefore$ $d = 2.0$ g mL$^{-1}$

**Note** *We can calculate density by the following relation,*

$$\text{Molality} = \frac{1000 \times \text{molarity } (M)}{[1000d - Mm_1 \text{ (molar mass)}]}$$

# DPP-4 Acid-Base Neutralisation Titration

**1.** (i) **Plan** Mass of metal = 0.24 g

Mass of oxide of metal = 0.40 g

Thus, oxygen combined = 0.40 − 0.24 = 0.16 g

Taking fact

1 equivalent metal = 1 equivalent of oxygen = 8 g oxygen

0.16 g oxygen combined with = 0.24 g metal

$$8 \text{ g oxygen combined with} = \frac{0.24}{0.16} \times 8 = 12 \text{ g equiv}^{-1}$$

(ii) **Plan** $W \text{ (weight of solute)} = \frac{NEV}{1000}$

$$\therefore \quad E = \frac{1000\,W}{NV}$$

where $N$ = Normality

$E$ = equivalent mass

$V$ = Volume in mL

$$E = \frac{1000 \times 6.0}{1 \times 100} = 60 \text{ g equiv}^{-1}$$

Molar mass = $E$ × basicity = 1 × 60 = 60 g mol$^{-1}$

**2.** (a) Molar mass of $H_2C_2O_4 \cdot 2H_2O$ (oxalic acid dihydrate) = 126 g mol$^{-1}$

It is a dibasic acid, hence,

$$\text{Equivalent mass} = \frac{126}{2} = 63 \text{ g equiv}^{-1}$$

Volume = 250 mL = 0.250 L

$$\text{Concentration} = \frac{6.3/63}{0.250} \text{ L} = 0.4 \text{ equiv L}^{-1}$$

For neutralisation $\underset{\text{NaOH}}{N_1V_1} = \underset{\text{Oxalic acid}}{N_2V_2}$

$$0.1 \times V_1 = 0.4 \times 10$$

$$V_1 = 40 \text{ mL}$$

**3.** (b) $\underset{\substack{1 \text{ mol} \\ \frac{0.265}{106} \text{ mol,}}}{Na_2CO_3} + \underset{\substack{2 \text{ mol} \\ \frac{2 \times 0.265}{106} \text{ mol}}}{2HCl} \longrightarrow 2NaCl + H_2O + CO_2]$

Molarity × volume = millimol ($MV$)

$$\frac{MV}{1000} = \text{mol}$$

$$V = 28.5 \text{ mL}$$

$$\therefore \quad \frac{M\text{ (HCl)} \times 28.5}{1000} = \frac{2 \times 0.265}{106}$$

$$\therefore \quad M\text{ (HCl)} = 0.175 \text{ M}$$

**Alternately**

In an acid-base/redox reaction, equivalent of species (1) = equivalent of species (2)

Equivalent of $Na_2CO_3$ = Equivlaent of HCl

$$\frac{0.265}{53} = 28.5 N_1/1000$$

$$\therefore \quad N_1 = 0.175$$

$$\therefore \quad M_1 = 0.175\text{M} \qquad \text{(being monobasic)}$$

**Note** 0.175 M = 0.175 N HCl *(being monobasic acid)*

**4.** (a) Molar mass of $H_2C_2O_4 \cdot 2H_2O$ = 126 g mol$^{-1}$

$$\text{Equivalent mass} = \frac{\text{molar mass}}{\text{basicity}} = 63 \text{ g equiv}^{-1}$$

**Method I In terms of moles**

$$\underset{1 \text{ mol}}{H_2C_2O_4} + \underset{2 \text{ mol}}{2NaOH} \longrightarrow Na_2C_2O_4 + 2H_2O$$

Also 1 N NaOH = 1 M NaOH (a monoacid base)

Let Volume of NaOH = $V$ mL

$$\therefore \quad MV = \text{millimol}$$

$$\therefore \quad \frac{MV}{1000} = \text{mol}$$

$$M = 1$$

$$\frac{V}{1000} = \text{moles of NaOH}$$

$$\therefore \text{ Moles of } H_2C_2O_4 \cdot 2H_2O = \frac{V}{2000} \qquad \text{(half of NaOH)}$$

$$\text{Moles of } H_2C_2O_4 \cdot 2H_2O = \frac{1.575}{126} = 0.0125$$

$$\therefore \quad \frac{V}{2000} = 0.0125$$

$$\therefore \quad V = 25 \text{ mL}$$

**Method II** In terms of equivalent

In a reaction

Equivalent of reactant I ≡ Equivalent of reactant II

$$NV = \text{milli equivalent}$$

$$\frac{NV}{1000} = \text{equivalent}$$

$$N = 1 \text{ (NaOH)}$$

$$\therefore \quad \frac{N}{1000} = \text{equivalent of NaOH}$$

$$\text{Also, } \frac{V}{1000} = \text{equivalent of } H_2C_2O_4 \cdot 2H_2O$$

$$\text{Equivalent of } H_2C_2O_4 \cdot 2H_2O = \frac{1.575}{63}$$

$$\therefore \quad \frac{1.575}{63} = \frac{V}{1000}$$

$$\therefore \quad V = 25 \text{ mL}$$

**5.** (a) $M$ is monovalent

$$M_2CO_3 \rightleftharpoons 2M^+ + CO_3^{2-}$$

$$M_2CO_3 + 2HCl \longrightarrow 2MCl + H_2O + CO_2$$

$$M_2CO_3 \equiv 2H^+$$

$$\therefore \text{ Equivalent weight of } M_2CO_3 = \frac{M}{2} = \frac{2M + 60}{2} = (M + 30)$$

Let, unused acid = $V$ mL

$\therefore$ $V$ mL of 1 N HCl ≡ 100 mL of 0.5 NaOH

$$\therefore \quad V = \frac{100 \times 0.5}{1} = 50 \text{ mL}$$

$$\therefore \quad \text{Used HCl} = 150 - 50 = 100 \text{ mL}$$

$$\therefore \quad \text{Equivalent of HCl} = \frac{100 \times 1}{1000} = 0.1$$

$$\text{Equivalent of } M_2CO_3 = \frac{100 \times 1}{1000} = 0.1$$

$$\text{Actual equivalent of } M_2CO_3 \text{ taken} = \frac{5.3}{M + 30}$$

$$\therefore \quad \frac{5.3}{M + 30} = 0.1$$

$$\therefore \quad M + 30 = 53$$

$$M = 23$$

**6.** (d) $\underset{\substack{1 \text{ mol} \\ 24 \text{ g}}}{Mg} + \underset{\substack{1 \text{ mol} \\ 1 \text{ mol}}}{H_2SO_4} \longrightarrow MgSO_4 + H_2$

Thus, 2.4 g = 0.1 mol mg

$$H_2SO_4 \text{ taken} = \frac{200 \times 1}{1000} = 0.2 \text{ mol}$$

$H_2SO_4$ reacted with Mg = 0.1 mol

Unreacted $H_2SO_4$ = 0.2 − 0.1 = 0.1 mol in 1 L dilute solution

Thus, $[H_2SO_4]$ = 0.1 M

7. (c) $CaCO_3 + 2HCl \longrightarrow CaCl_2 + H_2O + CO_2$
1 mol  2 mol

$0.10\ g = \frac{0.10}{100} = 0.001$ mol $CaCO_3$ required $= 0.002$ mol HCl

$[HCl] = 0.01\ M = 0.01\ mol\ L^{-1}$

0.01 mol HCl is in = 1000 mL

$\therefore$ 0.002 mol HCl is in $= \frac{1000}{0.010} \times 0.002 = 200$ mL

Thus, $x = 200$ mL

$CaCO_3 + H_2SO_4 \longrightarrow CaSO_4 + H_2O + CO_2$
1 mol  1 mol  0.001 mol

As above, 0.001 mol $H_2SO_4$ is in = 100 mL $H_2SO_4$ solution

Thus, $y = 100$ mL

8. (a) $H_2SO_4 + 2NaOH \longrightarrow Na_2SO_4 + 2H_2O$

$2H_3PO_4 + 2NaOH \longrightarrow 2NaH_2PO_4 + 2H_2O$
Dihydrogen orthophosphate

Thus, $H_2SO_4 : H_3PO_4 = 1:2$

9. (b) $H_2SO_4$ is a dibasic acid.

$H_2SO_4 \rightleftharpoons 2H^+ + SO_4^{2-}$.

$H_3PO_3$ is also a dibasic acid
(with only two ionisable $H^+$ ions)

$H_3PO_3 \rightleftharpoons 2H^+ + HPO_3^{2-}$

**Note** *In* $H_3PO_3$, *one* H *is attached to* P*-atom and is not ionised*

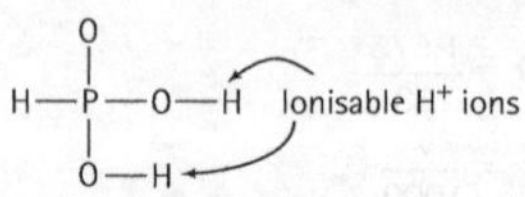


100 mL of 0.1 M $H_2SO_4$ = 100 mL of 0.2 N $H_2SO_4$
200 mL of 0.1 M $H_3PO_3$ = 200 mL of 0.2 N $H_3PO_3$
Resultant $[H^+]$ (in terms of normality)

$$= \frac{100 \times 0.2 + 200 \times 0.2}{100 + 200}$$

$= 0.2$ M or 0.2 N

10. (a) It is Kjeldahl's method of determination of percentage of N

$$N\% = \frac{1.4\ N_1(\text{normality})\ V_1\ (\text{Volume})}{W\ (\text{amount})}$$

$$= \frac{1.4 \times 0.1 \times 100}{0.31} = 45.16\%$$

11. (c) Solution Volume

KOH = 0.1 M = 0.1 N 100 mL
$Ca(OH)_2$ = 0.1 M = 0.2 N 200 mL
$H_2SO_4$ = 0.1 M = 0.2 N 300 mL

$$N'(OH^-) = \frac{N_1V_1 + N_2V_2}{V_1 + V_2} = \frac{0.1 \times 100 + 0.2 \times 200}{300} = 0.167$$

$N(H^+) = 0.2$ N

Thus, mixture now contains
(i) 300 mL of 0.167 N $[OH^-]$ and (ii) 300 mL of 0.200 N $[H^+]$
Since, $NV(H^+) > NV(OH^-)$

$(300 \times 0.200) > (300 \times 0.167)$

thus, resultant mixture is acidic.

$$\therefore \text{Resultant } [H^+] = \frac{(300 \times 0.200) - (300 \times 0.167)}{600}$$

$$= \frac{0.200 - 0.167}{2} = 0.017\ M$$

12. (d) Iron taken = 11.2 g
Iron sulphide formed = 17.6 g
Thus, sulphur combined = 17.6 − 11.2 = 6.4 g

Equivalent mass of sulphur $= \frac{\text{atomic mass}}{2} = 16$

6.4 g sulphur combined with = 11.2 g Fe

$\therefore$ 16 g sulphur combined with $= \frac{11.2}{6.4} \times 16 = 28$ g

Thus, equivalent mass of Fe = 28 g $equiv^{-1}$

Thus, 11.2 g Fe $= \frac{11.2}{28} = 0.4$ equivalent

13. (a, b) In terms of oxygen, equivalent weight is = 8

$8\ g\ O_2 = \frac{8}{32}$ mol $O_2 = 0.25$ mol

14. (b, c) $P_4O_{10} = 2.84\ gL^{-1} = \frac{2.84}{284}\ mol\ L^{-1} = 0.01$ M

$P_4O_{10} + 6H_2O \longrightarrow 4H_3PO_4$
Thus, 0.01 M  0.04 M

Also 0.04 M $H_3PO_4$ = 0.12 N $H_3PO_4$

(a) Resulting solution is 0.04 M $H_3PO_4$
Thus, (a) is incorrect.

(b) $H_3PO_4 + 3NaOH \longrightarrow Na_3PO_4 + 3H_2O$

$N_1V_1 = N_2V_2$
$NaOH = H_3PO_4$
$0.3 \times 400 = 0.12 \times 1000$

Thus, correct.

(c) $H_3PO_4 + Ca(OH)_2 \longrightarrow CaHPO_4 + 2H_2O$
In this case only two $(H^+)$ of $H_3PO_4$ are neutralised

Thus, 0.04 M $H_3PO_4$ ≡ 0.08 N $H_3PO_4$

Thus $N_1V_1 = N_2V_2$
$Ca(OH)_2$  $H_3PO_4$
$0.2 \times 400 = 0.08 \times 1000$

Thus, correct (d) incorrect.

15. (b, d)
(a) $H_3PO_2$ is a monobasic acid with only one ionisable $H^+$ attached to O-atom.

Two other H atoms are attached to P-atom and explain its reducing nature.

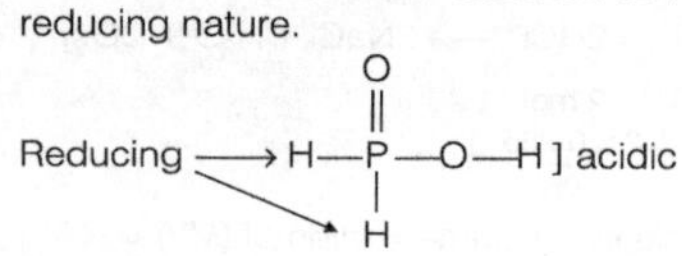


Thus, (a) is incorrect.

(b) $H_3PO_2 \rightleftharpoons H^+ + H_2PO_2^-$

! mol $H_3PO_2$ ≡ 1 mol $H^+$ ≡ 1 mol $(OH^-)$ ≡ 0.5 mol $Ca(OH)_2$

Thus, (b) is correct.

(c) $H_3PO_2 \rightleftharpoons H_2PO_2^- \rightarrow X$

Thus, $NaH_2PO_2$ is not acid salt.
Thus, (c) is incorrect.

(d) $3H_3PO_2 \longrightarrow 2H_3PO_3 + PH_3$
+1  +3  −3

Thus, $H_3PO_2$ disproportionates.

16. (a, c)

| | $BaF_2$ + | $H_2SO_4$ ⟶ | $BaSO_4$ + | $H_2F_2$ |
|---|---|---|---|---|
| Initial | 0.1 mol | 0.2 mol | 0 | 0 |
| After reaction | 0 | 0.1 mol | 0.1 | 0.1 |
| | | $0.2H^+$ | | $0.2H^+$ |

100 mL of 1 M, $BaF_2$ = 0.1 mol
100 mL of 2 M $H_2SO_4$ = 0.2 mol
Total $(H^+)$ = 0.2 + 0.2 = 0.4 mol in 200 mL solution

$(H^+)$ molarity $= \frac{0.4}{0.200} = 2$ M

Also, $BaSO_4$ formed = 0.1 mol

17. (b, d) $SO_2Cl_2 + 2H_2O \longrightarrow H_2SO_4 + 2HCl$
1 mol  1 mol  1  2 mol

$H_2SO_4 + 2NaOH \longrightarrow Na_2SO_4 + 2H_2O$
$2HCl + 2NaOH \longrightarrow 2NaCl + 2H_2O$

Thus, total moles of NaOH = 4

$$H_2SO_4 + Ca(OH)_2 \longrightarrow CaSO_4 + 2H_2O$$
$$2HCl + Ca(OH)_2 \longrightarrow CaCl_2 + 2H_2O$$

Total moles of $Ca(OH)_2 = 2$

**Explanations for Q. Nos. (18-19)**

1.0 L of 0.15 N NaOH = 0.15 mol

| | 2 NaOH | + | $CO_2$ | ⟶ | $Na_2CO_3$ | + | $H_2O$ |
|---|---|---|---|---|---|---|---|
| | 2 mol | | 1 mol | | 1 mol | | |
| Initial | 0.15 mol | | $11.2\times10^{-3}$ mol | | 0 | | 0 |
| After reaction (0.15 − 0.0112 × 2) | | | 0 | | 0.0112 mol | | |
| | 0.1276 mol | | | | | | – |

$\therefore$ $$[NaOH] = \frac{0.1276 \text{ mol}}{1 \text{ L}} = 0.1276 \text{ M}$$

$$[Na_2CO_3] = \frac{0.0112 \text{ mol}}{1 \text{ L}} = 0.0112 \text{ M}$$

**18.** (a) **19.** (b)

**20.** (6) $HCl \equiv NH_3$

$xHCl \equiv xNH_3 \equiv$ moles of $[Cr(NH_3)_xCl_3]$

$$\text{Moles of HCl} = \frac{24.26 \times 1.50}{1000} = 0.03639$$

$\therefore$ Moles of $[Cr(NH_3)_xCl_3] = \left(\frac{0.03639}{x}\right)$

Molar mass of $[Cr(NH_3)_xCl_3] = 52 + 17x + 106.5$
$= (158.5 + 17x)$

$$\text{Moles of } [Cr(NH_3)_xCl_3] \text{ taken} = \frac{1.580}{(158.5 + 17x)}$$

$\therefore$ $$\frac{1.580}{(158.5 + 17x)} = \frac{0.03639}{x}$$

$$1.580x = 0.03639\,[158.5 + 17x]$$

$\therefore$ $$x = \frac{5.7678}{0.96137} = 6$$

# DPP-5 Redox-Based Titration

**1.** (c) In acidic medium $MnO_4^-$ is reduced to $Mn^{2+}$ and $Cr^{3+}$ is oxidised to $Cr_2O_7^{2-}$

$$\underset{+7}{MnO_4^-} + 8H^+ + 5e^- \longrightarrow \underset{+2}{Mn^{2+}} + 4H_2O$$

Change in O.N. of Mn = 7 − 2 = 5 units

Thus, 0.200 M $MnO_4^-$ = 0.200 × 5
= 1.0 N $MnO_4^-$

$$\underset{+3}{Cr^{3+}} \longrightarrow \frac{1}{2}\underset{+6}{Cr_2O_7^{2-}} + 3e^-$$

Change in O.N. of Cr = 3 units

Thus, 0.125 M $Cr^{3+}$ = 0.125 × 3 = 0.375 N

$$\underset{MnO_4^-}{N_1V_1} = \underset{Cr^{3+}}{N_2V_2}$$
$$1.0 \times V_1 = 0.375 \times 32$$

$\therefore$ $$V_1 = \frac{0.375 \times 32}{1} = 12.0 \text{ mL}$$

**2.** (a) $Fe^{3+}(aq) + Sn^{2+}(aq) \longrightarrow Sn^{4+}(aq) + Fe^{2+}(aq)$ [$Sn^{2+} \to Sn^{4+}$: Oxidation; $Fe^{3+} \to Fe^{2+}$: Reduction]

**Caution** Take concentration in normality

$$Fe^{3+} + e^- \longrightarrow Fe^{2+}$$

Thus, equivalent mass of $Fe^{3+}$ = molar mass

Thus, 0.250 M $Fe^{3+}$ = 0.250 N $Fe^{3+}$

$$Sn^{2+} \longrightarrow Sn^{4+} + 2e^-$$

$Fe^{3+}$ $Sn^{2+}$ + $Sn^{4+}$ + $Fe^4$ $Fe^2$
Oxidity Reaction

Thus, equivalent mass of $Sn^{2+} = \frac{\text{molar mass}}{2}$

$\therefore$ 0.100 M $Sn^{2+}$ = 0.200 N $Sn^{2+}$

$$\underset{Fe^{3+}}{N_1V_1} = \underset{Sn^{2+}}{N_2V_2}$$
$$0.250 \times V_1 = 0.200 \times 20$$

$\therefore$ $$V_1 = \frac{0.200 \times 20}{0.250} = 16 \text{ mL}$$

**3.** (a) $$2CuSO_4 + 4KI \longrightarrow 2K_2SO_4 + Cu_2I_2 + I_2$$
$$I_2 + KI \rightleftharpoons KI_3$$
$$2Na_2S_2O_3 + I_2 \longrightarrow 2NaI + Na_2S_4O_6$$
$$2CuSO_4 \equiv I_2 \equiv 2Na_2S_2O_3$$
$$2CuSO_4 \equiv 2I \equiv 2Na_2S_2O_3$$

$\therefore$ $1\,CuSO_4 \equiv I$

$1\,Na_2S_2O_3 \equiv I$

$\therefore$ Equivalent weight of $CuSO_4 \cdot 5H_2O$ = molecular weight = 240.5

Equivalent weight of $Na_2S_2O_3$ = molecular weight

Thus, Molarity = Normality

$$\text{Moles of } CuSO_4 \cdot 5H_2O = \frac{0.2495}{249.5} = 1 \times 10^{-3}$$

Volume = 100 mL = 0.1 L

$\therefore$ $$M(CuSO_4) = \frac{1 \times 10^{-3}}{0.1} = 0.01 \text{ M}$$

$$M_1V_1 = M_2V_2$$
$$Na_2S_2O_3 = CuSO_4$$
$$M_1 \times 25 = 0.01 \times 100$$
$$M_1 = 0.04 \text{ M}$$

**4.** (c) $Fe^{2+}$ (Iron (II) in +2 oxidation state) is oxidised to $Fe^{3+}$ by $MnO_4^-/H^+$

$$\underset{+2}{5Fe^{2+}} + \underset{+7}{MnO_4^-} + 8H^+ \longrightarrow \underset{+3}{5Fe^{3+}} + \underset{+2}{Mn^{2+}} + 4H_2O$$

We express concentration in terms of equivalent

Change in oxidation number of Mn = 5 units

Change in oxidation number of Fe = 1 unit

$\therefore$ 0.02 M $KMnO_4$ = 0.02 × 5
= 0.1 N $KMnO_4$

$$\text{Equivalent of } KMnO_4 \text{ used} = \frac{NV}{1000} = \frac{0.1 \times 25}{1000}$$

$$\text{Equivalent of } Fe^{2+} \text{ used} = \frac{0.1 \times 25}{1000}$$

$\therefore$ $Fe^{2+}$ (pure) = Equivalent × equivalent mass

$= \frac{0.1 \times 25 \times 56}{1000} = 0.14$ g in 1 g impure iron are

$\therefore$ Percentage of Fe $= \frac{0.14 \times 100}{1} = 14\%$

**5.** (c) Mohr's salt is $FeSO_4 \cdot (NH_4)_2SO_4 \cdot 6H_2O$

Only $Fe^{2+}$ is oxidised

$$5Fe^{2+} + MnO_4^- + 8H^+ \longrightarrow 5Fe^{3+} + 4H_2O + Mn^{2+}$$
$$6Fe^{2+} + Cr_2O_7^{2-} + 14H^+ \longrightarrow 6Fe^{3+} + 7H_2O + 2Cr^{3+}$$
$$5Fe^{2+} \equiv 1\,MnO_4^-$$
$$1 \text{ mole } Fe^{2+} \equiv \frac{1}{5} \text{ mole } MnO_4^-$$
$$6Fe^{2+} \equiv 1\,Cr_2O_7^{2-}$$
$$1 \text{ mole } Fe^{2+} \equiv \frac{1}{6} \text{ mole } Cr_2O_7^{2-}$$
$$\text{Ratio} = \frac{1/5}{1/6} = \frac{6}{5} = 1.2$$

**6.** (d) $Cr_2O_7^{2-} \longrightarrow 2Cr^{3+}$ **Change in ON**
$\quad +12 \qquad\qquad +6 \qquad\qquad$ 6 units

$H_2SO_3 \longrightarrow HSO_4^-$
$\quad +4 \qquad\qquad +6 \qquad\qquad$ 2 units

Thus, 1 mole $Cr_2O_7^{2-}$ = 6 equivalents

3 moles of $H_2SO_3$ = 6 equivalents

Thus, ratio = 1 : 1

**7.** (a) $C_2H_5OH \longrightarrow 2CO_2$

ON $2x + 6 - 2 = 0 \quad 2(x - 4) = 0$

$2x = -4 \qquad 2x = 8$

Change in O.N. of C of $C_2H_5OH$ = 12 units

(Note : Take all the C-atoms)

$\therefore$ Equivalents of $C_2H_5OH$ in 5 moles = $12 \times 5 = 60$

$KMnO_4$ required = 12 moles

$\therefore$ Equivalents of $C_2H_5OH$ oxidised per mol of $MnO_4^- = \frac{60}{12} = 5$

**8.** (c) **Reaction** **Change in ON**

$Sn^{2+} \longrightarrow Sn^{4+} + 2e^-$ 2 units

$Fe^{2+} \longrightarrow Fe^{3+} + e^-$ 1 unit

$Cr_2O_7^{2-} + 6e^- \rightarrow 2Cr^{3+}$ 6 units

Thus, 0.04 M $Sn^{2+}$ = 0.08 N $Sn^{2+}$

$x$ M $Fe^{2+}$ = $x$ N $Fe^{2+}$

0.125 M $Cr_2O_7^{2-}$ = $0.125 \times 6$ = 0.750 N $Cr_2O_7^{2-}$

Equivalent of $Fe^{2+}$ and $Sn^{2+}$ = Equivalent of $Cr_2O_7^{2-}$

$$\frac{15 \times 0.08 + 15x}{1000} = \frac{18 \times 0.750}{1000}$$

$1.2 + 15x = 13.5$

$15x = 12.3$

$x = 0.82$ M

**9.** (a) $\underset{+7}{MnO_4^-} + \underset{+4}{SO_3^{2-}} + H^+ \longrightarrow \underset{+6}{SO_4^{2-}} + \underset{+2}{Mn^{2+}}$

$MnO_4^-$ is reduced to $Mn^{2+}$

$MnO_4^- + 8H^+ + 5e^- \longrightarrow Mn^{2+} + 4H_2O$

$SO_3^-$ is oxidised to $SO_4^{2-}$

$SO_3^{2-} + H_2O \longrightarrow SO_4^{2-} + 2H^+ + 2e^-$

To balance electrons

$2MnO_4^- \equiv 5SO_3^{2-}$

1 mol $SO_3^{2-} \equiv \frac{2}{5}$ mol $MnO_4^-$

= 0.4 mol $MnO_4^-$

**10.** (b) $H_2O_2 \longrightarrow O_2 + 2H^+ + 2e^-$

$\therefore$ 0.1 M $H_2O_2$ = 0.2 N $H_2O_2$

Let, change in ON of $MnO_4^-$ = $n$ unit

$\therefore$ 0.1 M $MnO_4^-$ = $0.1 \times n$ N $MnO_4^-$

$N_1V_1 = N_2V_2$

$H_2O_2 = MnO_4^-$

$0.2 \times 24 = 0.1n \times 16$

$\therefore \quad n = 3$

Thus, $\underset{+7}{MnO_4^-}$ is reduced to $\underset{+4}{MnO_2}$

**11.** (a) $MnO_4^- + 8H^+ + 5e^- \longrightarrow Mn^{2+} + 4H_2O$

$H_2O_2 \longrightarrow 2H^+ + O_2 + 2e^-$

To balance electrons

$5H_2O_2 \equiv 2MnO_4^-$

$H_2O_2$ taken = 1 g impure = $\frac{x}{100}$ g pure = $\frac{X}{3400}$ mol

$MnO_4^-$ = $X$ mL of molarity $M$ = $MX$ millimol = $\frac{MX}{1000}$ mol

5 mol $H_2O_2 \equiv 2$ mol $MnO_4^-$

$\therefore \quad \frac{x}{3400}$ mol $H_2O_2 = \frac{2X}{5 \times 3400}$ mol $MnO_4^-$

$\therefore \quad \frac{MX}{1000} = \frac{2X}{5 \times 3400}$

$\therefore \quad M = \frac{2 \times 1000}{5 \times 3400} = 0.12$ M

Molarity of $MnO_4^-$ = 0.12 M

**Alternately**

Equivalent of $MnO_4^-$ ≡ Equivalent of $H_2O_2$

$\frac{5 \times M}{1000} = \frac{X}{1700}$

$\therefore \quad M = 0.12$ M

**12.** (c) $NaHC_2O_4 \rightleftharpoons Na^+ + H^+ + C_2O_4^{2-}$

$MnO_4^- + 8H^+ + 5e^- \longrightarrow Mn^{2+} + 4H_2O$

$C_2O_4^{2-} \longrightarrow 2CO_2 + 2e^-$

$H^+ + NaOH \longrightarrow H_2O + Na^+$

Let, $NaHC_2O_4 = M_1$

$\therefore \quad N_{(NaHC_2O_4)} = M_1$ as an acid

$N_{(NaHC_2O_4)} = 2M_1$ as a reducing agent

**Titration I** $N_{(MnO_4^-)} = 0.5$ N

100 mL of $NaHC_2O_4$ ≡ 50 mL of 0.1 M $KMnO_4$

$\therefore \quad 100 \times 2M_1 = 50 \times 0.5$ N

$\therefore \quad M_1 = \frac{0.25}{2}$ N = 0.125

**Titration II** 100 mL of $NaHC_2O_4$ ≡ $V$ mL of 0.1 M NaOH

$100 \times 0.125 = V \times 0.1$

$\therefore \quad V = 125$ mL

**13.** (a) Iron (II) sulphate (VI) is $\underset{+2}{Fe}\underset{+6}{S}O_4$

Iron (III) sulphate (VI) is $\underset{+3}{Fe_2}(\underset{+6}{S}O_4)_3$

$Fe_2(SO_4)_3$ is already in $Fe^{3+}$ state hence, it does not require $KMnO_4$.

$FeSO_4$ is oxidised to $Fe^{3+}$ by $KMnO_4$

$5Fe^{2+} + MnO_4^- + 8H^+ \longrightarrow 5Fe^{3+} + Mn^{2+} + 4H_2O$

100 mL of 2 M $KMnO_4$ = $\frac{100 \times 2}{1000}$ = 0.2 mol $MnO_4^-$

1 mol $MnO_4^-$ ≡ 5 mol $Fe^{2+}$

$\therefore$ 0.2 mol $MnO_4^-$ ≡ 1 mol $Fe^{2+}$ or 1 mol $FeSO_4$

Total moles = 3

$\therefore \quad X_{FeSO_4} = \frac{1}{3}$

**14.** (b) $H_2O_2 \longrightarrow 2H^+ + O_2 + 2e^-$

$\underset{+7}{MnO_4^-} \xrightarrow{H^+} \underset{+2}{Mn^{2+}}$

Thus, 1 M $KMnO_4$ = 5 N $KMnO_4$

$\underset{+7}{MnO_4^-} \xrightarrow{OH^-} \underset{+4}{MnO_2}$

1 M $KMnO_4$ = 3 N $KMnO_4$

Thus, 100 mL of 5 N $KMnO_4$ = 100 mL of N′$H_2O_2$

and $V$ mL of 3 N $KMnO_4$ = 100 mL of N′$H_2O_2$

$\therefore \quad 100 \times 5 = 3V$

$\therefore \quad V = \frac{500}{3}$ mL

Let, $MnO_4^-$ = VmL

VmL of 0.01 M $MnO_4^-$ = VmL of 0.05 N $MnO_4^-$

$\therefore$ = $0.05 \times V$ milliequivalents

$\therefore \quad 0.05V = 4$

$V = 80$ mL

**15.** (b) $Fe^{2+}$ is 1.0 ppm

Thus, $10^6$ g solution has $Fe^{2+} = 1$ g

250 mL (= 250 g) solution has $Fe^{2+} = \frac{250}{10^6}$ g (**Note** $d = 1$ g/mL)

56 g of $Fe^{2+}$ is in = 392 g salt

$\therefore \quad \frac{250}{10^6}$ g of $Fe^{2+}$ is in = $\frac{392}{56} \times \frac{250}{10^6}$ g = $1.75 \times 10^{-3}$ g

**16.** (a, b)

(a) $FeSO_3 \rightleftharpoons Fe^{2+} + SO_3^{2-}$

$Fe^{2+}$ and $SO_3^{2-}$ both are oxidised by $Cr_2O_7^{2-}/H^+$

$$6Fe^{2+} + Cr_2O_7^{2-} \longrightarrow 6Fe^{3+} + 2Cr^{3+} + 7H_2O + 14H^+$$

$$6SO_3^{2-} + 2Cr_2O_7^{2-} + 16H^+ \longrightarrow 6SO_4^{2-} + 4Cr^{3+} + 8H_2O$$

$$6(FeSO_3) + 3Cr_2O_7^{2-} \longrightarrow \ldots$$

$$6FeSO_3 \equiv 3Cr_2O_7^{2-}$$

$$1\,FeSO_3 \equiv 0.5\,Cr_2O_7^{2-}$$

(b) $FeC_2O_7 \rightleftharpoons Fe^{2+} + C_2O_4^{2-}$

$Fe^{2+}$ and $C_2O_4^{2-}$ both are oxidised by $Cr_2O_7^{2-}/H^+$ as in, (a)

$$6Fe^{2+} + Cr_2O_7^{2-} \longrightarrow \ldots$$

$$6C_2O_4^{2-} + 2Cr_2O_7^{2-} \longrightarrow \ldots$$

$$6FeC_2O_4 + 3Cr_2O_7^{2-} + \ldots$$

$$6FeC_2O_4 \equiv 3Cr_2O_7^{2-}$$

$$1\,FeC_2O_4 \equiv 0.5\,Cr_2O_7^{2-}$$

(c) $Fe_2(SO_4)_3 \longrightarrow 2Fe^{2+} + 3SO_4^{2-}$

$Fe^{3+}$ and $SO_4^{2-}$ are in oxidised state

$Cr_2O_7^{2-}$ required = Nil

(d) $Al_2(SO_3)_3 \longrightarrow 2Al^{3+} + 3SO_3^{2-}$

$Al^{3+}$ is already oxidised

Only $SO_3^{2-}$ is oxidised to $SO_4^{2-}$

$$6SO_3^{2-} + 2Cr_2O_7^{2-} + \ldots \longrightarrow 6SO_4^{2-} + 4Cr^{3+} \quad 6SO_3^{2-} \rightleftharpoons 2Cr_2O_7^{2-}$$

$$1\,Al_2(SO_3)_3 \equiv 3SO_3^{2-} \equiv 1\,Cr_2O_7^{2-}$$

Thus, (a) and (b).

**17.** (a, d) $KHC_2O_4 \cdot H_2C_2O_4 \cdot 2H_2O$ has

(i) Three ionisable $[H^+]$

$\therefore$ Molarity (acid) = $\frac{\text{Normality}}{3}$

$\therefore$ Molarity (acid) = $\frac{0.2}{3} = 0.067$ M

Thus, (d) is correct.

(ii) It has two $C_2O_4^{2-}$, $\quad 2C_2O_4^{2-} \longrightarrow 4CO_2 + 4e^-$

Thus, normality (reducing agent) = $4 \times 0.067 = 0.267$ N, thus (a)

**Explanations for Q. Nos. (18-19)**

Mixture contains $Na_2C_2O_4$ and $KHC_2O_4$ in 1 : 1 molar ratio

Mixture required 100 mL of 0.01 M KOH

It is only required by $KHC_2O_4$

$$KHC_2O_4 \rightleftharpoons K^+ + H^+ + C_2O_4^{2-}$$

$$1\,(OH^-) \equiv 1\,(H^+) \equiv 1\,KHC_2O_4$$

Moles of $KHC_2O_4$ ≡ moles of KOH = $\frac{100 \times 0.01}{1000} = 0.001$ mol

Thus, $\quad Na_2C_2O_4 = 0.001$ mol

Total $C_2O_4^{2-} = 0.002$ mol

$\therefore$ Moles of oxalate = 0.002

$\therefore$ Equivalents of oxalate = $0.002 \times 2 = 0.004$

Milliequivalents of oxalate = $0.004 \times 1000 = 4$

Let $MnO_4^-$ = VmL, $\quad 0.01M\,MNO_4^- \equiv 0.05\,N\,MnO_4^-$

$\therefore 0.5V = 4, \quad V = 80$ mL

**18.** (d) **19.** (c)

**20.** (d)

(i) Iron (II) oxalate is $FeC_2O_4 \rightleftharpoons Fe^{2+} + C_2O_4^{2-}$

$Fe^{2+}$ and $C_2O_4^{2-}$ both are oxidised by $MnO_4^-$ in acidic medium

$$5Fe^{2+} + MnO_4^- + 8H^+ \longrightarrow 5Fe^{3+} + Mn^{2+} + 4H_2O$$

$$5C_2O_4^{2-} + 2MnO_4^- + 16H^+ \longrightarrow 10CO_2 + 2Mn^{2+} + 8H_2O$$

$$5FeC_2O_4 \equiv 3MnO_4^-$$

$$1\,FeC_2O_4 \equiv 0.6\text{ mol }MnO_4^-$$

**(i) → (r)**

(ii) Iron (II) sulphate (IV) is $\underset{+2}{Fe^{2+}}\,\underset{+4}{SO_3^{2-}}$

$Fe^{2+}$ is oxidised to $Fe^{3+}$

and $SO_3^{2-}$ is oxidised to $SO_4^{2-}$ by $MnO_4^-/H^+$

$$5Fe^{2+} + MnO_4^- + 8H^+ \longrightarrow 5Fe^{3+} + Mn^{2+} + 4H_2O$$

$$5SO_3^{2-} + 2MnO_4^- + H^+ \longrightarrow SO_4^{2-} + Mn^{2+} + H_2O$$

$$5FeSO_3 \equiv 3MnO_4^-$$

$$1\,FeSO_3 \equiv 0.6\text{mol }MnO_4^-$$

Thus, **(ii) → (r)**

(iii) Iron (III) sulphate (VI) is

$$\underset{+3\ \ +6}{Fe_2(SO_4)_3} \rightleftharpoons 2Fe^{3+} + 3SO_4^{2-}$$

No further oxidation

Thus, $MnO_4^-$ required = 0

Thus, **(iii) → (p)**

(iv) Iron (III) sulphate (IV) is

$$Fe_2(SO_3)_3 \rightleftharpoons 2Fe^{3+} + 3SO_3^{2-}$$

Only $SO_3^{2-}$ is oxidised.

$$5SO_3^{2-} + 2MnO_4^- + H^+ \longrightarrow 2Mn^{2+} + 5SO_4^{2-} + H_2O$$

5 mol $SO_3^{2-}$ ≡ 2 mol $MnO_4^-$

3 mol $SO_3^{2-} \equiv \frac{2}{5} \times 3 = 1.2$ mol (from 1 mol reactant)

Thus, **(iv) → (s)**

**21.** (6) $\quad \underset{+4}{SO_3^{2-}} + H_2O \longrightarrow \underset{+6}{SO_4^{2-}} + 2H^+ + 2e^-$

$\therefore$ 60 mL of 0.5 M $SO_3^{2-}$ ≡ 60 mL of 1.0 N $SO_3^{2-}$

$$\underset{+5}{IO_3^-} + (5-n)e^- \longrightarrow (I)^n$$

Equivalent weight of $KIO_3 = \frac{\text{Molar mass}}{(5-n)} = \frac{214}{(5-n)}$

Equivalents of $KIO_3 = \frac{2.14\,(5-n)}{214} = 0.01\,(5-n)$

Equivalents of $SO_3^{2-} = \frac{60 \times 1.0}{1000} = 0.06$

Equivalents of $KIO_3$ = Equivalents of $SO_3^{2-}$

$$0.01\,(5-n) = 0.06$$

$\therefore \quad 5 - n = 6$

$\therefore \quad n = -1$

Thus, change in oxidation state = 5 – (–1) = 6

**22.** (4)

In acidic medium

$$\underset{+7}{MnO_4^-} + 8H^+ + 5e^- \longrightarrow \underset{+2}{Mn^{2+}} + 4H_2O$$

In alkali medium

$$\underset{+7}{MnO_4^-} \rightarrow \underset{+n}{Mn^{n+}}$$

Change = $(7-n)$

20 of 1 M $KMnO_4$ in acidic medium

≡ 33.4 mL of 1 M $KMnO_4$ in alkali medium

$\therefore \quad 20 \times 5\,N = 33.4 \times (7-n)\,M$

$7 - n = 3$

$\therefore \quad n = 4$

**23.** (4) $MnO_4^-$ required = $100 \times 0.008$ millimol

$SO_2$ required = $5 \times (MnO_4^-)$

$= 100 \times 0.008 \times 5 = 4$

**24.** (6) Salt = $6.1\, g\, L^{-1}$

$$K_xH_y(C_2O_4)_z = xK^+ + yH^+ + zC_2O_4^{2-}$$

By change balance $(x + y) = 2z$

**Titration I** NaOH neutralises acid ($H^+$)

$25$ mL of $N_1(H^+) \equiv 18$ mL of 0.1 N NaOH

$N_1(H^+) \equiv 0.072$ N

$\therefore$ Equivalent weight of acid salt $= \dfrac{6.1}{0.072} = 84.72$ g equiv$^{-1}$

**Titration II** $KMnO_4$ oxidises $C_2O_4^{2-}$

25 mL $\times N_2(C_2O_4^{2-}) \equiv 24$ mL of 0.1 N $KMnO_4$

$\therefore \quad N_2(C_2O_4^{2-}) = \dfrac{24 \times 0.1}{25} = 0.096$ N

Equivalent weight of as $C_2O_4^{2-}$ part $= \dfrac{6.1}{0.096}$

$= 63.54$ g equiv$^{-1}$

Molar mass as $H^+ = 84.72y$

Molar mass as $C_2O_4^{2-} = 63.54 \times 2z$

$\therefore \quad 84.72y = 63.54 \times 2z$

$$\frac{y}{z} = \frac{63.54 \times 2}{84.72} = \frac{3}{2}$$

$y = 1.5z$

$x + y = 2z$

$x + 1.5z = 2z$

$\therefore \quad x = 0.5z$

$\dfrac{x}{z} = \dfrac{1}{2}$

$\therefore \quad x : y : z = 1 : 3 : 2$

$\therefore \quad x + y + z = 6$

## Revisal Problems (JEE Main)

**1.** (b) Let, $H_2 = x\,g = \dfrac{x}{2}$ mol

Then, $He = (2 - x)\,g = \dfrac{2 - x}{4}$ mol

Total moles $= \dfrac{x}{2} + \dfrac{2 - x}{4} = \left(\dfrac{x + 2}{4}\right)$

$$X_{H_2} = \frac{x/2}{\left(\frac{x+2}{4}\right)} = \frac{2x}{x + 2} = 0.5$$

$4x = x + 2$

$\Rightarrow \quad 3x = 2$

$\Rightarrow \quad x = \dfrac{2}{3}$

$\therefore \quad He = 2 - x = \left(2 - \dfrac{2}{3}\right) = \dfrac{4}{3}$ g

**2.** (c) Let, $CO_2 = x$ mol

Then, $CO = (1 - x)$ mol

$$\underset{x\text{ mol}}{CO_2} + C \longrightarrow \underset{2x\text{ mol}}{2CO}$$

Total CO after passing over hot graphite

$= \underset{\text{(in mixture)}}{(1 - x)} + \underset{\text{(from } CO_2)}{2x}$

$= (1 + x)$

22.4 L at STP = 1 mol

$\therefore$ 33.6 L at STP = 1.5 mol

$\therefore \quad 1 + x = 1.5$

$\Rightarrow \quad x = 0.5$

Thus, $CO_2 = x = 0.5$ mol

$CO = 0.5$ mol

$X_{CO_2} = 0.5$

**3.** (a) $H_2 = 1\,g = \dfrac{1}{2}$ mol $= \dfrac{8}{16}$

$He = 1\,g = \dfrac{1}{4}$ mol $= \dfrac{4}{16}$

$CH_4 = 1\,g = \dfrac{1}{16}$ mol $= \dfrac{1}{16}$

Total moles $= \dfrac{13}{16}$

$\therefore \quad X_{H_2} = \dfrac{8}{13} = 0.61, \; X_{He} = \dfrac{4}{13} = 0.31$

$X_{CH_4} = \dfrac{1}{13} = 0.08$

Thus, $X_{CH_4} < X_{He} < X_{H_2}$

Thus, $CH_4$ is not placed at correct position.

**4.** (b) Methanol = 1 g $= \dfrac{1}{0.75}$ mL $= \dfrac{100}{75}$ mL

$H_2O = 1\,g = \dfrac{1}{1}$ mL $= \dfrac{75}{75}$ mL

Total volume of mixture $= \dfrac{100}{75} + \dfrac{75}{75} = \dfrac{175}{75}$

$$\therefore \% \text{ of methanol} = \left(\frac{\frac{100}{75}}{\frac{175}{75}}\right) \times 100 = \frac{400}{7} = 57.14\%$$

**5.** (c) $H_2SO_4 = 100$ mL $= 100 \times 1.8$ g $= 180$ g

$H_2O = 100$ mL $= 100 \times 1.0$ g $= 100$ g

Total mass of solution = 280 g

$\therefore \; H_2SO_4 = \dfrac{180 \times 100}{280}\% = 64.28\% \approx 64.3\%$

**6.** (b) $10^6$ g of hard water has $CaCO_3 = 200$ g

100 g of hard water has

$$CaCO_3 = \frac{200}{10^6} \times 100 = 200 \times 10^{-4} = \frac{2}{100} = 0.02\%$$

**7.** (b) **Before heating** $w_1 + w_2 = 100$ g, $w_1$ = solute = 40 g

$w_2$ = water = 60 g

**After heating** Solute $= w_1 = 40$ g (no change in its amount)

Water $= w'_2$

Such that $\dfrac{40}{w'_2 + 40} = \dfrac{50}{100} \Rightarrow w'_2 + 40 = 80$ g

$\therefore \quad w'_2 = 40$ g

Thus, water lost = 60 − 40 = 20 g

**8.** (a) 100 mL of 0.40 M KCl $= \dfrac{100 \times 0.4}{1000}$ mol KCl = 0.04 mol KCl

100 mL of 0.50 KCl $= \dfrac{100 \times 0.5}{1000}$ mol KCl = 0.05 mol KCl

Thus, additional KCl required = 0.05 − 0.04 = 0.01 mol

$= 0.01 \times 74.5$ g

$= 0.745$ g KCl

**9.** (a) Solute in 800 g solution $= \dfrac{800 \times 40}{100} = 320$ g

Water in solution = 800 − 320 = 480 g

Solute removed = 100 g

Solute left in solution = 320 − 100 = 220 g in 480 g water

Total weight of solution = 220 + 480 = 700 g

Thus, % concentration $= \dfrac{220}{700} \times 100 = 31.4\%$

**10.** (b) $K_2CO_3$ is resistant to heat but $Li_2CO_3$ is easily converted to $CO_2$.

$$K_2CO_3 \xrightarrow{\Delta} X \text{ (no effect)}$$

$$\underset{\substack{2\text{ mol}\\1\text{ mol}}}{2KHCO_3} \longrightarrow \underset{\substack{1\text{ mol}\\0.5\text{ mol}}}{K_2CO_3} + CO_2 + H_2O$$

$$\underset{1\text{ mol}}{Li_2CO_3} \longrightarrow Li_2O + \underset{1\text{ mol}}{CO_2}$$

Total $CO_2 = 0.5 + 1.0 = 1.5$ mol

**11.** (b) Reaction is $\underset{1\text{ mol}}{Pb(NO_3)_2} + \underset{2\text{ mol}}{2NaCl} \longrightarrow PbCl_2 + 2NaNO_3$

Moles of NaCl taken $= MV = 2.25 \text{ mol L}^{-1} \times 1.0 \text{ L} = 2.25$ mol

Moles of $Pb(NO_3)_2$ taken $= 0.75 \text{ mol L}^{-1} \times V \text{ L} = 0.75\,V$

Moles of $Pb(NO_3)_2$ required based on NaCl $= \dfrac{2.25}{2} = 1.125$ mol

Thus, $0.75\,V = 1.125$

$\therefore$ $V = \dfrac{1.125}{0.75} = 1.5$ L

**12.** (a) $N_2 \equiv 3H_2 \equiv 6H$

Thus, equivalent mass of $N_2 = \dfrac{\text{Molar mass}}{6}$

Volume occupied by 1 equivalent $N_2$ at STP $= \dfrac{22.4}{6}$ L

$NH_3 \equiv 3H$

Thus, equivalent mass of $NH_3 = \dfrac{\text{Molar mass}}{3}$

Volume occupied by 1 equivalent $NH_3$ at STP $= \dfrac{22.4}{3}$ L

$H_2 \equiv 2H$

Thus, equivalent mass of $H_2 = \dfrac{\text{Molar mass}}{2}$

Volume occupied by 1 equivalent $H_2$ gas at STP $= \dfrac{22.4}{2}$ L

Thus, total volume $= V(N_2) + V(NH_3) + V(H_2)$

$$= \frac{22.4}{6} + \frac{22.4}{3} + \frac{22.4}{2}$$

$$= \frac{22.4}{6} + \frac{2 \times 22.4}{6} + \frac{3 \times 22.4}{6} = 22.4 \text{ L}$$

**13.** (b) Mass of metal combining with oxygen and halogen is same.

Metal $\equiv$ 4 g halogen = 0.4 g oxygen

0.4 g oxygen $\equiv$ 4 g halogen

Thus, 8 g oxygen $= \dfrac{4}{0.4} \times 8.0 = 80$ g

**14.** (d) $M + O_2 \longrightarrow M_xO_y$

Equivalent mass of oxygen = 8

Equivalent mass of metal = 16

Equivalent mass of metal oxide = 24

Thus, $\dfrac{E\text{ (metal oxide)}}{E\text{ (metal)}} = \dfrac{24}{16} = 1.5$

**15.** (b) $\underset{\substack{1\text{ mol}\\?}}{RH_2 + Ca^{2+}} \rightarrow RCa + \underset{\substack{2\text{ mol}\\0.01\text{ mol L}^{-1}}}{2H^+}$

When $[H^+]$ is 2 mol then, $[Ca^{2+}]$ is = 1 mol

When $[H^+]$ is 0.01 mol then $[Ca^{2+}]$ is $= \dfrac{0.01}{2}$ mol $L^{-1}$

$\therefore$ $[Ca^{2+}] = 0.005$ M $= 0.005 \times 40 \text{ gL}^{-1}$

$= 0.2 \text{ gL}^{-1}$

1000 mL of water has $[Ca^{2+}] = 0.2$ g

$10^6$ mL (parts per million) has $[Ca^{2+}] = \dfrac{0.2 \times 10^6}{10^3} = 200$

**16.** (a) Moles of $H_2O_2 = \dfrac{100X}{1000} = 0.1X$

$$\underset{2\text{ mol}}{2H_2O_2(l)} \longrightarrow 2H_2O(l) + \underset{1\text{ mol} = 24\text{ L}}{O_2(g)}$$

24 L $O_2$ is from = 2 mol $H_2O_2$

$\therefore$ 3 L $O_2$ is from $= \dfrac{2 \times 3}{24}$ mol $H_2O_2 = 0.25$ mol $H_2O_2$

$\therefore$ $0.1X = 0.25$

$\therefore$ $X = 2.5$ M

**17.** (c)

$$NH_2CONH_2 + 2NaOH \longrightarrow 2NH_3 + Na_2CO_3 + H_2O$$

$$2NH_3 + H_2SO_4 \longrightarrow (NH_4)_2SO_4$$

This method is called Kjeldahls' method for estimation of percentage of nitrogen

$$N\% = \frac{1.4N_1\text{(normality) } V_1\text{(volume)}}{W\text{ (amount)}}$$

Percentage of N in urea $= \dfrac{28}{60} \times 100$

$= 46.666$

$46.666 = \dfrac{1.4 \times 0.1 \times V_1\,(H_2SO_4)}{0.6}$

$\therefore$ $V_1(H_2SO_4) = 200$ mL

But, $H_2SO_4$ taken = 300 mL

Unreacted $H_2SO_4$ = 300 − 200 = 100 mL

100 mL of 0.1 N $H_2SO_4 \equiv V$ mL of 1 N NaOH

$\therefore$ $V = \dfrac{100 \times 0.1}{1} = 10$ mL

**18.** (a) 1 mole $O_2$ = 32 g

Equivalent mass of $O_2 = \dfrac{\text{Molecular weight}}{4} = 8$

Equivalents in one mole $= \dfrac{32}{8} = 4$

Thus, 1 mole $O_2 \equiv$ 4 equivalents $O_2$ = 30 L

$\therefore$ Volume of 1 equivalent $O_2$ gas $= \dfrac{30}{4} = 7.5$ L

**19.** (a) $H_2O_2 + 2H^+ + 2e^- \longrightarrow 2H_2O$

If $H_2O_2$ is = $x$ M (molarity).

Then, it is = $2x$ N (normality)

$$Cr^{3+} + 4H_2O \longrightarrow CrO_4^{2-} + 8H^+ + 3e^-$$

$Cr^{3+} = 0.05$ M $= 0.15$ N

$\therefore$ $N_1V_1 = N_2V_2$

$H_2O_2 = Cr^{3+}$

$2x \times 20 = 0.15 \times 40$

$\therefore$ $x = 0.15$ M

**20.** (b)

$$MnO_4^- + e^- \longrightarrow MnO_4^{2-}$$

Thus, molarity = normality

$KMnO_4 = \dfrac{12.64}{158} = 0.08$ mol

Volume = 0.500 L

$\therefore$ $M(KMnO_4) = N(KMnO_4)$

$= \dfrac{0.08\text{ mol}}{0.500\text{ L}} = 0.16$ N

**21.** (c) $NaHC_2O_4 \equiv Na^+ + H^+ + C_2O_4^{2-}, H^+ \equiv OH^-$

Thus, as an acid $NaHC_2O_4 \equiv$ NaOH = 0.1 M

$$C_2O_4^{2-} \longrightarrow 2CO_2 + 2e^-$$

Thus, 0.1 M $C_2O_4^{2-}$ = 0.2 N $C_2O_4^{2-}$

**22.** (d) Let, $X$ be $ACl_x \rightleftharpoons A + xCl$,

$Y$ be $ACl_y = A + yCl$

Molar mass of $x = 137$

$$\% \text{ of Cl in } X = \frac{35.5\,x}{137} = \frac{77.5}{100}$$

$$x = 3.0$$

$$\% \text{ of Cl in } Y = \frac{35.5y}{208} = \frac{85.1}{100}$$

$\therefore$ $y = 5$

Thus, $X$ is $ACl_3$. $Y$ is $ACl_5$.

**23.** (c) VD × 2 = Molecular weight

Molecular weight of mixture of $NO_2$ and $N_2O_4$

$= 38.3 \times 2 = 76.4 \text{ g mol}^{-1}$

Let, $NO_2$ be $= X\%$, $N_2O_4 = (100 - X)\%$

$$\overline{M} = \frac{M_1X_1 + M_2X_2}{X_1 + X_2}$$

$$\Rightarrow \quad 76.4 = \frac{46X + 92(100 - X)}{100}$$

$X = 33.913$ or $33.913\%\ NO_2$

In 100 g $NO_2 = 33.913$ g

$= 0.7372$ mol

**24.** (c) Sn taken = 1.056 g

Sn unreacted = 0.601 g

Sn used = 1.056 − 0.601 = 0.455 g

| | Mole | Ratio |
|---|---|---|
| Sn | $0.455/119 = 3.82 \times 10^{-3}$ | 1.0 |
| I | $1.947/127 = 1.53 \times 10^{-2}$ | 4.0 |

Thus, compound is $SnI_4$.

**25.** (a) $\underset{2\text{ mol}}{2Al} + 6HCl \longrightarrow 2AlCl_3 + \underset{3\text{ mol}}{3H_2(g)}$

Volume of foil $= 10.25 \text{ cm} \times 5.50 \text{ cm} \times \frac{0.600}{10}$ cm

$= 3.3825 \text{ cm}^3$

Density $= 2.70 \text{ g/cm}^3$

$\therefore$ Mass of Al $= 3.3825 \times 2.70$

$= 9.13275$ g

Moles of Al $= \frac{9.13275}{27}$

$= 0.33825$ mol

$H_2$ formed = 1.5 times of Al reacted

$= 1.5 \times 0.33825$

$= 0.51$ mol

## Revisal Problems (JEE Advanced)

**1.** (a) I. $\underset{1\text{ g}}{Zn} + 2NaOH \longrightarrow Na_2ZnO_2 + H_2$

II. $\underset{1\text{ g}}{Zn} + H_2SO_4 \longrightarrow ZnSO_4 + H_2$

**In each experiment**

1 mole Zn gives = 1 mole $H_2$

$= 22.4 \text{ L } H_2$ at STP

1 g Zn gives $= 22.4 \times \frac{1}{m}$ L $H_2$

Thus, ratio 1 : 1

**2.** (d) $[Co(NH_3)_5SO_4]Br$ and $[Co(NH_3)_5Br]SO_4$ are ionisation isomers which ionise to form cation and anion in such a way that anions are from outside coordinate parts.

$$\underset{0.02\text{ mol}}{[Co(NH_3)_5SO_4]Br} \rightleftharpoons [Co(NH_3)_5SO_4]^+ + \underset{0.02\text{ mol}}{Br^-}$$

$$Ag^+ + Br^- \longrightarrow \underset{0.02\,(P)}{AgBr\downarrow}$$

$SO_4^{2-}$ ion inside coordination sphere does not ionise.

$$\underset{0.02}{[Co(NH_3)_5Br]SO_4} \rightleftharpoons [Co(NH_3)_5Br]^{2+} + \underset{0.02}{SO_4^{2-}}$$

$$Ba^{2+} + SO_4^{2-} \longrightarrow \underset{0.02\,(Q)}{BaSO_4}$$

$Br^-$ ion inside sphere does not ionise.

**3.** (a) Given reaction is not balanced

Based on stoichiometry

1 mol $Na_2CO_3$ is from 2 mol $NaHCO_3$

2 mol $NaHCO_3$ is from 2 mol $NH_4HCO_3$

2 mol $NH_4HCO_3$ is from 2 mol $NH_3$

Thus, 1 mol $Na_2CO_3$ is from = 2 mol $NH_3$

106 g $Na_2CO_3$ is from $= 2 \times 17$ g $NH_3$

34 g $NH_3$ gives = 106 g $Na_2CO_3$

Thus, 1000 g $NH_3$ gives $= \frac{106 \times 1000}{34}$ g $Na_2CO_3$

$= 3118$ g $Na_2CO_3$

$= 3.118$ kg $Na_2CO_3$

**4.** (c) Sodium sulphate (IV) is $Na_2SO_3$ and sodium sulphate (VI) is $Na_2SO_4$

50 mL of 0.1 M $A^{3+}$ is reduced to 50 mL of 0.1 M $A^{n+}$

$$\underset{+3}{A^{3+}} + (3 - n)e^- \longrightarrow \underset{n}{A^n}$$

$$\underset{+4}{SO_3^{2-}} + H_2O \longrightarrow \underset{+6}{SO_4^{2-}} + 2H^+ + 2e^-$$

50 mL of 0.1 M $A^{3+} \equiv 25$ mL of 0.1 M $Na_2SO_3$

or 50 mL of $0.1 \times (3 - n)$ N $A^{3+} \equiv 25$ mL of 0.2 N $Na_2SO_3$

$\therefore$ $(3 - n) = 1 \Rightarrow n = 2$

**5.** (d) Based on stoichiometry

1 mol AgBr required 2 mol of $Na_2S_2O_3$

0.3760 g AgBr $= \frac{0.3760}{188}$ mol AgBr required

$= \frac{2 \times 0.3760}{188}$ mol of $Na_2S_2O_3$

Let, volume of 0.015 M $Na_2S_2O_3 = V$ mL

$\therefore$ $\frac{V \times 0.015}{1000}$ = moles of $Na_2S_2O_3$

$= \frac{2 \times 0.3760}{188}$

$\therefore$ $V = 267$ mL

**6.** (c) $\underset{1\text{ mol}}{K_2CrO_4} + \underset{2\text{ mol}}{2AgNO_3} \longrightarrow Ag_2CrO_4 + 2KNO_3$

Let, volume of $K_2CrO_4$ solution = $V$ mL

$\therefore$ Moles of $K_2CrO_4 = \frac{MV}{1000} = \frac{0.650V}{1000}$

Moles of $AgNO_3 = \frac{415 \times 0.186}{1000}$

$K_2CrO_4$ required for given moles of $AgNO_3 = \frac{415 \times 0.186}{1000 \times 2}$

$\therefore$ $\frac{0.650V}{1000} = \frac{415 \times 0.186}{1000 \times 2}$

$V = \frac{415 \times 0.186}{2 \times 0.650} = 59.4$ mL

**7.** (c) Let, $A = x$ mol, $B = y$ mol

Volume of $MnO_4^-$ = Volume of NaOH = $V$ mL

Thus, moles of $MnO_4^- = \frac{0.1V}{1000} = 1 \times 10^{-4}$ V

Moles of NaOH = $\frac{0.1V}{1000} = 1 \times 10^{-4}$ V

$A: Na_2C_2O_4 \rightleftharpoons 2Na^+ + C_2O_4^{2-}$

$B: KHC_2O_4 \cdot H_2C_2O_4 \rightleftharpoons K^+ + 3H^+ + 2C_2O_4^{2-}$

| | $C_2O_4^{2-}$ | $H^+$ |
|---|---|---|
| A. | $x$ | – |
| B. | $2y$ | $3y$ |
| Total | $x + 2y$ | $3y$ |

$$2MnO_4^- + 5C_2O_4^{2-} + 16H^+ \longrightarrow 2Mn^{2+} + 8H_2O + 10CO_2$$

$$H^+ + OH^- \longrightarrow H_2O$$

5 moles of $C_2O_4^{2-} \equiv 2$ moles $MnO_4^-$

$\therefore$ $(x + 2y)$ moles $C_2O_4^{2-} \equiv 0.4\,(x + 2y)$ moles $MnO_4^-$

1 mole $H^+ \equiv$ 1 mole of $OH^-$

$3y$ moles $H^+ \equiv 3y$ moles $OH^-$

$\therefore$ $0.4\,(x + 2y) = 1 \times 10^{-4}$

$3y = 1 \times 10^{-4}$

$y = \frac{1 \times 10^{-4}}{3}$

$0.4\left(x + \frac{2 \times 10^{-4}}{3}\right) = 1 \times 10^{-4}$

$x + \frac{2 \times 10^{-4}}{3} = \frac{1 \times 10^{-4}}{0.4}$

$x = \frac{1 \times 10^{-4}}{0.4} - \frac{2}{3} \times 10^{-4} = 1.83 \times 10^{-4}$

$\frac{x}{y} = \frac{1.83 \times 10^{-4} \times 3}{1 \times 10^{-4}} = 5.5$

Thus, ratio is = 5.5 : 1

**8.** (d) $Ca(HCO_3)_2 + CaO \longrightarrow 2CaCO_3 + H_2O$

| | 1 mol | 2 mol |
|---|---|---|
| | 56 g | 200 g |
| | 0.56 g | 2.00 g |

Thus, 10 L water (= $10^4$ g) water has hardness = 2 g

$\therefore$ $10^6$ g (ppm) water has hardness = $\frac{2 \times 10^6}{10^4} = 200$

**9.** (a) Mass of one atom of element $X = \frac{23}{N_0}$

Volume of one element = $\frac{23}{N_0 \times d} = \frac{23}{N_0 \times 6.2}$ cm$^3$

Thus, for a spherical molecule $V = \frac{4}{3}\pi r^3$

$\therefore$ $\frac{4}{3}\pi r^3 = \frac{23}{6.02 \times 10^{23} \times 6.2}$

$\therefore r^3 = \frac{23 \times 3}{6.02 \times 10^{23} \times 6.2 \times 4\pi} = 1.4711 \times 10^{-24}$ cm$^3$

$\therefore$ $r = 1.137 \times 10^{-8}$ cm = 1.137 Å

$AB = 2r = 2 \times 1.137 = 2.274$ Å

**10.** (a) Let, $MCl = x\,g = \left(\frac{x}{M + 35.5}\right)$ mol

NaCl = $(11.2 - x)$ g = $\frac{(11.2 - x)}{58.5}$ mol

$$AgNO_3 + MCl \longrightarrow AgCl + MNO_3$$

1 mol 1 mol

$\left(\frac{x}{M + 35.5}\right)$ mol, $\left(\frac{x}{M + 35.5}\right)$ mol

$$AgNO_3 + NaCl \longrightarrow AgCl$$

1 mol 1 mol

$\frac{(11.2 - x)}{58.5}$ mol, $\left(\frac{11.2 - x}{58.5}\right)$ mol

Total AgCl = 28.7 g = $\frac{28.7}{143.5}$ mol

$\therefore$ $\left(\frac{x}{M + 35.5}\right) + \frac{(11.2 - x)}{(58.5)} = \frac{28.7}{143.5}$ ...(i)

On heating only $M$Cl is vaporised thus AgCl is due to $M$Cl only.

Thus, $\left(\frac{x}{M + 35.5}\right) = \frac{14.35}{143.5}$ mol ...(ii)

Put value in Eq. (i)

$\therefore$ $\frac{14.35}{143.5} + \frac{(11.2 - x)}{58.5} = \frac{28.7}{143.5}$

$\Rightarrow$ $\left(\frac{11.2 - x}{58.5}\right) = \frac{14.35}{143.5} = 0.1$

$11.2 - x = 5.85$

$x = 11.20 - 5.85$

$= 5.35$ g

$\therefore$ $MCl = 5.35$ g

Also $\left(\frac{x}{M + 35.5}\right) = 0.1$

$\frac{5.35}{M + 35.5} = 0.1$

$\therefore$ $M = 18.0$

Thus, molar mass of $M$Cl = 53.5 g mol$^{-1}$

Thsu, $MCl = \frac{5.35}{53.5} = 0.1$ mol

NaCl = 11.2 − 5.35 = 5.85 g = $\frac{5.85}{58.5} = 0.1$ mol

Ratio = 1 : 1

**11.** (c, d) $NH_2CONH_2 + 2NaOH \xrightarrow{\Delta} Na_2CO_3 + H_2O + 2NH_3$

| | |
|---|---|
| 1 mol | 2 mol |
| 60 g | 2 mol |
| 0.6 g | 0.02 mol |

0.6 g of urea gives = 0.02 mol $NH_3$

$$NH_3 + HCl \longrightarrow NH_4Cl$$

1 mol 1 mol

0.02 mol 0.02 mol

Thus, HCl required = 0.02 mol

$[H^+] = 0.02$ mol

(a) 100 mL of 0.1 M HCl = $\frac{100 \times 0.1}{1000} = 0.01$ mol

(b) 200 mL of 0.2 M HCl = $\frac{200 \times 0.2}{1000} = 0.04$ mol

(c) 100 mL of 0.2 M HCl = $\frac{100 \times 0.2}{1000} = 0.02$ mol

(d) 100 mL of 0.1 M $H_2SO_4 = \frac{100 \times 0.1}{1000}$

$= 0.01$ mol $H_2SO_4$

$= 0.02$ mol $H^+$

**12.** (a, b, c, d)

(a) $H_3PO_3$ is a dibasic acid having only two ionisable $H^+$ ion

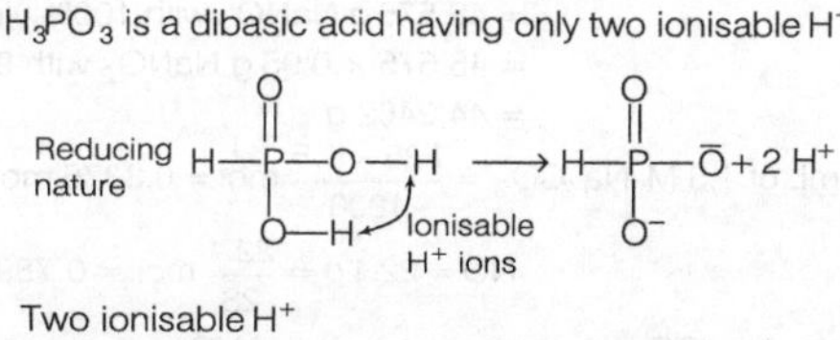


Two ionisable $H^+$

This is attached to is not ionised

Thus, $E = \frac{M}{2}$

(b) $H_3PO_3 \longrightarrow H_3PO_4$

O.N. +3 +5

$$E = \frac{M}{\text{Change in O.N}} = \frac{M}{2}$$

(c) $MnSO_4 \longrightarrow MnO_2$

O.N +2 +4

$$E = \frac{M}{\text{Change in O.N}} = \frac{M}{2}$$

(c) $C_2O_4^{2-} \longrightarrow 2CO_2 + 2e^-$

$$E = \frac{M}{\text{Exchange of electrons}} = \frac{M}{2}$$

**13.** (b) $C_6H_{12}O_6$

Molas mass = 180 g mol$^{-1}$

$$C = 6 \times 12 = 72 \text{ g} = \frac{72}{180} \times 100 = 40\%$$

$$H = 12 \times 1 = 12 \text{ g} = \frac{12}{180} \times 100 = 6.66\%$$

(a) $O = 6 \times 16 = 96 \text{ g} = \frac{96}{180} \times 100 = 53.33\%$

Molar mass of CO = 28 g

$$C = 12\text{g} = \frac{12}{28} \times 100 = 42.86\%$$

$$O = 16 \text{ g} = \frac{16}{28} \times 100 = 58.14\%$$

(b)

| | $C_6H_{12}O_6$ | | $CO(CH_2OH)_2$ | |
|---|---|---|---|---|
| C | 6 | 1 | 3 | 1 |
| H | 12 | 2 | 6 | 2 |
| O | 6 | 1 | 3 | 1 |

(a) Incorrect, (b) Correct, (c) Incorrect, (d) Incorrect

**14.** (b, d) $CaCO_3 \xrightarrow{\Delta} CaO + CO_2$

CaO = 5.6 g

$CO_2$ = 4.4 g

Total mass = 10.0 g

Residue = 20.0 − 10.0 = 10.0 g

Thus, % of pure $CaCO_3 = \frac{10}{20} \times 100 = 50\%$ pure

$$20 \text{ g impure } CaCO_3 = \frac{20}{1} \times \frac{50}{100} = 10 \text{ g pure } CaCO_3$$

$$10 \text{ g } CaCO_3 = \frac{10}{100} = 0.1 \text{ mol}$$

Molecules = 0.1 $N_0$

1 molecule has = 1 (Ca) + 1 (C) + 3 (O) atoms = 5 atoms

Thus, atoms = 0.5 $N_0$

## Explanations for Q. Nos. (15-16)

2 moles $Na_2CO_3$ give = 4 moles $NaNO_2$

0.3375 mole $Na_2CO_3$ gives

= 2 × 0.3375 = 0.6750 mol $NaNO_2$

= 0.6750 × 69 g $NaNO_2$

= 46.575 g $NaNO_2$ with 100% yield

= 46.575 × 0.95 g $NaNO_2$ with 95% yield

= 44.2462 g

$$225 \text{ mL of } 1.5 \text{ M } Na_2CO_3 = \frac{225 \times 1.5}{1000} \text{ mol} = 0.3375 \text{ mol}$$

$$NO = 22.1 \text{ g} = \frac{22.1}{28} \text{ mol} = 0.7893 \text{ mol}$$

2 moles $Na_2CO_3$ react with = 4 moles of NO

0.3375 moles $Na_2CO_3$ reacts with = 0.6750 mol NO

Thus, NO is in excess = 0.7893 − 0.6750 = 0.1143 mol

**15.** (a) **16.** (b)

## Explanations for Q. Nos. (17-18)

$3M + 2N \rightarrow M_3N_2$ compound

Molar mass of $M_3N_2 = 3M + 2N = (3M + 28)$

$$\text{Percentage of } M = \frac{3M \times 100}{(3M + 28)} = 72 \quad \text{(given)}$$

$$\therefore \frac{3M + 28}{3M} = \frac{100}{72} \Rightarrow 1 + \frac{28}{3M} = \frac{100}{72}$$

$$\therefore \frac{28}{3M} = \frac{28}{72} \Rightarrow M = 24$$

Molas mass of $M_3N_2 = (24 \times 3) + 2 \times 14 = 100$ g mol$^{-1}$

Atomic weight of $M$ = 24 g mol$^{-1}$

**17.** (c) **18.** (b)

**19.** (c) 100 g $CO_2 = \frac{100}{44}$ mol $CO_2$

2 mol $CO_2$ react with = 4 moles $KO_2$

$$\frac{100}{44} \text{ mol } CO_2 \text{ reacts with} = \frac{200}{44} \text{ mol } KO_2 = 4.54 \text{ mol } KO_2$$

**20.** (d) 176 g $CO_2 = \frac{176}{44}$ mol $CO_2$ = 4 moles $CO_2$

2 moles $CO_2$ give = 3 moles $O_2$

$$4 \text{ moles } CO_2 \text{ give} = \frac{3}{2} \times 4 = 6 \text{ moles } O_2$$

**21.** (d) 4 moles $KO_2$ give = 3 moles $O_2$

4 × 71 g $KO_2$ give = 3 moles $O_2$

4 × 71 × $10^3$ mg $KO_2$ give = 3 moles $O_2$

$$1 \text{ mg } KO_2 \text{ gives} = \frac{3}{4 \times 71 \times 10^3} \text{ mol } O_2 = 1.056 \times 10^{-5} \text{ mol}$$

= 10.56 × $10^{-6}$ mol = 10.56 micromol

## Explanations for Q. Nos. (22-23)

Pure ore = 100.0 g, Impure ore = 100 − 11 = 89 g

$$\text{Impure copper} = 75.4 \text{ g, Pure copper} = 75.4 \times \frac{89.5}{100} \text{ g} = 67.483 \text{ g}$$

Let, copper from $Cu_2S$ = $x$ g

and copper from CuS = (67.483 − $x$) g

From reaction (I), $2Cu \equiv 1 \, Cu_2S$

2 × 63.5 g ≡ 159 g

$$\therefore x \text{ g Cu} \equiv \frac{159x}{127} \text{ g } Cu_2S$$

From reaction (II), 1 Cu(s) ≡ 1 CuS

63.5 g Cu ≡ 95.5 g CuS

$$(67.483 - x) \text{ g Cu is from} = \frac{95.5}{63.5} \times (67.483 - x) \text{ g CuS}$$

$$\frac{159x}{127} + \frac{95.5\,(67.483 - x)}{63.5} = 89.0$$

$$\therefore x = 49.57 \text{ g}$$

$$Cu_2S = \frac{159x}{127} = \frac{159 \times 49.57}{127} = 62.06 \text{ g in 100 g impure ore}$$

∴ $Cu_2S$ = 62.06%

CuS = 89.0 − 62.06 = 26.94%

**22.** (b) **23.** (c)

**24.** (2) $x = n_{H_2} = \frac{pV}{RT} = \frac{5 \times 2}{0.0821 \times 400}$, $y = n_{N_2} = \frac{3 \times 5}{0.0821 \times 300}$

$$\therefore \frac{y}{x} = \frac{15}{300} \times \frac{400}{10} = 2$$

**25.** (2) $H_2SO_4$ solution is = 30.08% by mass

100 g $H_2SO_4$ solution has = 30.08 g $H_2SO_4$

$$\left(\frac{100}{1.303}\right) \text{mL } H_2SO_4 \text{ solution has} = 30.08 \text{ g } H_2SO_4$$

Thus, 500 mL of $H_2SO_4$ solution has

$$= \frac{30.08 \times 1.303 \times 500}{100} = 196 \text{ g} = \frac{196}{98} = 2 \text{ mol}$$

**26.** (8)

| Compound | Mass | Moles | Ratio | Empirical formula |
|---|---|---|---|---|
| C | 10.80 | 0.9 | 1.00=2 | |
| H | 1.36 | 1.39 | 1.5=3 | $C_2H_3Cl_3$ |
| Cl | 47.86 | 1.35 | 1.5=3 | |

Total atoms = 8

**Note** $C_2H_3Cl_3$ *can be* $CH_2(Cl)—CHCl_2 \cdot CH_3—CCl_3$

**27.** (d)

(i) Estimation of $S_2O_3^{2-}$ by $I_2$ : Iodimetric **(i) → (r)**

(ii) $MnO_4^- + Fe^{2+} + H^+ \rightarrow Fe^{3+} + Mn^{2+} + H_2O$

$MnO_4^-$ is self-indicator. Thus, **(ii) → (p)**

(iii) Estimation of N by Kjeldahl's method. Thus, **(iii) → (s)**

(S) $Cr_2O_7^{2-}$ is treated with $I^-$ and $I_2$ is estimated by hypo : Iodometric **(iv) → (q)**

**28.** (c)

(i) $\underset{\substack{2\text{ mol}\\1\text{ mol}}}{2CuSO_4} + 4KI \longrightarrow Cu_2I_2 + 2K_2SO_4 + \underset{\substack{1\text{ mol}\\0.5\text{ mol}}}{I_2}$

**(i) → (p)**

(ii) $\underset{1\text{ mol}}{Cr_2O_7^{2-}} + 14H^+ + 6I^- \longrightarrow 2Cr^{3+} + 7H_2O + \underset{3\text{mol}}{3I_2}$

**(ii) → (t)**

(iii) $\underset{1\text{ mol}}{H_2O_2} + 2I^- + 2H^+ \longrightarrow \underset{1\text{ mol}}{I_2} + 2H_2O$

Thus, **(iii) → q**

(iv) $\underset{1\text{ mol}}{O_3} + 6I^- + 6H^+ \longrightarrow 3H_2O + \underset{3\text{ mol}}{3I_2}$

Thus, **(iv) → (t)**

(v) $\underset{\substack{2\text{ mol}\\1\text{ mol}}}{2HNO_2} + 2I^- + 2H^+ \longrightarrow 2NO + \underset{\substack{1\text{ mol}\\0.5\text{ mol}}}{I_2} + 2H_2O$

Thus, **(v) → (p)**

**29.** (d)

(i) $N + P = 81$

$N = 31.7\%$ more than protons

Then $N = (P + 0.317P)$

$\therefore$ $(P + 0.317P + P) = 81$

$2.317P = 81$

$P = 35$

Thus, **(i) → (p)**

(ii) $M^{2+}$ has $= 6 + 16 = 22$ electrons

$M^{2+}$ has $= 24$ protons

$Z(M^{2+}) = 24$

$N = 24 + 2$

$= 26$

Thus, ionic mass $= 24 + 26 = 50$

Thus, **(ii) → (p)**

(iii) $X^-$ has protons $= 17$

Electrons $= 17 + 1 = 18$

$$\text{Neutrons} = \left(18 + \frac{18 \times 11.1}{100}\right)$$

$= 18 + 2$

$= 20$

Thus, ionic mass of $X = P + N - 37$

Thus, **(iii)→ (s)**

(iv) $N + P = 56$

Electrons in $M = P = Z$

Thus, electrons in $M^{3+} = (P - 3)$

Neutrons in

$$M^{3+} = (P-3) + (P-3) \times \frac{30.4}{100}$$

$N = (P - 3) + 0.304P - 0.912 + P = 56$

$P = 26$

Thus, **(iv) → (q)**

## JEE Main & AIEEE Archive

**1.** (b) The number of moles is given by

$$n = \frac{\text{weight } (w)}{\text{molecular weight } (M)}$$

Thus, ratio of moles of $O_2$ and $N_2$ is given by

$$\frac{n_{O_2}}{n_{N_2}} = \left(\frac{\dfrac{W_{O_2}}{M_{O_2}}}{\dfrac{W_{N_2}}{M_{N_2}}}\right) = \left(\frac{W_{O_2}}{W_{N_2}}\right)\left(\frac{M_{N_2}}{M_{O_2}}\right)$$

$$= \left(\frac{1}{4}\right) \times \left(\frac{28}{32}\right) = \frac{7}{32}$$

Hence, ratio of $n_{O_2}$ and $n_{N_2}$ is 7 : 32.

**2.** (a) The molarity of a resulting solution is given by

$$M = \frac{M_1V_1 + M_2V_2}{V_1 + V_2} = \frac{750 \times 0.5 + 250 \times 2}{750 + 250}$$

$$= \frac{875}{1000} = 0.875 \text{ M}$$

**3.** (d) 18 g $H_2O$ contain 2g H.

$\therefore$ 0.72 g $H_2O$ contain 0.08 g H

44 g $CO_2$ contain 12 g C.

$\therefore$ 3.08 g $CO_2$ contain 0.84 g C.

$$\therefore \text{C : H} = \frac{0.84}{12} : \frac{0.08}{1} = 0.07 : 0.08 = 7 : 8$$

$\therefore$ Empirical formula $= C_7H_8$

**4.** (d) Total mass of solution = 1000 g water + 120 g urea

= 1120 g

Density of solution = 1.15 g / mL

$$\text{Thus, volume of solution} = \frac{\text{mass}}{\text{density}} = \frac{1120 \text{ g}}{1.15 \text{ g / mL}}$$

= 973.91 mL = 0.974 L

$$\text{Moles of solute} = \frac{120}{60} = 2\text{mol}$$

$$\text{Molarity} = \frac{\text{moles of solute}}{\text{volume (L) of solution}}$$

$$= \frac{2 \text{ mol}}{0.974 \text{ L}} = 2.05 \text{ mol L}^{-1}$$

**5.** (c) Mohr's salt is $FeSO_4 \cdot (NH_4)_2SO_4 \cdot 6H_2O$.

Only oxidisable part is $Fe^{2+}$

$$[Fe^{2+} \longrightarrow Fe^{3+} + e^-] \times 6$$

$$Cr_2O_7^{2-} + 14H^+ + 6e^- \longrightarrow 2Cr^{3+} + 7H_2O$$

$$6Fe^{2+} + Cr_2O_7^{2-} + 14H^+ \longrightarrow 6Fe^{3+} + 2Cr^{3+} + 7H_2O$$

Millimoles of $Fe^{2+} = 750 \times 0.6 = 450$

$$\text{Moles of } Fe^{2+} = \frac{450}{1000} = 0.450 \text{ mol}$$

6 mol $Fe^{2+} \equiv$ 1 mol $Cr_2O_7^{2-}$

$$\therefore 0.450 \text{ mol } Fe^{2+} \equiv \frac{0.450}{6} = 0.075 \text{ mol } Cr_2O_7^{2-}$$

$= 0.075 \times 294$ g $= 22.05$ g

6. (a) Molality $= \dfrac{\text{moles of the solute}}{\text{mass of the solvent in kg}}$

Moles of urea ($n_{urea}$) $= \dfrac{0.0100 \text{ g}}{60 \text{ g mol}^{-1}}$

Mass of 1000 mL of solution = volume × density

$= 300 \text{ mL} \times \dfrac{1 \text{ g}}{\text{mL}}$ [$\because 1 \text{ dm}^3 = 1000 \text{ mL}$]

$= 300 \text{ g}$

Mass of solvent $= 0.300 \text{d m}^3 = 0.300 \text{kg}$

Molality $= \dfrac{0.0100}{60 \times 0.300}$

$= 5.55 \times 10^{-4} \text{ mol kg}^{-1}$

7. (c) Titration of oxalic acid by $KMnO_4$ in the presence of HCl gives unsatisfactory result because HCl is a better reducing agent than oxalic acid and HCl reduces preferably $MnO_4^-$ to $Mn^{2+}$.

8. (d) $2Al(s) + 6HCl(aq) \longrightarrow 2Al^{3+}(aq) + 6Cl^-(aq) + 3H_2(g)$

From the equation, it is clear that,

6 mol of HCl produces 3 mol of $H_2$

or 1 mole of HCl $= \dfrac{3 \times 22.4}{6}$ L of $H_2$

$= 11.2\text{L of } H_2$

9. (b) 8 moles of O-atom are contained by 1 mole $Mg_3(PO_4)_2$.

Hence, 0.25 moles of O-atom are contained by

$= \dfrac{1}{8} \times 0.25 \text{ mol } Mg_3(PO_4)_2$

$= 3.125 \times 10^{-2}$

10. (b) The mass of one mole of *a* substance will remain unchanged.

11. (b) $\underset{2\text{ mol}}{2BCl_3} + \underset{3\text{ mol}}{3H_2} \longrightarrow \underset{\substack{2\text{ mol}\\21.6\text{ g}}}{2B} + 6HCl$

$21.6 \text{ g B} = 2 \text{ mol B} \equiv 3 \text{ mol } H_2$

$pV = nRT$

$\therefore \quad V = \dfrac{nRT}{p} = \dfrac{3 \times 0.0821 \times 273}{1} = 67.2 \text{ L}$

12. (c) 560 g of Fe

Number of moles $= \dfrac{560 \text{ g}}{56 \text{ g mol}^{-1}} = 10 \text{ mol}$

For 70 g of N, 14 g N = 1 mol of N-atom

70 g N = 5 mol of N-atom

For 20 g of H, 1 g H = 1 mol of H-atom

20 g H ≡ 20 mol of H-atom

13. (a) Molar mass 108 g $\text{mol}^{-1}$

Total part by weight = 9 + 1 + 3.5 = 13.5

Weight of carbon $= \dfrac{9}{13.5} \times 108 = 72 \text{ g}$

Weight of hydrogen $= \dfrac{1}{13.5} \times 108 = 8 \text{ g}$

Weight of nitrogen $= \dfrac{3.5}{13.5} \times 108 = 28 \text{ g}$

| Atom | Amount | Mole |
|---|---|---|
| C | 72 g | 6 |
| H | 8 g | 8 |
| N | 28 g | 2 |

Hence, molecular formula = $C_6H_8N_2$.

## JEE Advanced & IIT JEE Archive

1. (8) There is no change in volume on dissolution.

Thus, 1000 mL solvent = 1000 mL solution

= 1000 × 0.4 g solvent = 400 g solvent = 0.4 kg solvent

Moles of solute = 3.2 mol in 1000 mL

or 0.4 kg solvent

Molality $= \dfrac{3.2}{0.4} = 8 \text{ mol kg}^{-1}$

2. (8) Mass of HCl in 1.0 mL stock solution

$= 1.25 \times \dfrac{29.2}{100}$

$= 0.365 \text{ g}$

Mass of HCl required for 200 mL 0.4 M HCl

$= \dfrac{200}{1000} \times 0.4 \times 36.5$

$= 0.08 \times 36.5 \text{ g}$

$\therefore$ 0.365 g of HCl is present in 1.0 mL stock solution.

$\therefore$ 0.08 × 36.5 g HCl will be present in $\dfrac{0.08 \times 36.5}{0.365} = 8.0$ mL

3. (c) Molarity $= \dfrac{\text{Moles of solute}}{\text{Volume of solution (L)}}$,

Moles of urea $= \dfrac{120}{60} = 2$

Weight of solution = weight of solvent + weight of solute

= 1000 + 120 = 1120g

Volume $= \dfrac{1120\text{g}}{1.15\text{g / mL}} \times \dfrac{1}{1000\text{mL / L}}$

= 0.974L

Molarity $= \dfrac{2.000}{0.974} = 2.05$ M

4. (b) From the given relative abundance, the average weight of Fe can be calculated as

$A = \dfrac{54 \times 5 + 56 \times 90 + 57 \times 5}{100}$

$= 55.95$

5. (d) $6\,Fe^{2+} + Cr_2O_7^{2-} + 14H^+ \longrightarrow 6\,Fe^{3+} + 2\,Cr^{3+} + 7H_2O$

Thus, $1\,Cr_2O_7^{2-} \equiv 6\,Fe^{2+}$

Thus, six (6)

6. (1008) The balanced reaction is

$6CaO + P_4O_{10} \longrightarrow 2Ca_3(PO_4)_2$

moles of $P_4O_{10} = \dfrac{852}{284} = 3$

moles of CaO required = 3 × 6 = 18

mass of CaO required = 18 × 56 = 1008 g

# Last 3 Years' Questions
# JEE Main & Advanced
# (2016-2018)

## JEE Main

### 2016

**1.** A stream of electrons from a heated filament was passed between two charged plates kept at a potential difference $V$ esu. If $e$ and $m$ are charge and mass of an electron, respectively, then the value of $h/\lambda$ (where, $\lambda$ is wavelength associated with electron wave) is given by **[Structure of Atom]**

(a) $2\,meV$ (b) $\sqrt{meV}$ (c) $\sqrt{2meV}$ (d) $meV$

### 2017

**2.** The radius of the second Bohr orbit for hydrogen atom is (Planck's constant $(h) = 6.6262 \times 10^{-34}$ Js; mass of electron $= 9.1091 \times 10^{-31}$ kg ; charge of electron $(e) = 1.60210 \times 10^{-19}$ C; permitivity of vacuum $(\varepsilon_0) = 8.854185 \times 10^{-12}\ \text{kg}^{-1}\text{m}^{-3}\text{A}^2$) **[Structure of Atom]**

(a) 1.65 Å (b) 4.76 Å (c) 0.529 Å (d) 2.12 Å

**3.** The most abundant elements by mass in the body of a healthy human adult are Oxygen (61.4%); Carbon (22.9%), Hydrogen (10.0 %); and Nitrogen (2.6%). The weight which a 75 kg person would gain if all $^1$H atoms are replaced by $^2$H atoms is **[Stoichiometry]**

(a) 15 kg (b) 37.5 kg (c) 7.5 kg (d) 10 kg

**4.** Which of the following reactions is an example of a redox reaction? **[Redox Reactions]**

(a) $XeF_4 + O_2F_2 \longrightarrow XeF_6 + O_2$ (b) $XeF_2 + PF_5 \longrightarrow [XeF]^+PF_6^-$

(c) $XeF_6 + H_2O \longrightarrow XeOF_4 + 2HF$ (d) $XeF_6 + 2H_2O \longrightarrow XeO_2F_2 + 4HF$

**5.** Which of the following species is not paramagnetic? **[Chemical Bonding]**

(a) NO (b) CO (c) $O_2$ (d) $B_2$

**6.** The group having isoelectronic species is **[Chemical Bonding]**

(a) $O^{2-}, F^-, Na^+, Mg^{2+}$ (b) $O^-, F^-, Na, Mg^+$

(c) $O^{2-}, F^-, Na, Mg^{2+}$ (d) $O^-, F^-, Na^+, Mg^{2+}$

### 2018

**7.** The ratio of mass per cent of C and H of an organic compound $(C_xH_yO_z)$ is 6 : 1. If one molecule of the above compound $(C_xH_yO_z)$ contains half as much oxygen as required to burn one molecule of compound $C_xH_y$ completely to $CO_2$ and $H_2O$. The empirical formula of compound $C_xH_yO_z$ is **[Stoichiometry]**

(a) $C_3H_6O_3$ (b) $C_2H_4O$ (c) $C_3H_4O_2$ (d) $C_2H_4O_3$

**8.** According to molecular orbital theory, which of the following will not be a viable molecule? **[Chemical Bonding]**

(a) $He_2^{2+}$ (b) $He_2^+$ (c) $H_2^-$ (d) $H_2^{2-}$

**9.** Which of the following compounds contain(s) no covalent bond(s)?
$KCl, PH_3, O_2, B_2H_6, H_2SO_4$ **[Chemical Bonding]**

(a) $KCl, B_2H_6, PH_3$ (b) $KCl, H_2SO_4$ (c) KCl (d) $KCl, B_2H_6$

**10.** Total number of lone pair of electron in $I_3^-$ ion is **[Chemical Bonding]**

(a) 3 (b) 6 (c) 9 (d) 12

# JEE Advanced

## 2016

**1.** $P$ is the probability of finding the $1s$ electron of hydrogen atom in a spherical shell of infinitesimal thickness, $dr$, at a distance $r$ from the nucleus. The volume of this shell is $4\pi r^2 dr$. The qualitative sketch of the dependence of $P$ on $r$ is

(a)
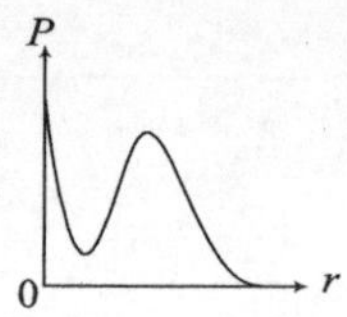


(b)
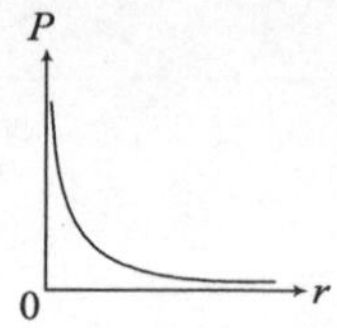


(c)
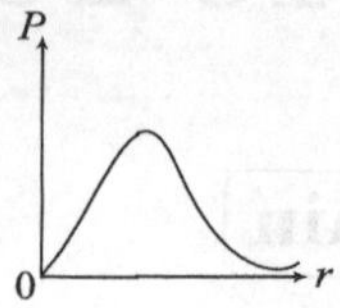


(d)
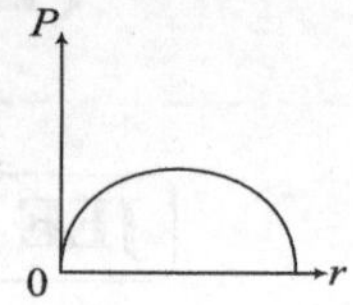


**[Structure of Atoms, Type 1])**

**2.** The compound(s) with two lone pairs of electrons on the central atom is (are) **[Chemical Bonding, Type 2]**

(a) $BrF_5$ (b) $ClF_3$ (c) $XeF_4$ (d) $SF_4$

**3.** According to molecular orbital theory, which of the following statements is(are) correct? **[Chemical Bonding, Type 2]**

(a) $C_2^{2-}$ is expected to be diamagnetic

(b) $O_2^{2+}$ is expected to have a longer bond length than $O_2$

(c) $N_2^+$ and $N_2^-$ have the same bond order

(d) $He_2^+$ has the same energy as two isolated He atoms

## 2017

**4.** The sum of the number of lone pairs of electrons on each central atom in the following species is **[Chemical Bonding, Type 3]**

$[TeBr_6]^{2-}$, $[BrF_2]^+$, $SNF_3$ and $[XeF_3]^-$

(Atomic numbers : N = 7, F = 9, S = 16, Br = 35, Te= 52, Xe = 54)

**5.** Among $H_2$, $He_2^+$, $Li_2$, $Be_2$, $B_2$, $C_2$, $N_2$, and $F_2O_2^-$, the number of diamagnetic species is (Atomic numbers : H = 1, He = 2, Li = 3, Be = 4, B = 5, C = 6, N = 7, O = 8, F = 9) **[Chemical Bonding, Type 3]**

**Answer Q. 6, Q. 7 and Q. 8 by appropriately matching the information given in the three columns of the following table.**

The wave function, $\psi_{n,l,m_l}$ is a mathematical function whose value depends upon spherical polar coordinates $(r, \theta, \phi)$ of the electron and characterised by the quantum number $n$, $l$ and $m_l$. Here $r$ is distance from nucleus, $\theta$ is colatitude and $\phi$ is azimuth. In the mathematical functions given in the Table, $Z$ is atomic number and $a_0$ is Bohr radius. **[Structure of Atom]**

| | Column 1 | | Column 2 | | Column 3 |
|---|---|---|---|---|---|
| (I) | 1s-orbital | (i) | $\psi_{n,l,m_l} \propto \left(\frac{Z}{a_0}\right)^{\frac{3}{2}} e^{-\left(\frac{Zr}{a_0}\right)}$ | (P) | 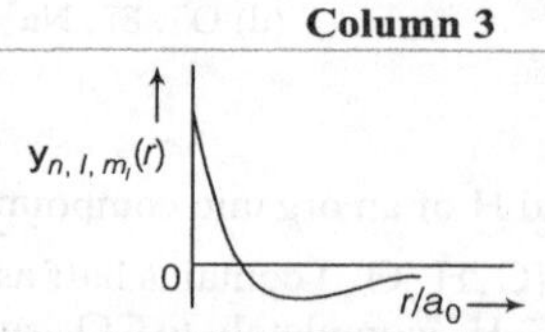  |
| (II) | 2s-orbital | (ii) | One radial node | (Q) | Probability density at nucleus $\propto \frac{1}{a_0^3}$ |
| (III) | $2p_z$-orbital | (iii) | $\psi_{n,l\,m_l} \propto \left(\frac{Z}{a_0}\right)^{5/2} re^{-\left(\frac{Zr}{a_0}\right)} \cos\theta$ | (R) | Probability density is maximum at nucleus |
| (IV) | $3\,d_z^2$-orbital | | xy-plane is a nodal plane | (S) | Energy needed to excite electron from $n = 2$ state to $n = 4$ state is $\frac{27}{32}$ times the energy needed to excite electron from $n = 2$ state to $n = 6$ state |

***Note*** ***Type-1*** *Only one correct option,* ***Type-2*** *More then one correct option,* ***Type-3*** *Single digit Answer,* ***Type-4*** *Numerical value based*

**6.** For $He^+$ ion, the only **INCORRECT** combination is
(a) (I) (i) (S) (b) (II) (ii) (Q) (c) (I) (iii) (R) (d) (I) (i) (R

**7.** For the given orbital in Column 1, the Only **CORRECT** combination for any hydrogen-like species is
(a) (II) (ii) (P) (b) (I) (ii) (S) (c) (IV) (iv) (R) (d) (III) (iii) (P)

**8.** For hydrogen atom, the only **CORRECT** combination is
(a) (I) (i) (P) (b) (I) (iv) (R) (c) (II) (i) (Q) (d) (I) (i) (S)

## 2018

**9.** Among the species given below, the total number of diamagnetic species is____
H atom, $NO_2$ monomer, $O_2^-$ (superoxide), dimeric sulphur in vapour phase, $Mn_3O_4$, $(NH_4)_2[FeCl_4]$, $(NH_4)_2[NiCl_4]$, $K_2MnO_4$, $K_2CrO_4$ **[Chemical Bonding, Type 4]**

**10.** Galena (an ore) is partially oxidised by passing air through it at high temperature. After some time, the passage of air is stopped, but the heating is continued in a closed furnace such that the content undergo self-reduction. The weight (in kg) of Pb produced per kg of $O_2$ consumed is .... (Atomic weights in g $mol^{-1}$ : O = 16, S = 32, Pb = 207) **[Stoichiometry, Type 4]**

**11.** To measure the quantity of $MnCl_2$ dissolved in an aqueous solution, it was completely converted to $KMnO_4$ using the reaction,

$$MnCl_2 + K_2S_2O_8 + H_2O \longrightarrow KMnO_4 + H_2SO_4 + HCl$$

(equation not balanced).
Few drops of concentrated HCl were added to this solution and gently warmed. Further, oxalic acid
(225 mg) was added in portions till the colour of the permanganate ion disappeared. The quantity of $MnCl_2$ (in mg) present in the initial solution is ......... .
(Atomic weights in g $mol^{-1}$: Mn = 55, Cl = 35.5) **[Stoichiometry, Type 4]**

**12.** The ammonia prepared by treating ammonium sulphate with calcium hydroxide is completely used by $NiCl_2 \cdot 6H_2O$ to form a stable coordination compound. Assume that both the reactions are 100% complete. If 1584 g of ammonium sulphate and 952 g of $NiCl_2 \cdot 6H_2O$ are used in the preparation, the combined weight (in grams) of gypsum and the nickel-ammonia coordination compound thus produced is____
(Atomic weights in g $mol^{-1}$ : H = 1, N = 14, O = 16, S = 32, Cl = 35.5, Ca = 40, Ni = 59) **[Stoichiometry, Type 4]**

# Answer *with* Explanations

## JEE Main

**1.** *(c) Plan* As you can see in options, energy term is mentioned hence, we have to find out relation between $h/\lambda$ and energy. For this, we shall use de-Broglie wavelength and kinetic energy term in eV.

de-Broglie wavelength for an electron

$$(\lambda) = \frac{h}{p} \Rightarrow p = \frac{h}{\lambda} \quad ...(i)$$

Kinetic energy of an electron = eV

As we know that, $KE = \frac{p^2}{2m}$

$$\therefore \quad eV = \frac{p^2}{2m} \text{ or } p = \sqrt{2meV} \quad ...(ii)$$

From Eqs. (i) and (ii), we get

$$\frac{h}{\lambda} = \sqrt{2meV}$$

**2.** *(d)* Bohr radius $(r_n) = \epsilon_0 n^2h^2$

$$r_n = \frac{n^2h^2}{4\pi^2 me^2 kZ} \Rightarrow k = \frac{1}{4\pi \epsilon_0}$$

$$\therefore \quad r_n = \frac{n^2h^2 \epsilon_0}{\pi me^2 Z} = n^2 \frac{a_0}{Z}$$

where, $m$ = mass of electron, $e$ = charge of electron
$h$ = Planck's constant, $k$ = Coulomb constant

$$r_n = \frac{n^2 \times 0.53}{Z} \text{ Å}$$

Radius of $n^{th}$ Bohr orbit for H-atom = $0.53\, n^2$ Å
[Z = 1 for H-atom]

$\therefore$ Radius of $2^{nd}$ Bohr orbit for H-atom
$= 0.53 \times (2)^2 = 2.12$ Å

3. *(c)* Given, abundance of elements by mass oxygen = 61.4%, carbon = 22.9%, hydrogen = 10% and nitrogen = 2.6%

Total weight of person = 75 kg

$$\text{Mass due to } {}^1\text{H} = \frac{75 \times 10}{100} = 7.5 \text{ kg}$$

${}^1$H atoms are replaced by ${}^2$H atoms,

Mass due to ${}^2$H = (7.5 × 2) kg

∴ Mass gain by person = 7.5 kg

4. *(a)* The reaction in which oxidation and reduction occur simultaneously are termed as redox reaction.

$$\overset{+4}{(\text{Xe})}\ F_4 + \overset{+1}{O_2}\ (F_2) \longrightarrow \overset{+6}{X}\ eF_6 + \overset{0}{O}_2$$

Since, Xe undergoes oxidation while O undergoes reduction. So, it is an example of redox reaction.

5. *(b)* To identify the magnetic nature we need to check the molecular orbital configuration. If all orbitals are fully occupied, species is diamagnetic while when one or more molecular orbitals is/are singly occupied, species is paramagnetic.

(a) NO (7 + 8 = 15) $-\sigma 1s^2, \sigma^* 1s^2, \sigma 2s^2,$

$\sigma^* 2s^2, \pi 2p_x^2 = \pi 2p_y^2, \pi 2p_z^2, \pi^* 2p_x^1 = \pi^* 2p_y^0$

One unpaired electron is present.
Hence, it is paramagnetic.

(b) CO (6 + 8 = 14) $-\sigma 1s^2, \sigma^* 1s^2, \sigma 2s^2,$

$\sigma^* 2s^2, \pi 2p_x^2 = \pi 2p_y^2, \sigma 2p_z^2$

No unpaired electron is present.
Hence, it is diamagnetic.

(c) $O_2$ (8 + 8 = 16) $-\sigma 1s^2, \sigma^* 1s^2, \sigma 2s^2,$

$\sigma^* 2s^2, \sigma 2p_z^2, \pi 2p_x^2 = \pi 2p_z^2, \pi^* 2p_x^1 = \pi^* 2p_y^1$

Two unpaired electrons are present.
Hence, it is paramagnetic.

(d) $B_2$(5 + 5) $-\sigma 1s^2, \sigma^* 1s^2, \sigma 2s^2, \sigma^* 2s^2, \pi 2p_x^1 = \pi 2p_y^1$

Two unpaired electrons are present.
Hence, it is paramagnetic.

6. *(a)* Isoelectronic species are those which contains same number of electrons.

| Species | Atomic number | Number of electrons |
|---|---|---|
| $O^{2-}$ | 8 | 10 |
| $F^-$ | 9 | 10 |
| $Na^+$ | 11 | 10 |
| $Mg^{2+}$ | 12 | 10 |
| $O^-$ | 8 | 9 |
| Na | 11 | 11 |
| $Mg^+$ | 12 | 11 |

∴ Option (a) is correct which contains isoelectronic species $O^{2-}, F^-, Na^+, Mg^{2+}$.

7. *(d)* We can calculate the simplest whole number ratio of C and H from the data given, as

| Element | Relative mass | Molar mass | Relative mole | Simplest whole number ratio |
|---|---|---|---|---|
| C | 6 | 12 | $\frac{6}{12} = 0.5$ | $\frac{0.5}{0.5} = 1$ |
| H | 1 | 1 | $\frac{1}{1} = 1$ | $\frac{1}{0.5} = 2$ |

Alternatively this ratio can also be calculated directly in the terms of $x$ and $y$ as

$$\frac{12x}{y} = \frac{6}{1} \text{ (given and molar mass of C = 12, H = 1)}$$

Now, after calculating this ratio look for condition 2 given in the question i.e. quantity of oxygen is half of the quantity required to burn one molecule of compound $C_xH_y$ completely to $CO_2$ and $H_2O$. We can calculate number of oxygen atoms from this as consider the equation.

$$C_xH_y + \left[x + \frac{y}{4}\right]O_2 \longrightarrow xCO_2 + \frac{y}{2}H_2O$$

$$\text{Number of oxygen atoms required} = 2 \times \left[x + \frac{y}{4}\right] = \left[2x + \frac{y}{2}\right]$$

$$\text{Now given, } z = \frac{1}{2}\left[2x + \frac{y}{2}\right] = \left[x + \frac{y}{4}\right]$$

Here we consider $x$ and $y$ as simplest ratios for C and H so now putting the values of $x$ and $y$ in the above equation.

$$z = \left[x + \frac{y}{4}\right] = \left[1 + \frac{2}{4}\right] = 1.5$$

Thus, the simplest ratio figures for $x$, $y$ and $z$ are $x = 1$, $y = 2$ and $z = 1.5$

Now, put these values in the formula given i.e.

$C_xH_yO_z = C_1H_2O_{1.5}$

So, empirical formula will be $[C_1H_2O_{1.5}] \times 2 = C_2H_4O_3$

8. *(d)* **Key Idea** *According to M.O.T, the viability of any molecule can be judged through the calculation of bond order.*

| Electronic | Configuration | Bond order |
|---|---|---|
| $He_2^+$ | $\sigma_{1s^2}\sigma^*_{1s^1}$ | $\frac{2-1}{2} = 0.5$ |
| $H_2^-$ | $\sigma_{1s^2}\sigma^*_{1s^1}$ | $\frac{2-1}{2} = 0.5$ |
| $H_2^{2-}$ | $\sigma_{1s^2}\sigma^*_{1s^2}$ | $\frac{2-2}{2} = 0$ |
| $He_2^{2+}$ | $\sigma_{1s^2}$ | $\frac{2-0}{2} = 1$ |

The molecule having zero bond order will not be viable hence, $H_2^{2-}$ (option d) is the correct answer.

9. *(c)* KCl is the only ionic compound. The structure of $PH_3$, $O_2$, $B_2H_6$ and $H_2SO_4$ are given below

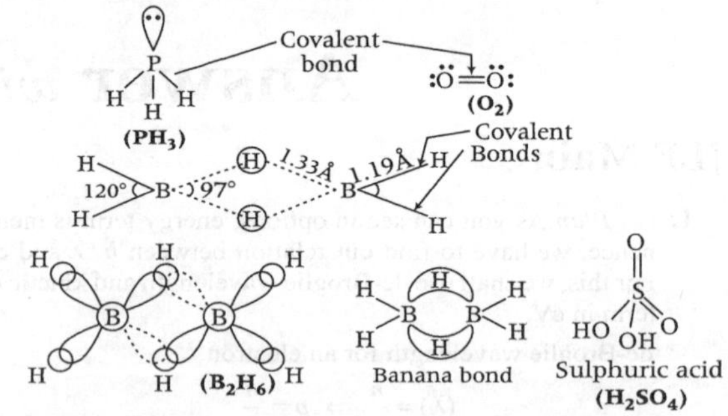


All bond between S and O atom are covalent bonds.

10. *(c)* The structure of $I_3^-$ ion is

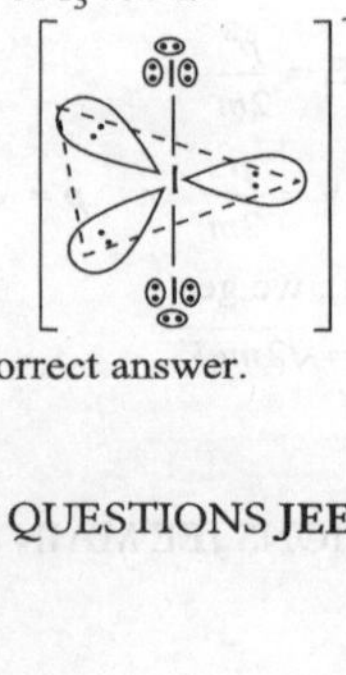

Hence, 9 is the correct answer.

# JEE Advanced

1. *(c)* This graph shows the probability of finding the electron within shell at various distances from the nucleus (radial probability). The curve shows the maximum, which means that the radial probability is greatest for a given distance from the nucleus. This distance is equal to Bohr's radius = $a_0$

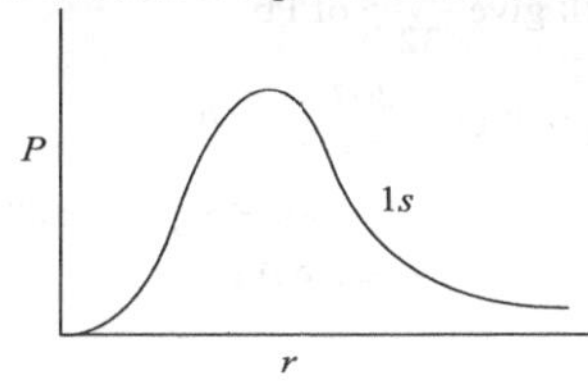


(a) It is for 2*s*-orbital.
(b) It is radial wave function for 1*s*.
(c) Correct
(d) Probability cannot be zero at a certain distance from nucleus.

2. *(b,c)*

| Compounds | Hydridisation | Structures | Lone pair on central atom |
|---|---|---|---|
| $BrF_5$ | $sp^3d^2$ | 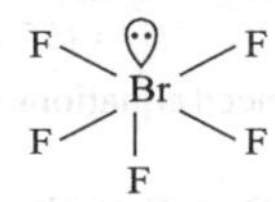 | 1 |
| $ClF_3$ | $sp^3d$ | 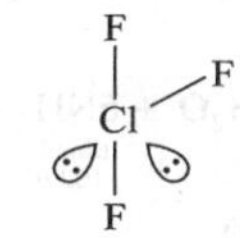 | 2 |
| $XeF_4$ | $sp^3d^2$ | | 2 |
| $SF_4$ | $sp^3d$ | 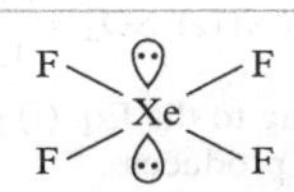 | 1 |

3. *(a, c)*

| Species | Electrons | MOEC | $N_B$ | $N_A$ | BO | Magnetic character |
|---|---|---|---|---|---|---|
| $C_2^{2-}$ | 14 | $\sigma 1s^2, \sigma^* 1s^2, \sigma 2s^2, \sigma^* 2s^2, \pi 2p_x^2 \approx -\pi 2p_y^2, \sigma 2p_z^2$ | 10 | 4 | 3 | Diamagnetic |
| $O_2^{2+}$ | 14 | As above according to number of electrons | 10 | 4 | 3 | Diamagnetic |
| $O_2$ | 16 | | 10 | 6 | 2 | Paramagnetic |
| $N_2^+$ | 13 | | 9 | 4 | 2.5 | Paramagnetic |
| $N_2^-$ | 15 | | 10 | 5 | 2.5 | Paramagnetic |
| $He_2^+$ | 3 | | 2 | 1 | 0.5 | Paramagnetic |

Thus, (a) is correct.
(b) Bond order $O_2^{2+} > O_2$ thus,
Bond length of $O_2^{2+} < O_2$ thus, incorrect.
(c) $N_2^+$ and $N_2^-$ have same bond order thus correct.
(d) $He_2^+$ with bond order = 0.5 is more stable thus, less energy than isolated He atoms. Thus, (d) is incorrect.

4. *(6)*

| S.N. | Species | No. of σ-bonds with central atom | No. of L.P at central atom |
|---|---|---|---|
| (i) | In $[TeBr_6]^{2-}$ | 6 | 1 |
| (ii) | In $[BrF_2]^+$ | 2 | 2 |
| (iii) | In $SNF_3$ | 4 | 0 |
| (iv) | In $[XeF_3]^-$ | 3 | 3 |

5. *(6)* $H_2$, $Li_2$, $Be_2$, $C_2$, $N_2$ and $F_2$ are diamagnetic according to molecular orbital theory.

6. *(d)* In the wave function ($\psi$) expression for 1*s*-orbital of $He^+$, there should be no angular part. Hence, (iii) can't be true for $\psi_{1s}$ of $He^+$.

7. *(b)* Correct : 2s orbital has one radial node.
No of radial node = $n - l - 1 = 2 - 0 - 1 = 1$
Also, when radial part of wave function ($\psi$) is plotted against "*r*", wave function changes its sign at node.

8. *(d)* *i* is the correct expression of wave function for 1*s*-orbital of hydrogenic system.

$$\text{Also, } E(2-4) = kZ^2\left(\frac{1}{4} - \frac{1}{16}\right) = kZ^2\left(\frac{3}{16}\right) \quad \ldots\text{(i)}$$

$$E(2-6) = kZ^2\left(\frac{1}{4} - \frac{1}{36}\right) = kZ^2\left(\frac{2}{9}\right) \quad \ldots\text{(ii)}$$

$$\Rightarrow \quad E(2-6) \times \frac{27}{32} = kZ^2 \times \frac{2}{9} \times \frac{27}{32}$$

$$= kZ^2\left(\frac{3}{16}\right) = E(2-4)$$

9. *(1)* Among the given species only $K_2CrO_4$ is diamagnetic as central metal atom Cr in it has $[Ar]3d^0$ electronic configuration i.e., all paired electrons. The structure and oxidation state of central metal atom of this compound are as follows

**Structure** $K^+$ 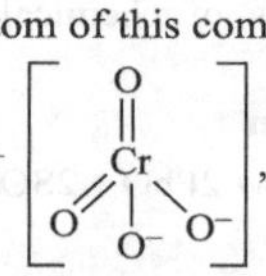

**Oxidation state** $Cr^{6+}$

Rest all the compounds are **paramagnetic**. Reasons for their paramagnetism are given below

(i) H-atom have $1s^1$ electronic configuration i.e., 1 unpaired electron.

(ii) $NO_2$ i.e. 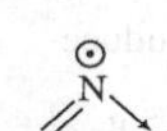 in itself is **an odd electron species.**

(iii) $O_2^-$ (Superoxide) has one unpaired electron in $\pi^*$ molecular orbital.

(iv) $S_2$ in vapour phase has $O_2$ like electronic configuration i.e., have 2 unpaired electrons in $\pi^*$ molecular orbitals.

(v) $Mn_3O_4$ has following structure

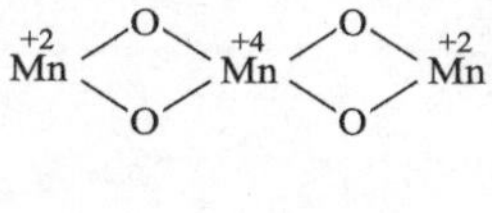

Thus, Mn is showing +2 and +4 oxidation states. The outermost electronic configuration of elemental Mn is $3d^5 4s^2$. Hence, in both the above oxidation states it has unpaired electrons as

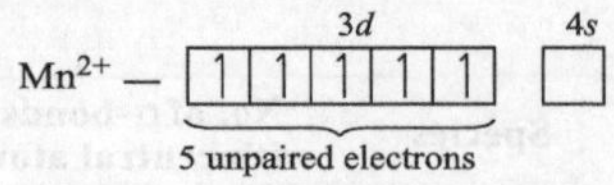


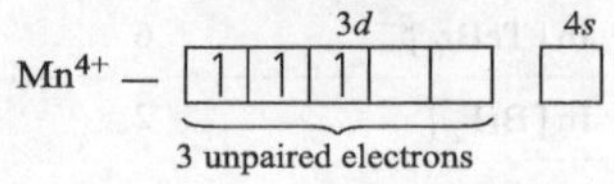


(vi) $(NH_4)_2FeCl_4$ has Fe as central metal atom with +2 oxidation state. The electronic configuration of $Fe^{2+}$ in the complex is

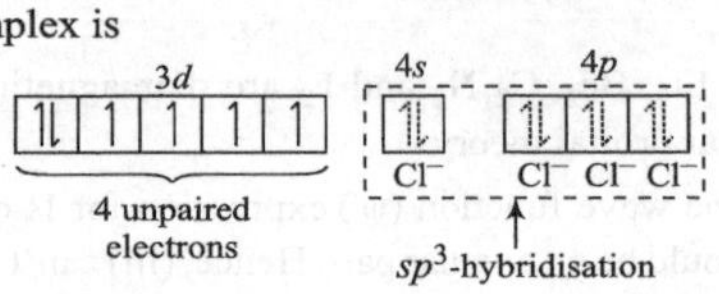


(vii) $(NH_4)_2NiCl_4$ has Ni as central metal atom with +2 oxidation state. The electronic configuration of $Ni^{2+}$ in the complex is

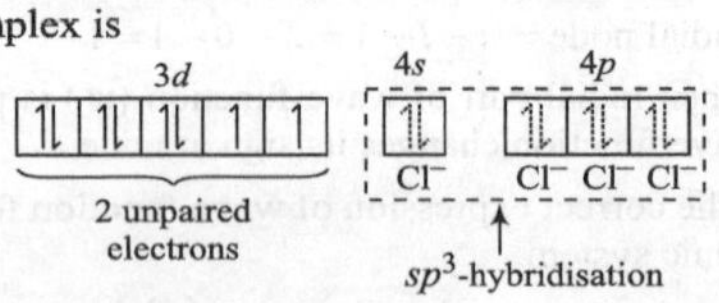


(viii) In $K_2MnO_4$ central metal atom Mn has +6 oxidation state with following structure

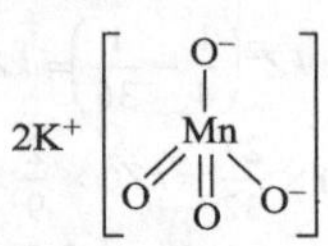


Electronic configuration of $Mn^{6+}$ is

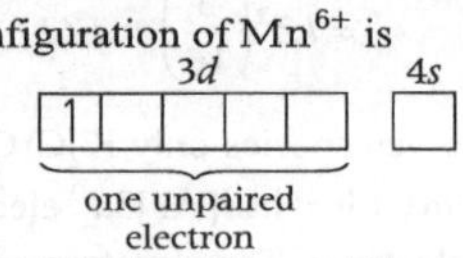


**10.** *(6.47 kg)* The equations of chemical reactions occurring during the process are—

**In the presence of oxygen**

$$2\,PbS + 3O_2 \longrightarrow 2PbO + 2SO_2 \qquad ...(i)$$

By self reduction

$$2PbO + PbS \longrightarrow 3Pb + SO_2$$

Thus 3 moles of $O_2$ produces 3 moles of Pb i.e.,

$32 \times 3 = 96$ g of $O_2$ produces

$3 \times 207 = 621$ g of Pb

So 1000 g (1kg) of oxygen will produce

$$\frac{621}{96} \times 1000 = 6468.15 \text{ g}$$

$$= 6.4687 \text{ kg} \approx 6.47 \text{ kg}$$

**Alternatively**

From the direct equation,

$$\underset{32\text{ g}}{PbS + O_2} \longrightarrow \underset{207\text{ g}}{Pb} + SO_2$$

So, 32 g of $O_2$ gives 207 g of Pb

1 g of $O_2$ will give $\frac{207}{32}$ g of Pb

1000g of $O_2$ will give $\frac{207}{32} \times 1000$

$$= 6468.75 \text{ g} = 6.46875 \text{ kg}$$

$$\approx 6.47 \text{kg}$$

**11.** *(126 mg)*

$$Mn^{2+} \longrightarrow MnO_4^- + 5e^-$$

$$S_2O_8^{2-} + 2e^- \longrightarrow 2SO_4^{2-}$$

$$\therefore \quad 2Mn^{2+} + 5S_2O_8^{2-} \longrightarrow 2MnO_4^- + 10SO_4^{2-}$$

Also, $2MnO_4^- + 5C_2O_4^{2-} \longrightarrow 2Mn^{2+} + 10CO_2$

Hence, $2Mn^{2+} \equiv 5C_2O_4^{2-}$

$$1\ MnCl_2 \equiv 2.5\ H_2C_2O_4$$

Oxalic acid taken = 225 mg

$$= \frac{225}{90} = 2.5 \text{ millimoles}$$

Hence, $MnCl_2 = 1$ millimole

$$= (55 + 71) = 126 \text{ mg}$$

**12.** *(2992)* Balanced equations of reactions used in the problem are as follows

(i) $\underset{\substack{1\text{ mol}\\132\text{ g}}}{(NH_4)_2SO_4} + Ca(OH)_2 \longrightarrow \underset{\substack{1\text{ mol}\\172\text{ g}}}{CaSO_4 \cdot 2H_2O} + \underset{\substack{2\text{ mol}\\(2\times17)=34\text{g}}}{2NH_3}$

(ii) $\underset{\substack{1\text{ mol}\\238\text{ g}}}{NiCl_2 \cdot 6H_2O} + \underset{\substack{6\text{ mol}\\102\text{ g}}}{6NH_3} \longrightarrow \underset{\substack{1\text{ mol}\\232\text{ g}}}{[Ni(NH_3)_6]Cl_2} + 6H_2O$

Now, in Eq. (i)

if, 1584 g of ammonium sulphate is used.

i.e., 1584 g $(NH_4)_2SO_4 = \frac{1584}{132} = 12$ mol

So, according to the Eq. (i) given above 12 moles of $(NH_4)_2SO_4$ produces

(a) 12 moles of gypsum

(b) 24 moles of ammonia

Here, 12 moles of gypsum = $12 \times 172$

$= 2064$ g

and 24 moles of $NH_3 = 24 \times 17 = 408$ g

Further, as given in question, 24 moles of $NH_3$ produced in reaction (i) is completly utilised by 952g or 4 moles of $NiCl_2 \cdot 6H_2O$ to produce 4 moles of $[Ni(NH_3)_6]Cl_2$.

So, 4 moles of $[Ni(NH_3)_6]Cl_2 = 4 \times 232 = 928$gms

Hence, total mass of gypsum and nickel ammonia coordination compound $[Ni(NH_3)_6]Cl_2$

$$= 2064 + 928 = 2992$$